COMETS

A
Descriptive
Catalog

D1236140

by

Gary W. Kronk

Foreword by Brian G. Marsden
Smithsonian Astrophysical Observatory

ENSLOW PUBLISHERS, INC.

Bloy St. & Ramsey Ave.   P.O. Box 38
Box 777   Aldershot
Hillside, N.J. 07205   Hants GU12 6BP
U.S.A.   U.K.

This book is dedicated to my wonderful wife Karen and to the memory of my father, Richard Kronk, Jr.

Library of Congress Cataloging in Publication Data

Kronk, Gary.
   Comets: a descriptive catalog.

   1. Comets—Catalogs. I. Title.

QB722.K76   1984   523.6   82-20971
   ISBN 0-89490-071-4

Printed in the United States of America

10 9 8 7 6 5 4 3 2 1

CONTENTS

iii

PREFACE

I have been interested in general astronomy since age eight, but it was Comet Kohoutek in 1973 that turned me toward comets. Ever since seeing Kohoutek I have been collecting all the descriptive material I could find on the comets recorded since antiquity. I have read at least 7000 references to compile information, mostly from the general sources listed in the back of this book. Along the way I corresponded with Dr. Brian Marsden of the Smithsonian Astrophysical Observatory, who thought that my information should be published so as to be generally available. This book presents useful data and readable descriptions of every comet observed from the years −371 to 1982 in a manner useful to both the armchair buff and the active comet observer.

The long-period and non-periodic comets are listed chronologically in the first section. Every effort has been made to include the brightest magnitude, longest tail length, and largest coma diameter achieved by each comet. In addition, as many details as possible have been included about each comet's discovery.

The second section deals with the short-period comets. Since these comets are always referred to in the literature by their proper names, the arrangement is alphabetical. Included are details of each comet's discovery, orbital investigations, and subsequent recoveries, as well as other pertinent information.

There were two problems concerning the handling of the data. The first involved measurement of the tail length and coma diameters of early comets. Modern observations are usually recorded in degrees, but Oriental records before 1600 were measured in "chi," which means "feet." In an article in 1972, T. Kiang matched stars in ancient Chinese literature to those in modern catalogs and concluded that one chi was equal to about 1.50 degrees, so this conversion was used when handling data for ancient comets.

The second problem concerned the brightness of comets. Although the Greek astronomer Hipparchus was the first to classify each star observed with the naked eye in terms of its brightness and magnitude, it was not until 1856 that the English astronomer Norman R. Pogson proposed the current system of comparing the nucleus to neighboring stars. Unfortunately, due to the extent of the comas of most comets, the total magnitude still could not be made reliably using Pogson's system, with variations of two or more magnitudes occurring between observers.

In the mid-1890s a new method of determining comet total magnitudes was developed by J. Holetschek. He tried to duplicate with telescope observations what other astronomers had done with the naked eye while observing bright comets— compare a comet to a star of known brightness while the apparent size was small. Telescopes generally enlarge comets to sizes greatly exceeding the apparent sizes of neighboring stars, but by using as little magnification as possible in an observation, Holetschek tried to keep the apparent size small to allow an easier comparison to local stars. Other observers adopted this method. The current method of defocusing a star's image until it equals the size of the coma of the comet before measuring the magnitude was begun during the second decade of the 1900s and was led by George van Biesbroeck of Yerkes Observatory.

This brief history of comet magnitude determinations shows how new the art is. As a matter of fact, observing conditions, telescope size, and the acuity of an observer's eyes still prevent this from being an exact science.

The following three kinds of magnitude data can be distinguished in the descriptions in the book:

A. Raw data. They can be identified because I have attributed them directly to the persons who made the observations.

B. Calculated magnitudes. When there was no magnitude recorded for a particular date or period of time, I sometimes calculated the magnitude using a formula. For these calculations I used several observations from one or more observers to make what I consider to be an accurate projection. These calculated magnitudes are indicated by an asterisk. An advanced researcher may not want to use these figures but to go back to the original reports (see the references) and derive figures for himself.

C. Averaged magnitudes. Magnitudes neither attributed to a particular observer nor marked by an asterisk were averaged from observations available for the particular date or period of time specified.

I have used the following customary formulas, originally developed by Orlov in 1911, to translate an observed magnitude (m) into an absolute magnitude (Ho) and vice versa:

$$(1) \quad Ho = m - 5 \log d - [2.5n(\log r)]$$

$$(2) \quad m = Ho + 5 \log d + [2.5n(\log r)]$$

where,

Ho = absolute magnitude
m = apparent magnitude
d = distance of the comet from Earth
r = distance of the comet from the sun
n = a factor chosen to match the pattern of the comet's brightening and dimming

Most magnitudes for comets observed prior to the 1890s had to be computed from very rough descriptions. Phrases such as "as bright as Saturn" or "nearly as bright as Antares" and descriptions of whether a comet was faint or bright in certain-sized telescopes were converted into rough magnitudes and then used in equation (1) above to determine the comet's absolute magnitude. Once an absolute magnitude of sufficient accuracy was obtained, equation (2) was utilized to project the comet's probable apparent magnitude for a particular date. If the number of rough observations was too small to statistically obtain a reliable absolute magnitude, my values were compared to the extensive absolute magnitude lists of Holetschek and Vsekhsvyatskii to determine consistency before putting computed apparent magnitudes into this book. These values fit into category B. above (and are designated by an asterisk). For comets observed during the twentieth century, this type of calculation is rare, since a greater number of observations are available; however, there are some instances, primarily when no observations could be made at all, where the calculation was used (as before, these are designated by an asterisk).

I hope my book will prove generally useful and enjoyable to read and will stimulate interest in those marvels known as comets.

FOREWORD

At the 1967 meeting of the International Astronomical Union we considered that an up-to-date descriptive catalog of comets—a new "cometography" as we called it— should be prepared. An ad hoc committee was formed to look into this. In presenting its report in 1970, the committee concluded that it would be very nice to have a cometography, but that the effort required to produce such a catalog was greater than the members of the committee could devote to it. Therefore, it gives me pleasure to know that Gary Kronk has taken up the challenge and has produced a book that meets many of the aims of the committee.

The book does not contain orbital information, but the physical descriptions and documentation of the interval of time during which a comet could be observed can be put together with orbital information from elsewhere. This would enable one to examine such questions as, for example, why that comet had not been seen earlier. Gary Kronk's book is thus a natural companion to my *Catalog of Cometary Orbits.** One can read the description of a comet in *Comets,* and then look up the orbital figures in the *Catalog.* The procedure for intercomparing observational and orbital data is conveniently contained in the article, "Basic Information and References."**

Prior to *Comets: A Descriptive Catalog* one leaned mainly on two older sources for descriptions of the comets. *Cométographie ou Traité Historique et Théorique des Comètes* by A.-G. Pingré, 2 vols., Paris, 1783-84, is still very authoritative but, of course, much out of date. (It may be one of the oldest scientific books never reprinted that is still useful.) *Physical Characteristics of Comets* by S.K. Vsekhsvyatskii (a Russian book published in 1958, translated into English by NASA) is also now dated and contains a rather large number of errors; supplements were published in the U.S.S.R. but, unfortunately, they have not been translated into English. Other earlier sources include the series of papers by J. Holetschek, *Untersuchen über die Grösse und Helligkeit der Kometen und ihrer Schweife,* Vienna, 1896-1917, and the annual reports in the *Monthly Notices* and *Quarterly Journal* of the Royal Astronomical Society.

The observation of comets is one of the areas of astronomy where amateurs still can make major contributions. This book should be useful to them and to professional astronomers actively working in the field. It also should find a place in science libraries and in the offices of popular publications that need a reference to comets in the news. I recommend it.

Brian G. Marsden

Astronomer, Smithsonian Astrophysical Observatory.

Lecturer on Astronomy, Harvard University.

Director, Central Bureau for Astronomical Telegrams and Minor Planet Center,
*   International Astronomical Union.*

---

*   Marsden, Brian G., *Catalog of Cometary Orbits,* Enslow Publishers, Box 777, Hillside, NJ 07205.

** Marsden, Brian G. and Elizabeth Roemer, "Basic Information and References," in Wilkening, Laurel L., editor, *Comets,* Tuscon, AZ, The University of Arizona Press, 1982.

ABBREVIATIONS

arcmin–arc minutes or an angular measure representing 1/60th of a degree.

arcsec–arc seconds or an angular measure representing 1/60th of an arcmin.

AU–Astronomical Unit or the average distance between the sun and Earth (149,597,870 kilometers).

cm–centimeters.

Max. Mag.–maximum magnitude.

Max. Tail–maximum length of the tail.

P–orbital period in years.

q–perihelion distance in AU or the point in the orbit when closest to the sun.

r–the distance from the sun in AU.

T–the date perihelion is reached.

-371   Discovered: Winter
    Aristotle described this comet as first appearing near the sun in the west
and said the tail later extended over one-third of the sky. According to the
Roman philosopher Seneca, the Greek historian Ephorus wrote about this comet's
division into two parts. No formal orbit can be calculated, but Alexandre-Guy
Pingré estimated a very small perihelion in the direction of either Virgo or
Libra and an inclination of about 150 degrees. This rough estimate, as well
as the comet's brightness and visibility near the sun, has made it a possible
candidate for the family of sungrazing comets.

-146   Discovered: August 6
    Discovered in China with a tail nearly 15 degrees long. By the 8th it
was nearly 80 degrees in length and white in color. Although last observed
by the Chinese on August 16, the Europeans kept it under observation for a
total of 32 days. They described it as the size of the sun with a red color,
like fire, "spreading sufficient light to dissipate the darkness of night."
The size and brightness steadily decreased each day as it moved from Scorpius
through Sagittarius to near Zeta Ophiuchi. The perihelion date has been
determined as June 28 (r= 0.38 AU).   ·

-68   Discovered: July 23
    The Chinese discovered this comet in Virgo with a white tail extending
3 degrees to the southeast. It remained visible for several days. The American mathematician Benjamin Peirce estimated a rough orbit indicating the
perihelion occurred in July (r= 0.8 AU).

178   Discovered: September
    This comet first appeared north of Virgo and was observed in China as
it moved into the Hercules-Ophiuchus region. The tail was several degrees
long at discovery and later grew to nearly 70 degrees, as well as taking on
a reddish color. After over 80 days of visibility, it disappeared in Eridanus. The English mathematician John Russell Hind estimated the perihelion
date as occurring in early September (r= 0.5 AU). The inclination was about
18 degrees.

240   Discovered: November 10
    Discovered by the Chinese in southern Scorpius with a tail nearly 30
degrees long. On December 19 it passed through Aquarius and disappeared before month's end. Johann Karl Burckhardt, a German mathematician, calculated
an approximate orbit which gave the perihelion date as November 9 (r= 0.37 AU).
The inclination was 44 degrees.

539   Discovered: November 17
    This comet was discovered in Sagittarius with a tail 2 degrees long.
It was observed in China and Europe and traveled primarily along the ecliptic. At maximum, the tail extended 15 degrees. An approximate orbit by
Burckhardt gave the perihelion date as October 20 (r= 0.34 AU). The inclination was only 10 degrees.

565   Discovered: July 22

This comet was observed by the Chinese in Ursa Major and after 100 days it vanished in the Aquarius region. The tail reached a maximum length of 15 degrees. Approximate orbits by Burckhardt indicated a perihelion date between July 9 and 14 (r= 0.72 to 0.83 AU). The motion was retrograde and on September 10 the comet passed less than 0.5 AU from Earth.

568   Discovered: September 3

This comet was discovered by the Chinese in the "claws" of Scorpius, at which time it was white in color and resembled "loose cotton." It gradually increased in size and at the end of the month appeared "like a piece of cloth" as it passed closest to Earth at a distance possibly as small as 0.13 AU. By October 16 the comet crossed into the Andromeda- Pisces region and had faded and grown considerably smaller. On November 5 it was located 2 degrees north of the Alpha and Beta Arietis region and by the 11th it was no longer visible. There is a possibility that this comet is identical to the "nova" reported by the Chinese in Libra on July 28, since an orbit by E. Laugier in 1846 shows the comet near the position indicated. The orbit also gives a perihelion date of August 29 (r= 0.91 AU). The inclination of only 4 degrees makes this comet a candidate for the family of short-period comets.

574   Discovered: April 4

This bluish-white tailless comet was first seen in Auriga by the Chinese. On May 8 it was in Ursa Major and after June 9 it was no longer visible. At maximum the tail extended 15 degrees. An uncertain orbit by Hind in 1844 gave a perihelion date of April 7 (r= 0.96 AU).

770   Discovered: May 26

This comet was discovered when only 22 degrees from the sun. It was then situated in Auriga in the evening sky and the Chinese described the tail as white and nearly 80 degrees long. The comet's motion was eastward and it was last observed on July 25. Calculations by Hind in 1846 gave the date of perihelion as June 7 (r= 0.64 AU).

868   Discovered: January 21

This comet was discovered in Ursa Major and observed in China, Japan and Europe. During the next few weeks it moved through Triangulum and by Beta Arietis before disappearing after Feburary 15. S. Kanda cal- culated an orbit in 1932, which gave the perihelion date as March 4 (r= 0.42 AU).

961   Discovered: January 28, 962

The Chinese discovered this "guest star" in Ophiuchus and described it as possessing a tail and emitting faint rays. On February 19 it passed near Alpha Librae and upon reaching the Alpha Hydrae region on April 2, it disappeared. Orbital calculations indicate a perihelion date of December 30 (r= 0.54 AU).

1092   Discovered: January 8
     This object's lack of a tail brought it the designation of "guest star"
when discovered by the Chinese in Orion; however, it was decidedly cometary
as it traveled northeastward through Eridanus, Taurus and Aries to reach
the Andromeda-Pisces border on January 30. It was last seen on May 8. Hind
calculated a perihelion date of February 15 (r= 0.93 AU). The inclination
was 29 degrees.

1097   Discovered: October 6
     Between October 6 and 25 a bright comet appeared near Alpha and Gamma
Librae. It was observed in China, Korea and Japan and possessed a white
tail nearly 5 degrees long. On the 9th of October the tail had grown to
nearly 10 degrees as the comet entered the Hercules-Ophiuchus region. On
the 13th the tail extended 50 degrees. Three days later, the comet was
near Iota and Kappa Ophiuchi and on the 17th it was near Alpha Herculi. An
orbital calculation gave the perihelion date as September 22 (r= 0.74 AU).
It also indicated an approach to within 0.17 AU of Earth between October 17
and 25.

1106   Discovered: February 4
     This comet was first seen in Europe as a bright star nearly 2 degrees
from the sun. Apparently then near perihelion, it remained visible nearly
all day, but faded enough to remain lost in the sun's glare in the following
days. On February 7 observers in Constantinople and Palestine saw it in
the southwestern sky after sunset. By February 10 the comet had become
widely visible in China, Korea and Japan in the western sky with an east-
ward-pointing tail nearly 90 degrees long and 5 degrees wide. By February
11 the tail was still over 15 degrees long and on the 12th it extended 10
degrees. The comet remained visible nearly until the middle of March. No
reliable orbit has ever been calculated due to conflicts in observations.
European observations of February 7 indicate the comet's head to have been
south of the ecliptic, and elsewhere in Europe, observers described the
motion in the following days to have been to the northwest. On the other
hand, observers in the Orient indicated a movement consistent with that
expected from a sungrazing comet--an identification that is strengthened
by the comet's physical appearance. Some researchers now believe the
European observations refer to the movement of the comet's tail across
the sky. There is a possibility that this comet may be a previous appearance
of comet 1965 VIII.

1132   Discovered: October 5
     The Japanese first recorded this comet near Lambda Orionis with a white
tail pointing westward for 5 degrees. On the 7th it was located in Aries and
the tail consisted of intense rays extending 50 degrees to the northwest. The
tail had decreased to a length of 15 degrees by October 8. Observations were
made in China, Korea and Japan on the 9th, when it was located near Beta Ceti
and possessed a tail nearly 5 degrees long. The final observation was made on
October 27. An orbit by Ogura in 1917 estimated a perihelion date of August
30 (r= 0.74 AU).

1231   Discovered: December 15, 1230
     This comet first appeared on the Ophiuchus-Serpens border as an "excep-
tional star" to the Chinese. Further observations are scarce until February
1231, when on the 6th it approached to within 0.08 AU of Earth and appeared
as bright as Saturn while situated in Cygnus. Observations were made in China,
Japan and Europe, with the comet being last seen on March 30. An orbit by
Pingré in 1783 linked the December 15 observation with those of February and
March. It also gave the perihelion date as January 30 (r= 0.95 AU). The low
inclination of 6 degrees makes this comet a possible candidate for the family
of short-period comets.

1240   Discovered: January 27
     The Japanese discovered this comet in Pegasus as a reddish-white object
nearly 5 degrees long and pointing southeast. On the 29th it was the "same
size as Saturn" with a tail 6 degrees long. The Chinese first observed it on
January 31 and the next day estimated it as being "as large as Venus" with a
tail 8 degrees long. The comet passed near Alpha and Gamma Pegasi on February
5 and was near the Andromeda-Pisces border on the 13th. It passed into Cassiopeia on the 23rd and finally disappeared on March 31. An orbital calculation by Ogura in 1917 gave the perihelion date as January 21 (r= 0.67 AU).

1264   Discovered: July 14
     This comet was first detected in Europe with a tail nearly 100 degrees
long. After one week it had moved from the evening sky into the morning sky
and was situated in Cancer. Shortly thereafter, it retrograded into Gemini and
remained visible until November. A survey of the orbits calculated for this
comet in the 18th and 19th centuries reveals the difficulty astronomers have
faced when interpreting the observations made in 1264. The perihelion date
varies from July 15 to 20 (r= 0.41 to 0.82 AU), and the inclination falls between 16 and 30 degrees.

1299   Discovered: January 24
     First seen in China and, later, in Europe, this comet was of moderate
size and dark blue in color when it first appeared in the evening sky in
Columba. Its motion was primarily northwestward as it moved into Taurus and
passed Aldebaran and the Pleiades. The comet was last seen in Europe on March
5, but the Chinese managed to continue observations until April 10. Pingré
calculated the perihelion date to be March 31 (r= 0.32 AU).

1337   Discovered: June 26
     This comet was discovered in both China and Europe as it was making its
way through Taurus. It was not considered a very prominent object and its tail
reached a maximum length of only 5 degrees. The comet faded as it moved northwestward through Perseus and Auriga, and between July 15 and 27 it was situated
within 20 degrees of the north celestial pole. It was last detected on August
28. Orbital calculations reveal a perihelion date between June 15 and 22 (r=
0.83 to 0.94 AU).

1351   Discovered: November 24
     Discovered in Andromeda and observed only until the 30th by the Chinese

and Europeans. This comet was not brighter than magnitude 3 and was lost in
moonlight after November. A rough orbit by Burckhardt gave a perihelion date
of November 26 (r= 1.0 AU). The orbit also indicated that the comet was only
0.1 AU from Earth at discovery and when last seen had moved within this distance.

1362   Discovered: March 4
     When first seen by the Chinese near Alpha Aquarii, the tail was about
2 degrees long with a bluish-white color. On March 17 it had moved to a
point near Nu, Tau and Eta Pegasi. By late March, observations were also
being made in Europe where the tail was described as extending 30 degrees to
the west. On April 1 the comet had become involved with morning twilight
causing the tail to become invisible; however, the comet still appeared very
large and possessed a dull, faint color. It was last observed on April 7.
Orbital calculations have been somewhat discordant, with the perihelion date
ranging from February 23 to March 11 (r= 0.27 to 0.46 AU).

1385   Discovered: October 23
     The Chinese and Japanese discovered this comet near the Leo-Virgo border.
On October 30 it entered Crater and possessed a tail over 15 degrees long.
By November 4 it was in Hydra with a tail nearly 30 degrees in length which
swept through the southern constellation Vela. Thereafter, the comet's southern declination rendered it invisible to Northern Hemisphere observers. An
orbit by Hind in 1844 gave a perihelion date of October 16 (r= 0.77 AU).

1402   Discovered: February 8
     This comet first appeared in the southwest after sunset with a short
tail. On February 20 the Koreans reported a tail nearly 8 degrees long
pointing eastward and on the 22nd it was over 15 degrees long. Observations
also came from China, Japan and Europe. Between March 22 and 29 the comet
was visible in broad daylight--a duration matched by no other comet in history. It remained visible until the end of April. In 1877, Hind found an
orbit which explained the comet's track across the sky, but not its extraordinary brightness. The perihelion date was given as March 21 (r= 0.38 AU).
The inclination was near 55 degrees.

1433   Discovered: September 15
     This comet was discovered in Boötes by the Chinese and possessed a tail
nearly 15 degrees long. Soon afterwards, it was found by observers in Korea,
Japan and Poland. It passed near Alpha Ophiuchi on October 12 and was last
seen on November 5 as it neared conjunction with the sun. An orbit calculated by Celoria in 1921 gave the perihelion date as November 8 (r= 0.49 AU).

1449   Discovered: December 20
     Discovered in Ophiuchus by the Chinese when 22 degrees from the sun,
this comet was found less than a week later by observers in Korea and Europe
when the tail was extending nearly 5 degrees. On January 19, 1450 it had
moved into Serpens. The final observation was made by Paolo dal Pozzo
Toscanelli (Florence, Italy) on February 13. In 1921, Celoria estimated the
perihelion date as December 9 (r=0.33 AU).

1457 I   Discovered: January 14
    This comet was observed for only 10 days. It was discovered when close
to Alpha and Gamma Tauri and possessed a tail 1 degree long. Its movement
was to the southeast. Observations were made in China, Japan and Europe.
In 1921 Celoria calculated the comet's perihelion date to be January 18
(r= 0.70 AU) and found the inclination to be near 13 degrees. There has
been much debate as to whether this comet is actually the periodic comet
Crommelin, but the current consensus is that this identity is not possible.

1457 II   Discovered: June 15
    This comet was discovered in Aquarius with a tail 1 degree long. By
June 22 the tail had increased to 15 degrees in length and observers in China,
Korea, Japan and Europe continued observations until August 30. An orbit
by Celoria in 1921 gave the perihelion date as August 8 (r= 0.77 AU). The
inclination was only 7 degrees.

1462   Discovered: June 29
    This "star" was discovered by the Chinese near the Cepheus-Cygnus bor-
der. Its color was darkish white. During the following days it moved
through Draco, Ursa Minor and Camelopardalis as it passed within 10 degrees
of the north celestial pole. On July 16 it was located below Lambda and
Mu Ursae Majoris and gradually faded from sight thereafter. Hind calculated
a rough orbit which gave the perihelion date as August 6 (r= 0.31 AU).

1468   Discovered: September 18
    First discovered about 5 degrees from Alpha Hydrae, observations came
from China, Korea, Japan and Europe. The comet had a bluish tail 30 degrees
long on September 23, which pointed to the southwest. The comet passed near
Eta Ursae Majoris and through Boötes before disappearing after December 8.
An orbit by Laugier in 1846 gave the perihelion date as October 7 (r= 0.85
AU).

1472   Discovered: December 25, 1471
    This celebrated comet was discovered just south of the Virgo-Libra bor-
der and, shortly thereafter, passed near Spica. The comet was observed in
China, Korea, Japan and Europe and by mid-January it possessed a tail over
30 degrees long while located in the limbs of Boötes. On January 21, it
passed within 0.07 AU of Earth and was then traveling nearly 40 degrees per
day across the sky. This close proximity to Earth made it visible in broad
daylight. Thereafter, it moved through Cepheus, Cassiopeia and Andromeda.
On February 17 it was near Alpha and Delta Piscium and it was last observed
on the 21st. Celoria calculated the perihelion date as March 1 (r= 0.49 AU).

1490   Discovered: December 31, 1490
    This comet was first observed by the Chinese in northeastern Cygnus with
a tail pointing to the northeast. By January 2, 1491 it was located near
Iota and Kappa Pegasi. When first observed by the Japanese on January 4 it
was located west of Alpha and Beta Pegasi and possessed a tail nearly 8
degrees long. On January 22 it was situated in the middle of Cetus. The
comet was last observed on February 14 by observers in Korea. Calculations

of the orbit have been somewhat discordant with the date of perihelion varying
from December 24, 1490 to January 5, 1491. The perihelion distance seems
firmly established as near 0.75 AU, but it is unclear whether the comet was
moving direct or retrograde since the inclination was either 51 or 105
degrees.

1499   Discovered: August 16
    This comet was discovered in Ophiuchus and was widely observed in China
and Korea. In late August it moved through the north polar region as it
passed close to Earth. It was last observed on September 6 in Draco. An
orbit by Hind indicated a perihelion date of September 6 (r= 0.95 AU). The
inclination was 21 degrees.

1500   Discovered: April
    First observed in Europe, this comet was widely seen in China, Korea and
Japan beginning in early May. On May 8 the Chinese first observed it near
Beta Piscium and the comet soon afterward passed through Pegasus. On May 18
the Koreans estimated the tail length as 10 degrees and by the 24th the
Japanese estimated it as only 3 degrees. The comet later passed through
Cepheus and Draco before its final observation on July 10. Hind calculated
an approximate orbit in 1861, which gave the perihelion date as May 17 (r=
1.40 AU).

1506   Discovered: July 31
    This comet was discovered in China as a "large sphere" without a tail.
Its northern declination coupled with a right ascension nearly equal to the
sun's made it visible shortly before sunrise and shortly after sunset. At
discovery it was also visible in twilight and had a bluish tint. After several days it developed a tail and passed above Ursa Major. On August 7 the
Japanese recorded the tail at its longest length of nearly 15 degrees and
observers last viewed it on the 14th. An orbit by Laugier in 1846 gave the
perihelion date as September 4 (r= 0.39 AU).

1532   Discovered: September 2
    The Chinese discovered this comet in southern Gemini (near Zeta and
Gamma) with a tail 2 degrees long. It traveled southeast and moved into
Hydra by September 14. On October 2 the tail was 15 degrees long. Observations continued until December 30. Orbits by three mathematicians indicate a perihelion date of either October 18 or 19 (r= 0.51 to 0.61 AU). The
inclination was nearly 30 degrees.

1533   Discovered: July 1
    The Chinese discovered this comet in Auriga and described it as poss-
essing a tail 8 degrees long. On July 7 European observers reported it as
brighter than Jupiter (magnitude -2) and by the 21st the tail extended 15
degrees. The comet was last observed on September 16 in China. Heinrich
Olbers calculated the perihelion date to be June 15 (r= 0.33 AU). The
inclination was 28 degrees.

1556   Discovered: February 27
    Discovered by Heller (Nuremberg) as a "terrifying, prodigious, extra-

ordinary star" in Virgo three days after full moon. It was independently
discovered in China on March 1 when it possessed a tail 2 degrees long.
During mid-March it passed less than 0.08 AU from Earth and possessed a
tail nearly 4 degrees long. The comet was moving through the north polar
region at that time. It was last observed on May 10 by the Chinese. Orbital calculations indicate the perihelion date occurred on either April 21 or
22 (r= 0.50 AU).

1558   Discovered: July 14
     First observed by the Europeans in Leo, this comet became bright, but,
overall, remained fainter and smaller than comet 1556--according to several
observers. In early August the comet passed less than 0.23 AU from Earth
and was independently discovered in Korea on the 8th with a white tail
measuring nearly 8 degrees in length. In September it was visible near
the horizon in twilight and was last seen on the 6th. Orbital calculations
are somewhat discordant, but reveal a perihelion date falling between August
10 and September 14 (r= 0.28 to 0.58 AU). The inclination was about 70
degrees.

1577   Discovered: November 1
     Discovered in Peru nearly 5 days after perihelion (r= 0.18 AU) as an
exceptionally brilliant object. On November 3 observers in London noted
a tail nearly 7 degrees long and on the 8th observers in Japan independently discovered the comet with a white tail nearly 80 degrees long and a
prominent nucleus. The total magnitude was then called "as bright as the
moon." Tycho Brahe observed it on November 13 shortly before sunset at a
magnitude comparable to Venus (-4) and after sunset he reported a tail 22
degrees long with a coma 7 arcmin across. The coma was yellow and the tail
was reddish and curved. On the 14th the Chinese estimated the tail as
nearly 50 degrees long. The comet moved away from the sun at nearly 4
degrees per day when first seen, but later began approaching the sun at
up to one-half degree per day. The magnitude in December was still about
0*and the comet remained visible until January 26, 1578.

1580   Discovered: October 1
     This comet was discovered by the Chinese in the southeastern sky when
less than 0.3 AU from Earth. It was independently discovered on the 2nd,
by Michael Moestlin (Batang), who described the comet as round and brighter
than Venus. Brahe made numerous observations in the evening until November 25 and recovered the comet after the November 28 perihelion (r= 0.60
AU) in the morning sky on December 13. In the last half of November he
estimated the tail as extending over 10 degrees. The comet was last seen
on January 14, 1581.

1582   Discovered: May 12
     This comet was discovered in Gemini by Brahe, who described it as
possessing a tail 13 degrees long. Though he last observed it on May 17,
the Chinese independently discovered it on the 20th and continued observations until barely visible to the naked eye (magnitude 6) on June 9. Orbital calculations indicate a perihelion date of May 6 (r= 0.17 AU).

1585   Discovered: October 13
     Discovered in Aquarius by the Chinese with a tail nearly 2 degrees long,
this comet passed within 0.14 AU of Earth in the next few days and was widely
observed at that time at a magnitude which equalled Venus. As the comet receded from the Earth and sun, its size rapidly decreased. Brahe described
it as resembling the open star cluster M44 in early November and a thin
cloud on November 24. It was last seen on November 27. A parabolic orbit
gives the perihelion date as October 8 (r= 1.09 AU). The inclination was
only 6 degrees. Elliptical orbits were calculated by Laugier and Hind and
produced periods of 5.2 and 15.5 years, respectively, with the latter being
closer to explaining the comet's motion.

1590   Discovered: March 5
     This comet was discovered by Brahe in Pisces and was described as having
a coma 3 arcmin in diameter and a tail nearly 8 degrees long. On the 6th
the tail increased to 10 degrees. Chinese observers also saw this comet for
a short time; however, Brahe made the final observation on March 16. On
March 7 the total magnitude equalled Capella (magnitude 0) and by the 11th
the tail was nearly gone. Orbital calculations have established the perihelion date as February 8 (r= 0.57 AU).

1593   Discovered: July 30   Discovery Magnitude: 3[*]
     This comet was discovered in Gemini by the Chinese, but was observed
extensively in Europe. On August 3 the Japanese estimated the tail as
nearly 15 degrees long. On the 4th Brahe's assistant Christiernus Joannes
Ripensis estimated the comet's brightness as comparable to a 3rd-magnitude
star and the tail was said to extend nearly 5 degrees. The comet was described as fainter then 3rd magnitude on the 16th, but by the 23rd it was
reported to match Alpha Cephei in "light, size and color" (magnitude 2.5).
Although the comet was near its closest approach to Earth (0.49 AU) in late
August it was last seen on September 3 by Ripensis--observations thereafter
being affected by moonlight. In 1747 Nicholas Louis de Lacaille established
the perihelion date as July 19 (r= 0.09 AU).

1596   Discovered: July 19
     This comet may have been discovered by Moestlin on July 11, though no
"hard" data are available to positively identify the object he saw with the
comet. The Koreans first saw the comet on July 19 at a magnitude comparable
to Capella and it was independently found by Brahe on the 24th with a tail
7 degrees long. On August 5 the Chinese described the comet's appearance
as "a large sphere without tail" and the final observation was made during
twilight and moonlight on August 12. Slightly discordant orbital elements
give the perihelion date as being between July 23 and 25 (r= 0.57 AU).

1618 I   Discovered: August 25
     Discovered at Kaschau, Hungary, and independently found two days later
by Johannes Kepler (Linz, Austria) where it rose in the morning sky to the
southeast. On August 28 the Koreans reported a blue-white tail nearly 15
degrees long and by September 1 Kepler described it as 5 degrees in length.
By September 6 the tail had vanished, but the comet remained visible in a

telescope until September 25. An orbit by Pingre in 1784 indicated a perihelion date of August 17 (r= 0.51 AU). The inclination was 21 degrees.

1618 II   Discovered: November 16

   This comet was discovered in the morning sky in Libra by several observers
as a tail projecting above the horizon. As the northerly motion continued, the
comet became visible in the Northern Hemisphere in late November. On
December 10 the Danish astronomer Christianus Longomontanus estimated the tail
to extend 104 degrees and on the 12th a Jesuit mathematician,Horatio Grassi,
estimated it as 60 degrees long. On the 13th, the comet passed into Boötes.
A reddish hue was present in early December, when the comet passed closest to
Earth (0.3 AU), and at one time it was visible in broad daylight. The final
observation came on January 22, 1619. In 1805 Friedrich Wilhelm Bessel
established the perihelion date as November 8 (r= 0.39 AU). The inclination
was 37 degrees.

1652   Discovered: December 16   Discovery Magnitude: 2*

   First seen in Orion by observers in Brazil and detected soon afterward
in Europe and China. The comet was closest to Earth (0.13 AU) on December
19. On the 20th Johannes Hevelius (Danzig) described the comet as the
size of the moon (30 arcmin) and composed of minute stars. After discovery
the comet faded, while the tail reached a maximum of 8 degrees in mid-December. On December 23 it was located one degree from the Pleiades. By
January 1, 1653 the comet had faded to a magnitude near 4*and the final
observation came on the 10th. In 1705 Edmund Halley calculated the perihelion date as November 13 (r= 0.85 AU).

1661   Discovered: February 3

   The comet was discovered with a tail 6 degrees long in the morning
twilight in Europe. Hevelius later observed this comet extensively with his
telescope and reported multiple structure in the nucleus up until February 20.
Afterwards, the comet's decreasing size and brightness made detailed observations difficult, especially with a full moon on March 16. The comet was last
seen on March 28. In 1785, Pierre F. A. Mechain established the perihelion
date as January 27 (r= 0.44 AU). The inclination was 33 degrees.

1664   Discovered: November 18   Discovery Magnitude: 4*

   This comet was discovered in Spain nearly three weeks before its December 4 perihelion passage (r= 1.03 AU); however, continuous observation did
not begin until the first days of December, when it was independently discovered in northern Europe at about magnitude 2*. During the comet's closest
approach to Earth in mid-December, the total magnitude reached 1 and the tail
extended nearly 20 degrees. By late December the tail reached a maximum
length of 37 degrees. Hevelius made numerous observations as the comet
exited the sun's vicinity and by mid-January, he reported a magnitude near
3.5 and a tail 20 degrees long. By February 18 the comet was no longer
visible to the naked eye, but telescopic observations continued through
March 20, 1665.

1665   Discovered: March 27
    At least four independent discoveries occurred within five days during
late March as this comet appeared in the morning sky. Hevelius began observations on April 6, when the comet was in Pegasus, and reported it being
visible to the naked eye in moonlight with a tail nearly 20 degrees long.
Between April 10 and 13 observers reported a tail 30 degrees in length. The
comet was last seen on April 20 low over the horizon with the sun just minutes away from rising. The magnitude must then have equalled Venus. In
1705 Halley calculated the perihelion date as April 24 (r= 0.11 AU).

1668   Discovered: March 3
    First detected by observers at the Cape of Good Hope (South Africa),
this comet was extensively observed in the Southern Hemisphere. Six days
after discovery a tail 30 degrees long was reported in Rome and between the
10th and 12th of March it reached a maximum length of 37 degrees. The
most accurate observations came from Goa, India, where Gottignies studied
the comet daily between March 9 and 17. The comet was last seen on March
30. In 1901 Carl Heinrich Friedrich Kreutz calculated the perihelion date
as February 28 (r= 0.07 AU). There is a possibility that this comet was
a member of the sungrazer family, in which case the perihelion date would
be March 1 (r= 0.01 AU).

1672   Discovered: March 2
    Discovered only 19 degrees above the horizon, in strong twilight, by
Hevelius, this comet was described as possessing a tail 2 degrees long.
Fading came slowly and by March 26 observers in Paris still estimated the
magnitude as near 3. The final observation was made by Hevelius on April 21
when he viewed it with his telescope near magnitude 5.5[*] Orbital calculations
revealed a perihelion date of March 1 (r= 0.70 AU).

1677   Discovered: April 27
    Hevelius discovered this comet a short distance above the horizon in
morning twilight and estimated the tail length as 2 degrees. During the
next several days, observers in Paris, as well as Danzig, caught only glimpses
of the comet due to cloudiness, but these observations indicated the comet was
growing brighter as it neared the sun. On May 2 it became widely observed in
Europe in both the morning and evening sky due to its northern declination and
a right ascension which nearly equalled the sun's. Paris observers saw it 4
degrees above the horizon that morning and, in the evening, Hevelius saw it
3 degrees above the horizon with a thin tail extending into Cassiopeia. On
May 3 Hevelius reported a tail 3 degrees long and the next day John Flamsteed
estimated it as 6 degrees in length. The comet was last seen on May 8 with
the aid of a telescope after the stars and Jupiter had vanished in the morning twilight (magnitude -3). An orbit by Halley in 1705 gave the perihelion
date as May 6 (r= 0.25 AU).

1678   Discovered: September 11
    Philippe de La Hire (Paris) discovered this comet and continued observa-
tions until October 7, when the magnitude had faded to near 6. Parabolic
orbital calculations indicate a perihelion date of August 27 (r= 1.24 AU).

The inclination was only 3 degrees. The comet also was near its closest
approach to Earth (0.29 AU) when discovered. There has been much debate on
this comet's identity with the periodic comet De Vico-Swift; however, current
studies disprove this link.

1680   Discovered: November 14   Discovery Magnitude: 4
     During observation of the moon and Mars in the morning sky, the German
astronomer Gottfried Kirch made the first telescopic comet discovery in
history. This comet then possessed no tail, but two days later, Kirch noted
the first traces emanating from the coma. On November 21 the tail extended
one-half degree and the coma was comparable in magnitude to Regulus (magnitude 1.5). On December 4 the tail was clearly 15 degrees long and the
total magnitude was near 2*. Soon afterwards, the comet was lost in the sun's
glare. On December 18 it was at perihelion (r= 0.006 AU) and in late December
the comet reappeared in the evening sky with a tail 70 to 90 degrees long and
a magnitude near 2*. During this time, P. J. de Fontaney studied the comet
through a telescope, noting the transparent nature and commenting on how the
nucleus seemed to be absent at times. He added that when the nucleus was
present it occasionally appeared multiple. By January 10, 1681 the tail had
shrunk to 55 degrees and 13 days later, when the comet had reached magnitude
4*, the tail was 30 degrees long. After February 11 observations of the tail
were restricted to observers with telescopes and by the 18th the tail was
seen for the final time. The last observation of the comet was made on March
19, when located 2.2 AU and 2.5 AU from the sun and Earth, respectively.

1683   Discovered: July 21   Discovery Magnitude: 3
     Discovered by John Flamsteed, first Astronomer Royal of England, and
shortly thereafter by Francesco Bianchini (Verona), in the north polar region.
Hevelius began observations on the 30th and continued them until September
4. He noted that the head of the comet became fainter and larger during its
time of visibility and that no nucleus was ever observed. During August the
tail reached a maximum length of 4 degrees, while the brightness was near
magnitude 4. The comet was last seen on September 5 by Halley. Orbital
calculations indicate a perihelion date of July 13 (r= 0.56 AU). The comet
drew closer to Earth until nearest (0.3 AU) on about the final day of observation.

1684   Discovered: June 30
     Discovered in Corvus by Bianchini (Rome) when at perigee (0.20 AU) at
a magnitude probably between 2 and 3.* This comet never possessed a visible
tail and by July 19 it had faded to magnitude 5*. Telescopic observations continued until early August. The perihelion date has been established as June
8 (r= 0.96 AU).

1686   Discovered: August
     This comet was a very conspicuous object to observers in the southern
latitudes during August. It was first seen in Para, Brazil, and shortly
thereafter, by French Jesuit missionaries in Siam. The comet was closest to
Earth on August 17 (0.32 AU) and then possessed a nucleus equal to a first-
magnitude star with a tail 18 degrees long. Though the comet moved rapidly

away from Earth it moved towards the sun and in early September was still near magnitude 1[*] and was white in color. On September 17 the magnitude was slightly fainter than 2[*] and by the 20th it was equal to Delta Leonis (magnitude 2.6). It was last observed on September 22. Orbital calculations place the perihelion date at September 16 (r= 0.34 AU). The inclination was 32 degrees.

1689   Discovered: December

Discovered at sea in the first days of December, Pere Richaud (Pondicherry, India) made his first observation on December 9. By December 14 the total magnitude was between 3 and 4 and the strongly curved tail extended 68 degrees. The comet was also observed in Japan during this time. The final observation came in the first days of January, 1690. A parabolic orbit gives the perihelion date as November 30 (r= 0.06 AU). An association with the sungrazing family is questionable, but gives the perihelion date as December 2 (r= 0.01 AU).

1695   Discovered: October 28

Jacob (Brazil) first saw the tail of this comet rising above the horizon on October 28. Two days later, Bouvet (Surat, India) independently discovered the comet and described the tail as 30 degrees long. Observations began in China on November 2 when observers saw a tail extending 15 degrees above the horizon. The comet rose higher into northern latitude skies each day and by November 6 it was seen with a tail 40 degrees long. The comet faded noticeably thereafter and was last seen on November 19--shortly before full moon. Orbital calculations vary considerably with the perihelion dates ranging from October 24 to November 19 (r= 0.04 to 0.84 AU). An assumption that this comet belongs to the Kreutz sungrazing family produces a perihelion date of October 23 (r= 0.01 AU).

1698   Discovered: September 2   Discovery Magnitude: 3

This short-tailed comet was discovered by La Hire (Paris) between Beta and Kappa Cassiopeiae. It was thereafter observed exclusively by La Hire and Giovanni Cassini as it reached a maximum brightness of 2nd magnitude on September 8, when in Cepheus. The comet moved rapidly to the south and passed only 0.23 AU from Earth on the 9th. It was seen for the final time between Xi and Psi Scorpii on September 28. An accurate orbit by Hind in 1876 gave the perihelion date as October 17 (r= 0.73 AU).

1699   Discovered: February 17

Discovered by Fontaney (Paris) in the evening sky during the comet's closest approach to Earth (0.17 AU). The comet faded thereafter, but remained visible to the naked eye until March 2. Overall, the comet remained tailless, though Cassini did observe a short tail at one time. The comet was last seen on March 6. An orbit by Hind in 1879 gave the perihelion date as January 13 (r= 0.75 AU).

1701   Discovered: October 28

First seen in Crater before sunrise by Pallu (Pau, France) and independently by A. Thomas (Peking, China). In the first days of November the magnitude was near 3 and the tail was at least 4 degrees long. The comet was

last seen on November 11 by Thomas. An orbit by Burckhardt in 1809 gave
the perihelion date to be October 17 (r= 0.59 AU).

1702 I   Discovered: February 20
     The tail of this comet was first seen in the evening sky at the Cape
of Good Hope. Observations thereafter were almost exclusively made in the
Southern Hemisphere, the only exception being an observation of the tail by
Jacques Maraldi (Rome) at the end of February and in early March. The tail
reached maximum on February 28 when observers in the Gulf of Bengal estimated it as 43 degrees long. The final observation of the comet came shortly
after March 2. Accurate observations are few and no mathematician has
succeeded in calculating a definitive orbit. Kreutz found that the typical
sungrazer orbit fit the rough positions obtained on February 27, 28 and
March 2 reasonably well, if the perihelion date is assumed to be February
15.1 (r= 0.01 AU).

1702 II   Discovered: April 20
     Within a few hours on the morning of April 20 this comet was independ-
ently discovered by three observers: La Hire (Paris), Bianchini (Rome) and
Maraldi (Rome). The next morning G. Kirch (Berlin) also found the comet.
All these discoveries were made with the naked eye and each reported a short
tail. On May 5 the waning moon caused observations of this comet to cease
as it had rapidly faded below naked-eye visibility. In 1807, Burckhardt gave
the perihelion date as March 14 (r= 0.65 AU). His orbit also showed that
the comet had passed only 0.04 AU from Earth on April 19. The orbital inclination of only 4 degrees has caused some astronomers to suggest a possible
short-period orbit.

1706   Discovered: March 18
     This comet was discovered by Cassini in Corona Borealis when 0.31 AU
from Earth. It was then described as resembling the Andromeda Nebula to the
naked eye. The comet was lost in the moonlight between March 28 and 30, but
naked-eye observations resumed on the 31st and continued until April 13. The
comet was last seen with the aid of a telescope on April 16. In 1753, Struyck
calculated the perihelion date to be January 30 (r= 0.43 AU).

1707   Discovered: November 25
     This comet was discovered with the naked eye by Manfredi as "a nebulous
star of Jupiter's size" with a small tail. On the 28th Cassini and Maraldi
independently discovered the comet from Paris. It remained visible to the
naked eye throughout December and was last observed on January 23, 1708.
In 1753, Struyck calculated the perihelion date to be December 12 (r= 0.86 AU).
At discovery the comet had been only 0.19 AU from Earth.

1718   Discovered: January 18
     This comet was discovered by Christfried Kirch when only 0.11 AU from
Earth as it moved nearly 2 degrees per hour across Ursa Minor. Kirch compared the comet's brightness to Beta Ursae Minoris (magnitude 2.1) and found
the comet brighter by nearly one magnitude. In his telescope he could find
no nucleus or tail, but estimated the coma to be nearly 11 arcmin across.

Bad weather prevented Kirch from further observing this comet until January 21, at which time he found the magnitude to be hardly brighter than 4 and the coma diameter to have decreased to 7 arcmin. The rate of motion was then estimated as only 20 arcmin per hour. Two days later Kirch found a magnitude of 5 and a coma diameter of 5 arcmin. The comet was last observed on February 5, when the magnitude was near 8.* In 1829 Friedrich Wilhelm Argelander computed the perihelion date to have been January 15 (r= 1.03 AU).

1723   Discovered: October 10
    Kogler and Gian-Priamo, both of Peking, independently discovered this comet during its closest approach to Earth (0.17 AU) and described it as possessing a tail 4 degrees long. On October 15 the tail had increased to a length of nearly 8 degrees. Independent discoveries were made on October 16 by observers in Lisbon, and on October 17 by Sanderson (Bombay, India). The comet was visible to the naked eye in the morning sky up until November 16, and thereafter it could only be observed using telescopes. It was last seen in England on December 18. Orbital calculations gave the perihelion date as September 28 (r=1.00 AU).

1729   Discovered: August 1   Discovery Magnitude: 5
    This comet was first detected by Sarabat (Nismes, France), an ecclesiastic of the Jesuit Order, when barely visible to the naked eye. Due to increasing moonlight, the comet was not again seen by Sarabat until a total eclipse of the moon on August 8. Soon afterward, Jacques Cassini (Paris) was notified and he began observations on the 26th, when he reported a diameter of 1.5 arcmin. The comet slowly faded and was observed until it disappeared in evening twilight on January 21, 1730--the magnitude then being near 8.* Orbital calculations gave the perihelion date as June 16, 1729 (r= 4.05 AU). The comet's motion was so slow, due to its great distance, that it traversed only 15 degrees during 6 months of observation. The inclination was 77 degrees.

1737 I   Discovered: February 6
    First detected in Jamaica, independent discoveries were made February 7 in Philadelphia and on the 9th in Lisbon and Gibraltar. The comet's bright tail was then 7 degrees in length. Observers in Paris on the 16th estimated a tail nearly 3 degrees long with a total magnitude of 2. By the 22nd the comet appeared as a nebulous star "of Venus' size" and possessed a tail 2 degrees long. Naked-eye observations ceased on March 18 and the comet was last seen on April 6 by observers in Peking. Orbital calculations revealed a perihelion passage on January 30 (r= 0.22 AU). The inclination was 18 degrees.

1737 II   Discovered: July 2   Discovery Magnitude: 3
    This comet was discovered by Kogler (Peking) with the naked eye and independently on July 7 elsewhere in China. In a 7.5-cm telescope, Kogler described the comet as round, tailless and "larger than Jupiter." He last saw the comet on July 10 when moonlight and subsequent bad weather prevented further observations; however, the comet continued to be observed by other observers in China until July 16, at which time the comet should have

been at a declination of about -12 degrees. There is a possibility that this
comet was a previous return of comet 1862 III, though an investigation of the
orbit fails to decide the matter fully. It is likely that the perihelion fell
between June 3 and 16 (r= 0.84 to 1.02 AU). The inclination was between 36
and 113 degrees.

1739   Discovered: May 27
    Discovered with the naked eye by Zanotti (Bologna, Italy), this comet was
described as a nebulous star with a tail 2 degrees long. During July the tail
was only 1 degree long. By August 18 the comet could not be seen with the
naked eye, but was detected with a telescope. Observations thereafter were
prevented by moonlight. Orbital calculations gave the perihelion date as
June 17 (r= 0.67 AU).

1742   Discovered: February 6
    This comet was first seen at the Cape of Good Hope as it neared its
February 8 perihelion (r= 0.77 AU). In late February sailors in the Indian
Ocean independently discovered the comet, and on March 2 Pereyra (Peking,
China) made the first discovery of this comet in the Northern Hemisphere.
Shortly thereafter, Grant, an amateur astronomer in Ireland, discovered the
comet as it sat low over the horizon and described it as about magnitude 1,
with a tail nearly 5 degrees long. In late March the comet was still an easy
naked-eye object at a magnitude of 3*. The comet was last seen on May 9 with
the aid of a telescope. For a long time, astronomers believed this comet
was an earlier return of 1907 II, but in 1974 Brain G. Marsden showed that
this identity "seems to be highly improbable." Perihelion occurred on
February 8 (r= 0.77 AU).

1743 II   Discovered: August 18
    This comet was discovered by Klinkenberg (Haarlem, Netherlands), who
continued to observe the comet until its final observation on September 13.
Struyck (Amsterdam, Netherlands) observed the comet on August 21 with the
naked eye and indicated it as being between magnitude 3 and 4, with a short
tail. Klinkenberg and, later, other mathematicians established the perihelion
date as September 21 (r= 0.52 AU).

1744   Discovered: November 29, 1743   Discovery Magnitude: 4
    Though it now appears that this comet was observed before December,
the official discoverers for many years were Klinkenberg, on December 9, and
Philippe Loys de Cheseaux (Lausanne, Switzerland), on December 13. The comet
was found in the evening sky with the naked eye, and then possessed a short
tail. By mid-January 1744 the comet was widely observed as a 1st-magnitude*
object with a 7-degree tail when still over a month from perihelion. On
February 1 it was brighter than Sirius (hence, a magnitude of about -2) and
possessed a curved tail 15 degrees long, and on the 18th it matched Venus in
brightness (magnitude -4) and possessed a double tail which extended 7 and
24 degrees across the sky. The comet attained a magnitude of near -7* on
February 27, and observers in Europe reported it visible in broad daylight
when only 12 degrees from the sun. Perihelion came on March 1 (r= 0.22 AU).
The observers reported that it reappeared on March 6 with a spectacular tail

system, which persisted until the 9th.   The broad tail then consisted of from
5 to 11 distinct rays and the depicted appearance resembled a Japanese hand fan.
On March 18, the tail extended nearly 90 degrees in length, and thereafter,
fading was rapid as the comet moved quickly away from both the sun and Earth.
It was last seen on April 22.

1747   Discovered: August 13, 1746
     Discovered by de Cheseaux as a faint object with a coma nearly 5 arcmin
across and a tail 24 arcmin long, this comet had a large perihelion distance
of 2.2 AU and changed little in appearance during 4 months of observations.
When last seen in Paris on December 5 it was still barely visible to the
naked eye.   The comet passed perihelion on March 3, but was then located on
the opposite side of the sun from Earth.

1748 I   Discovered: April 25
     This comet was discovered in Peking by Hallerstein and Gaubil in the
morning sky with a tail over 1 degree long.   It passed perihelion on April
29 (r= 0.84 AU) and was independently discovered the next day with the
naked eye by J. D. Maraldi (Paris), who described the tail as 2 degrees
long.   Maraldi reported the comet to be brighter than the Andromeda Nebula
on May 9 (hence, a magnitude near 4).   During the early days of observation
a nucleus could be seen with a telescope, but as the comet faded, the nucleus
could no longer be detected.   In late June the comet dropped below naked-eye
visibility and observations finally ceased on June 30.

1748 II   Discovered: May 19
     This comet was discovered and observed only by Klinkenberg and, besides
the date of discovery, was seen only on May 20 and 22.   During the time of
observation, the comet was between 0.45 and 0.47 AU from Earth and was heading
towards a perihelion date of June 19 (r= 0.63 AU).

1750   Discovered: February 1   Discovery Magnitude: 2.5
     Discovered by Wargentin (Uppsala, Sweden) to the west-southwest soon
after sunset, this comet was described as appearing as bright as Epsilon
Pegasi (magnitude 2.5) with a tail 2 degrees long.   Wargentin and a colleague
were the only observers of the comet during the short time it was seen.   After
a second observation on the 2nd the comet was lost to cloudy skies until a
partial clearing on the 5th gave the observers their final glimpse.   When
skies cleared on the 8th and 9th, the comet's motion of nearly 5 degrees
per day had taken it too close to the sun for further observation.   Marsden
calculated a very uncertain orbit in 1973, which gave the perihelion date as
February 23 (r= 0.20 AU), and indicated a close approach of 0.25 AU from
Earth near the time of observation.

1757   Discovered: September 11   Discovery Magnitude: 5.5
     First discovered by Gärtner (Dresden, Germany) in the morning sky, this
comet was independently found by James Bradley (Greenwich, England) on September 12 and Klinkenberg on the 16th.   Bradley described the comet as a dull
star of magnitude 5 or 6 when first seen, with a short tail and a nucleus
visible in a telescope.   In October the comet reached magnitude 2[*]and poss-

essed a bright tail. The comet was visible 20 degrees from the sun at
Greenwich on the 18th of October, but was last seen at Marseilles on the
27th. An orbital calculation by Bradley soon afterward, gave the perihelion
date as October 21 (r= 0.34 AU). The inclination was 12 degrees.

1758   Discovered: May 26   Discovery Magnitude: 2.5
     This comet was discovered by de la Nux (Bourbon Island, Indian Ocean)
with the naked eye and was described as possessing a tail 1.5 degrees long
and a bright nucleus. Observations steadily became more difficult thereafter
as the comet approached the sun. On June 7 it was observed in bright
twilight when only 5 degrees above the horizon and the brightness was comparable to Alpha Orionis (magnitude 0). On June 8 the comet was seen when
only 7.5 degrees from the sun and must then have been at least near magnitude
-3. Perihelion came on June 11 (r= 0.22 AU), and the comet was recovered on
the 18th, when it was seen with the naked eye when only 18 degrees from the
sun. The comet remained at naked-eye visibility throughout July, and by
August 28, it was said to equal the Crab Nebula in brightness (magnitude
8.5). The comet was last observed on November 2 by Charles Messier.

1759 II   Discovered: January 26, 1760   Discovery Magnitude: 5.5
     Discovered in the morning sky by Messier (Paris) using a telescope, this
comet was visible, with difficulty, to the naked eye and possessed a tail
3 degrees long. The comet was near magnitude 4*when observed on February
5, and by the 10th the tail extended nearly 5 degrees. Naked-eye visibility
continued until early March and the comet was last seen on March 18 by
Messier. The comet had passed perihelion on November 27, 1759 (r= 0.80 AU)
and was discovered when only 0.46 AU from Earth.
(Messier)

1759 III   Discovered: January 7, 1760
     This rapidly moving comet was first seen by observers in Lisbon with the
naked eye. The motion was greatest (1 degree per hour) on the 8th, as the
comet passed only 0.07 AU from Earth and independent discoveries were then
made in Paris by Messier, Cesar Francois Cassini de Thury, J. D. Maraldi and
Jean Chappe d'Auteroche, and in Chelsea, England, by Samuel Dunn. These
observers described the comet as a nebulous object 30 arcmin across with a
nucleus of magnitude 2 and a tail 4 degrees long. Fading was rapid and by
January 9 the nucleus was no brighter than magnitude 4. Naked-eye visibility ended on January 17 and the comet was last seen on February 11. Orbital calculations indicated a perihelion date of December 17 (r= 0.97 AU).
(Great Comet)

1762   Discovered: May 17
     This comet was discovered by Klinkenberg (The Hague, Netherlands), who
described it as possessing a tail 15 arcmin long. Perihelion came on May
28 (r= 1.01 AU) and observers said it then resembled a star of the 4th magnitude. Naked-eye visibility persisted through the first half of June and
the comet was observed telescopically until July 12.
(Klinkenberg)

1763   Discovered: September 28
     This small, tailless comet was discovered when only 0.16 AU from Earth
and was first seen by Messier, who described the comet as bright, but appar-

ently not visible to the naked eye. Although the comet rapidly moved away from Earth after discovery, fading was slow since perihelion was due on November 2 (r= 0.50 AU). On October 28 the total magnitude was near 6, while the nuclear magnitude was close to 7. The comet was last seen on November 25 at a magnitude between 8 and 9.
(Messier)

1764   Discovered: January 3   Discovery Magnitude: 3.0
     This comet was at its closest point to Earth (0.29 AU) when discovered by Messier with the naked eye. He then described the comet as 14 arcmin in diameter with a tail 2.5 degrees long and a brightness which equalled Delta Draconis. When observed on January 11 the magnitude had faded to about 4 and the tail had decreased to 2 degrees. The comet experienced an increase in brightness between January 20 and 29, but afterwards the magnitude returned to normal and the comet was last seen in evening twilight on February 11 when the nucleus appeared very bright with almost no coma. Perihelion came on February 13 (r= 0.56 AU).
(Messier)

1766 I   Discovered: March 8   Discovery Magnitude: 6.0
     Having already passed perihelion on February 17 (r= 0.51 AU) this comet was discovered by Messier on March 8, in twilight as it was approaching conjunction with the sun. An observation on the 11th allowed the magnitude to be estimated as near 6. The comet was last observed in twilight as a very difficult object low over the horizon.
(Messier)

1769   Discovered: August 8   Discovery Magnitude: 5.5
     While sweeping for comets with a small telescope during the evening hours of August 8, Messier (Paris) discovered this comet as a nebulosity several arcmin in diameter between the stars 24, 29 and 31 Arietis. On August 15 it was visible to the naked eye and possessed a tail 6 degrees long. Messier measured the tail as 34 degrees long on August 31, and on September 5 he described it as strongly curved with a length of 43 degrees. Pingré was at sea between Teneriffe and Cadiz on September 11 and reported the tail as 90 degrees long, "but so faint at its extremity that when Venus rose above the horizon, its light shortened the tail by several degrees." Another observer, de la Nux, estimated the tail as 98 degrees long on this same date, which was very near the time of the comet's closest approach to Earth (0.35 AU). The tail shrank to 60 degrees on the 13th and to almost nothing when last seen in twilight near the horizon on the 26th. Perihelion came on October 8 (r= 0.12 AU) and the comet was recovered to the left of Arcturus in the evening sky by Messier on October 24. It was then faintly visible to the naked eye and had a tail 2 degrees long. On the 26th the nucleus was of magnitude 3 and the tail was 3 degrees long. On November 3 the tail extended over 6 degrees. The comet's appearance on November 20 reflected that at discovery and on December 1, telescopes still revealed a tail 1.5 degrees in length. The comet was last seen on December 3.
(Messier)

1770 II   Discovered: January 9, 1771 Discovery Magnitude: 5.0
     This comet was discovered by de la Nux (Bourbon Island) and de la Grand (Milan) while it was passing only 0.19 AU from Earth. The next evening independent discoveries were made by Messier and R. G. Boscovich (Milan). The

comet was then visible to the naked eye, with a tail nearly 6 degrees long and a coma 18 arcmin in diameter. By January 16 the comet was no longer visible to the naked eye but observers still described the tail as 4 degrees long and the nucleus as magnitude 7 using telescopes. The final observation was made on January 20. An orbit by Pingré in 1784 gave the perihelion date as November 22, 1770 (r= 0.53 AU).
(Great Comet)

1771 Discovered: April 1

Messier discovered this comet with the naked eye and then noted that the nucleus was as bright at Epsilon Arietis (magnitude 4.5). The comet was then located low over the horizon, but climbed higher into the sky each day thereafter. On April 7 the tail was estimated as 3 degrees long. The comet varied little in brightness as it approached its April 19 perihelion (r= 0.90 AU) and due to a decreasing comet-Earth distance it was still near magnitude 4.5* when observed with the naked eye during the full moon of April 28. The comet was followed until July 20. Though the orbital elements are well established, mathematicians are still undecided as to whether the comet is traveling in a hyperbolic orbit or an orbit which is slightly elliptical.
(Messier)

1773 Discovered: October 13 Discovery Magnitude: 4.5*

This comet passed perihelion on September 5 (r= 1.13 AU), but remained in conjunction with the sun until shortly before Messier's discovery on October 13. On that evening Messier found the comet in bright twilight and described its appearance in his telescope as 4 arcmin in diameter, with a tail 8 arcmin long. During the last half of October, observers obtained twilight-free views of the comet, which showed the tail to be at least one-half degree long. Thereafter, the comet faded slowly, with the magnitude dropping from 5.5* on November 14 to 7.5* on January 11, 1774. The comet remained under observation until April 14 when the distances from the sun and Earth were 3.26 and 2.92 AU, respectively.
(Messier)

1774 Discovered: August 11 Discovery Magnitude: 6.5*

During a telescopic search for comets, Jacques Leibax Montaigne (Limoges, France) discovered this object when only 4 days from perihelion (r= 1.43 AU). After perihelion the comet brightened slowly as it approached the Earth to within 0.6 AU in late September. It reached a maximum magnitude of 5.5* at that time and by October 9 it had already faded to nearly 6.5*. The comet was last observed on November 8.
(Montaigne)

1779 Discovered: January 6 Discovery Magnitude: 5.5*

This comet passed perihelion on January 4 (r= 0.71 AU) and was discovered two days later by Johann Elert Bode (Berlin), who described it as a bright telescopic object with a tail 20 arcmin long. When independently found by Messier on the 19th, the tail had increased to a length of 33 arcmin. The comet reached naked-eye visibility for a few days in late January as it passed within 0.5 AU of the Earth, but during most of February, observers reported it visible only in telescopes. On March 21 Messier said the comet's brightness equalled that of the globular cluster M3 (magnitude 6.4) and by April 22 he said it equalled the elliptical galaxy M49 in both size and brightness (4 arcmin across and of magnitude 8.6). Messier discovered many

new objects for his famous catalog of clusters and nebulae during the comet's
trek through Virgo in May. In particular, on the 5th and 6th, the galaxy M61
was mistaken for the comet and it was not until the 11th that Messier
realized his mistake and listed the galaxy as a new discovery. The comet's
magnitude at that time was near 10*. The comet was last seen on May 17.
(Bode)

1780 I   Discovered: October 27   Discovery Magnitude: 7.0
     This comet passed perihelion on October 1 (r= 0.10 AU), but remained
behind the sun until discovered by Messier on October 27. At that time it
was visible only with a telescope in the morning sky. Although still in
twilight on November 7, the comet could be seen with the naked eye and possessed a diameter of 8.5 arcmin. Thereafter, Messier noted an unusually rapid
decrease in brightness, with the comet being described as very faint on November 26 and invisible after December 4.
(Messier)

1780 II   Discovered: October 18   Discovery Magnitude: 6.0
     This comet was discovered independently by Montaigne (Limoges) and Olbers
(Göttingen) when only 15 degrees above the horizon. Observing conditions
worsened each day thereafter as the comet moved closer to the sun and obser-
vations ceased after October 26. Olbers calculated the perihelion date as
November 28 (r= 0.52 AU).
(Montaigne-Olbers)

1781 I   Discovered: June 28   Discovery Magnitude: 7.0*
     This comet was discovered by Mechain (Paris) in Ursa Major and was then
described as a tailless coma 3 arcmin across. By the 30th, more precise estimates of brightness gave it a magnitude comparable to nearby M92 (magnitude
6.1). The comet continued to brighten and even developed a tail which was
visible in moonlight on July 5, but after the comet reached perihelion on
July 7 (r= 0.78 AU), it became a difficult object as it entered the sun's
glare. The southern motion caused it to be lost after July 17.
(Mechain)

1781 II   Discovered: October 9   Discovery Magnitude: 8.0
     Observing near Delta Cancri, Mechain discovered this comet and described
it as slightly fainter than the galaxy M81. On October 11, Messier estimated
the brightness as equal to M79 (magnitude 8.4) and 4 days later, the first
traces of tail had become visible. As the comet neared the Earth, a rapid
brightening as well as an increase in size became apparent. On October 23,
the comet was equal in size (12 arcmin) and brightness (magnitude 6.3) to the
globular cluster M2, according to Messier, and on November 7, when the comet
was only 0.25 AU from Earth, the coma was estimated as near 20 arcmin across.
Also, on the 7th, observers described the comet as an easy naked-eye object
(magnitude 3 or 4) with a tail 4 degrees long. Fading was rapid thereafter
and the comet was last seen on December 26. Orbital calculations revealed a
perihelion on November 29 (r= 0.96 AU).
(Mechain)

1784   Discovered: December 15, 1783   Discovery Magnitude: 5.0
     This comet was first detected by de la Nux (Bourbon Island) and was
visible to the naked eye in Paris a short time later. Perihelion came on
January 21 (r= 0.71 AU), and around that time observers in Brazil estimated
the tail length as 6 degrees. Bright moonlight prevented naked-eye observa-

tions in early February, but on the 10th, the tail was still seen extending 2.5 degrees. Fading was rapid in late February, and by May, the comet was described as a diffuse nebulosity with a diameter of 3 arcmin. It was last seen on May 26.
(Great Comet)

1785 I   Discovered: January 7   Discovery Magnitude: 6.5

This telescopic comet was independently discovered by Messier and Mechain on the same day. Both described the comet as a faint object (in telescopes with apertures of only 3 cm) with central condensation, but no trace of tail. Messier's observations ceased on the 16th, but Mechain continued to study the comet until February 8, when it was only 9 degrees above the horizon. Orbital calculations reveal a perihelion date of January 27 (r= 1.14 AU).
(Messier-Mechain)

1785 II   Discovered: March 11   Discovery Magnitude: 7.0

Discovered in the evening sky by Mechain, this comet was lost in the sun's glare after the 22nd, only to reappear in the morning sky on the 30th. On April 4, the tail extended 5 degrees and observers remarked on the very prominent nucleus. The next morning, the tail was 8 degrees long. The comet was last seen on April 17, when low over the horizon, but very bright despite the dawn light. Perihelion came on April 8 (r= 0.43 AU). A long-period orbit of near 1,300 years is likely.
(Mechain)

1786 II   Discovered: August 1   Discovery Magnitude: 7.5

During a routine search for comets, Caroline Herschel came upon a round, diffuse object similar in brightness to the planetary nebula M27 (magnitude 7.6). A haziness across the sky prevented her from firmly establishing its cometary nature until the next evening. The comet became visible to the naked eye on the 17th, and the next evening, Messier detected a tail 1.5 degrees long using a telescope. On August 19, William Herschel (Caroline's brother) described the comet as "considerably brighter" than M3, thus establishing a magnitude between 5 and 6. He also detected a "very faint, scattered light towards the north" which extended nearly 4 arcmin. The comet was last seen on October 26. Perihelion had occurred on July 8 (r= 0.41 AU), and the comet possesses an orbit with a period near 9,000 years.
(Herschel)

1787   Discovered: April 10   Discovery Magnitude: 6.0

Mechain discovered this comet in the evening sky and, with other observers, continued observations until it was lost in the sun's glare after the 26th. The comet arrived at perihelion on May 10 (r= 0.35 AU), and seven days later, was recovered in the morning sky moving southward. When closest to Earth in early June (0.3 AU), the comet was easily seen with the naked eye by observers in the Southern Hemisphere. It was then described as bright (magnitude 3) with traces of a tail. The comet was last seen on July 26 (magnitude 7), by observers on Bourbon Island.
(Mechain)

1788 I   Discovered: November 26   Discovery Magnitude: 6.0

This comet possessed a bright nucleus and a faint tail nearly 3 degrees long when discovered by Messier in Ursa Major. Although the comet brightened to naked-eye visibility by the 30th, moonlight thereafter allowed only tele-

scopic observations until mid-December. On December 14, the brightness seemed
to have sharply increased, but by the 16th, it had returned to normal.
Mechain made the final observation on December 30, and his orbital calculations indicated a perihelion passage on November 10 (r= 1.06 AU).
(Messier)

1790 I   Discovered: January 7   Discovery Magnitude: 7.0*
       After being discovered by C. Herschel (Slough), this comet was observed
on only 3 other days before being lost in the sun's glare. On January 19,
Messier described the comet as nebulous with a bright condensation. He added
that the comet appeared equal in brightness to the globular cluster M15 (magnitude 6.3). The comet was last observed on January 21. Perihelion came on
January 15 (r= 0.76 AU).
(Herschel)

1790 III   Discovered: April 18   Discovery Magnitude: 7.0
       This comet was discovered in Andromeda by C. Herschel as a small, faint
object without a tail. After being contacted by Miss Herschel, Alexander
Aubert (London) searched with binoculars and described the comet as like "a
star of 7th magnitude" with no discernible coma. At the beginning of May,
the comet was described as slightly fainter than the Andromeda Galaxy (hence,
a magnitude near 4.5) and on the 4th, the first traces of a tail were detected.
With perihelion coming on May 21 (r= 0.80 AU), the tail rapidly increased in
length to 4 degrees by the 20th. The comet was closest to Earth (0.7 AU)
during the first days in June, and the comet was then slightly fainter than
magnitude 4*with a tail 1 degree long. By the middle of June, the tail had
vanished and the comet was last detected on the 29th.
(Herschel)

1792 I   Discovered: December 15, 1791   Discovery Magnitude: 6.0
       This comet was discovered in Lacerta by C. Herschel and was described as
a "pretty large, telescopic comet." W. Herschel observed the comet in more
detail that same evening and described it as 6 arcmin in diameter with a
central condensation 5 or 6 arcsec across. To the north there extended a
faint, ill-defined ray with a length of nearly 15 arcmin. The comet was then
moving at a rate of nearly 2 degrees per day and was later recognized as having been only a few days past its closest approach to Earth (0.8 AU). With
a rapidly increasing distance from Earth thereafter, the comet faded slowly
after discovery, despite an approaching perihelion (January 14--r= 1.29 AU).
The comet was last seen on January 28, by Messier.
(Herschel)

1792 II   Discovered: January 8, 1793   Discovery Magnitude: 2.0
       The Reverend Edward Gregory (Nottingham, England) first noticed this
comet during a naked-eye inspection of the sky near the Draco-Hercules border.
It then appeared as a "nebulous star" with a faint hint of tail. On January
11, with a brightness still near 2nd magnitude*, the coma measured 30 arcmin
in diameter as the comet was then less than 24 hours from its closest approach
to Earth (0.14 AU). After perigee, the retrograde orbit caused the comet to
move rapidly away from Earth and a rapid decrease in brightness, as well as
size, followed. When last seen near the moon on February 19, the comet was
near magnitude 8*, with a coma diameter of 1.5 arcmin and only a trace of tail.
Orbital calculations revealed a perihelion date of December 27, 1792 (r= 0.97
AU).
(Gregory)

1793 I   Discovered: September 27   Discovery Magnitude: 6.0
     Messier discovered this comet using a small telescope and described it as
resembling the globular cluster M13 (magnitude 5.8). At the end of September,
it was found to be "easily visible through a simple telescope" (magnitude 5)
and possessed a very bright nucleus, but no tail. The comet was independently
discovered near Delta Ophiuchi by C. Herschel on October 7, and less than a
week later, it was lost in evening twilight. Perihelion came on November
5 (r= 0.40 AU), and Messier was able to recover the comet on December 29, with
the help of Jean Baptiste de Saron's orbit. The comet was followed until January 8, 1794, when located in Canis Minor at a magnitude slightly fainter than
6.*
(Messier)

1793 II   Discovered: September 24   Discovery Magnitude: 7.5
     This comet was discovered by Perny (Paris) and was described soon after-
ward as being too faint to be seen in binoculars. The comet varied little in
brightness during its 2½ months of visibility, as the direct orbit (inclination
equalled 51 degrees) and the large perihelion (r= 1.50 AU) caused the distances
from both Earth and the sun to change by only 0.2 AU. Perihelion came on November 20, and the comet was last seen on December 8. Though a parabolic orbit
is generally listed in comet orbit catalogs, some mathematicians have calculated elliptical orbits with periods ranging from 12 years (Burckhardt) to
420 years (H. L. d'Arrest). The longer-period orbits are considered most likely.
(Perny)

1796   Discovered: March 31   Discovery Magnitude: 8.0
     Using a comet-seeker, Heinrich Wilhelm Matthias Olbers discovered this
comet in the region south of Alpha Virginis. Perihelion came on April 3 (r=
1.58 AU), and between the 4th and 11th, observers detected faint traces of a
small tail pointing to the northeast. The comet was last observed on April
14, when the magnitude was estimated as between 8 and 9.
(Olbers)

1797   Discovered: August 14   Discovery Magnitude: 3.0
     This comet was independently discovered by Eugene Bouvard (Paris), C.
Herschel (Slough) and St. Lee (Hackney, England) within a few hours of each
other. The comet was then located only 0.17 AU from Earth and was plainly
visible to the naked eye. During the next two days, the comet was independently discovered by several other observers and was described as possessing
a coma 10 arcmin in diameter. The comet passed closest to Earth on August
16.5, at a distance of only 0.088 AU. Thereafter, fading was rapid and the
comet dropped below naked-eye visibility on the 19th. By the 20th, the coma
measured nearly 7 arcmin in diameter. No tail was ever reported for this
comet, which was last observed on August 31, by Olbers. Perihelion had
occurred on July 9 (r= 0.53 AU).
(Bouvard-Herschel-Lee)

1798 I   Discovered: April 12   Discovery Magnitude: 6.0
     This comet was apparently just below naked-eye visibility when discovered
by Messier near the Pleiades, since he reported it as bright, but only visible
in a telescope. An observation by Burckhardt on April 25, when the comet was
only 5 degrees above the horizon, indicates a magnitude brighter than 7. The
comet faded considerably in May, and was last observed on the 24th. Perihel-

ion came on April 5 (r= 0.48 AU).
(Messier)

1801 II   Discovered: December 7   Discovery Magnitude: 7.0
     Bouvard discovered this comet one day before it reached its closest
approach to Earth (0.12 AU). It was then 5 arcmin in diameter and possessed
no tail. An independent discovery was made by Olbers on the 9th. When last
seen on the 12th, the magnitude was near 7.0*and traces of a tail were noted
to the southeast. Bad weather in Europe is generally accepted as the reason
for the comet's not being observed after the 12th. Perihelion occurred between December 31, 1798, and January 1, 1799 (r= 0.77 AU).
(Bouvard)

1799 I   Discovered: August 7   Discovery Magnitude: 6.5
     Mechain (Paris) discovered this comet as a diffuse, bright and tailless
nebulosity in the north polar region. In late August, the comet became visible
to the naked eye and possessed a tail 1 degree long. It was then independently discovered by Bode (Berlin), Baron Francis Xaver von Zach (Seeburg) and
Olbers (Bremen). Between September 6 and 8, observers reported a coma 10
arcmin in diameter with a tail 4 degrees long. Large telescopes revealed
jet emissions from the nucleus. On September 23 and 25, Nevil Maskelyne
(Greenwich) observed the comet with binoculars and found its brightness equi-
valent to nearby Rho Serpentis (magnitude 4.8). The comet was last seen on
October 25, by Messier. Perihelion came on September 7 (r= 0.84 AU).
(Mechain)

1799 II   Discovered: December 26   Discovery Magnitude: 4.5
     This comet was discovered in the morning sky by Mechain (Paris). He
then reported it as visible to the naked eye as a star near magnitude 4.5,
but added that in a telescope there was a nucleus and a tail 1 degree long.
On January 5, 1800, observers generally described the tail as 3 degrees long
and curved, though some estimates were as high as 7 degrees. The comet was
last seen on January 6, at a magnitude near 3.5* Observations thereafter
were ended as the comet entered the sun's glare. Perihelion occurred on the
day of the comet's discovery (r= 0.63 AU).
(Mechain)

1801   Discovered: July 11   Discovery Magnitude: 6.5
     This comet was first observed by Jean Louis Pons (Marseilles, France) in
Ursa Major. Independent discoveries were made the next evening by Messier,
Mechain and Bouvard. The comet was described as small, round and tailless.
Pons received an award of 600 francs which had been offered by Joseph Jerome
Lalande for the first comet discovered in the 19th Century. A German amateur
named Reissig later claimed to have seen the comet through a brief cloud open-
ing on June 30, near the Ursa Major-Camelopardalis border. In 1969, however,
Joseph Ashbrook calculated the comet's position for the date Reissig claimed
to have seen it and found a position in Andromeda, thus solving the 168-year-
old mystery. No tail was visible during the comet's appearance. After discovery the comet headed southwestward and became lost in the sun's rays after
July 23. Perihelion came on August 9 (r= 0.26 AU).
(Pons)

1802   Discovered: August 26   Discovery Magnitude: 7.0
     First discovered by Pons as a diffuse object 2 to 3 arcmin in diameter,
independent discoveries were later made by Mechain (August 28) and Olbers

(September 2). Mechain described the comet as equaling the globular clusters M10 and M12 in brightness (magnitude 6.5). During the first week of September, observers noticed the first traces of a tail pointing eastward, as well as a stellar nucleus. During the rest of the month, the brightness slowly decreased from magnitude 7 to 8*and when last observed on October 5, Messier found it near 9th magnitude*. Perihelion came on September 10 (r= 1.09 AU).
(Pons)

1804   Discovered: March 7   Discovery Magnitude: 6.5
     Pons discovered this comet in Libra during the time of its closest approach to Earth (0.22 AU). He then described the comet as tailless and below naked-eye visibility. Independent discoveries were made by Bouvard (March 10) and Olbers (March 12). The latter observer described the comet as larger (15 arc-min) and brighter (magnitude 6) than the globular cluster M5. Afterwards, the comet headed northwest and by month's end, had faded to magnitude 8*. The final observation was made by Olbers on April 1, with an attempted observation on April 8, ending in failure. Perihelion had occurred on February 14 (r= 1.07 AU).
(Pons)

1806 II   Discovered: November 10   Discovery Magnitude: 8.0
     This comet was discovered by Pons as a faint object slowly moving towards the sun. Bessel saw it through clouds on December 8, detecting a faint tail and indicating a magnitude near 6.5. The comet was lost in the sun's glare after December 20, and passed perihelion on the 29th (r= 1.08 AU). After reappearing on January 17, the comet remained low over the horizon, though an observation on the 25th indicated it was quite bright (magnitude 5) in a telescope. W. Herschel estimated the coma as 7 arcmin in diameter on February 1, and the comet was followed until the 12th.
(Pons)

1807   Discovered: September 9   Discovery Magnitude: 1.0
     First detected by Giovanni (Sicily), this comet was independently dis-covered by numerous people (including Pons) after the comet passed perihelion on September 19 (r= 0.65 AU). On the 26th, Flaugergues (Vivires) found the comet's nucleus to be near magnitude 2, while the coma measured 6 arcmin in diameter. He also estimated the tail to be 1.5 degrees long, which increased to nearly 8 degrees on the 27th. During the first days of October, observers indicated the comet's total magnitude was brighter than Alpha Serpentis (hence, magnitude 2) and the tail had become split with a long, straight tail and a short, curved one. On the 11th, the long tail measured 2.5 degrees long, while the shorter one measured only 1 degree. The longer tail reached maximum on the 22nd, at a length of 10 degrees. Bessel made several estimates of the comet's brightness during November and December. He gave an estimate of 3 for November 3, 4 for November 20 and 5.5 for mid-December. The comet was followed until March 27, when Wisniewsky (St. Petersburg) saw it for the final time in a telescope. Using observations spanning 5 months, W. Herschel calculated the diameter of the "round and well-defined disk" inside the coma to be 538 miles.
(Great Comet)

1808 I   Discovered: March 25   Discovery Magnitude: 7.0
     This comet was discovered by Pons when less than 10 degrees from the north celestial pole. On March 29, the comet was independently found by Wisniewsky with a small telescope and was then described as a tailless coma 3 arcmin in

diameter. When seen on April 3, the comet was described as considerably
fainter because of moonlight. The comet was not seen thereafter, and after
reaching perihelion on May 13 (r= 0.39 AU), remained in the sun's glare until
too distant for observation.
(Pons)

1808 II   Discovered: June 24   Discovery Magnitude: 7.0
     Pons discovered this comet in Camelopardalis as a very small, dim object.
In the next days it rapidly moved eastward through Lynx. Pons remained the
only observer and followed the comet until July 4, when it was situated less
than 15 degrees above the horizon. This low position and moonlight prevented
further observations. Perihelion came on July 12 (r= 0.61 AU).
(Pons)

1810   Discovered: August 23   Discovery Magnitude: 6.5
     This comet was discovered by Pons when less than 5 degrees from the north
celestial pole and was described as a tailless nebulosity. A waning moon temporarily stopped observations between September 9 and 16, but, thereafter,
the comet was followed until October 8, with the magnitude fading from near
6* in mid-September, to 7*. Perihelion came on October 6 (r= 0.97 AU).
(Pons)

1811 I   Discovered: March 25   Discovery Magnitude: 5.0
     This comet not only became a prominent object, but its 512 days of vis-
ibility was not surpassed until 1889, when much larger telescopes and photography were available. It was discovered by Flaugergues on March 25, and
was independently found on April 11, by Pons. On the latter date, the comet
was noticeable to the naked eye, though still over 2 AU from the sun and
Earth. Brightening was slow and the comet was lost in the sun's glare after
June 16. After conjunction with the sun, the comet reappeared on August 20,
and passed perihelion on September 12 (r= 1.04 AU). At the time of perihelion,
the comet was still located nearly 1.6 AU from Earth and was near magnitude 5*;
however, during the next month, it continued to approach Earth and brightened
from a magnitude of 4.5* on September 22, to nearly zero magnitude* by October
20. On the latter date, the comet was nearest Earth at a distance of about
1.1 AU. The comet had become circumpolar in early October, and soon afterward
began displaying twin tails nearly 16 degrees long. On October 15, W. Herschel
described the straight tail as 24 degrees in length, while the curved tail had
a width of nearly 7 degrees. The coma reached a maximum diameter of 20 arcmin
soon afterward. The tail continued to grow into December, and was measured as
nearly 70 degrees long, but at the beginning of 1812, the comet rapidly became
an inconspicuous object. By mid-January, it dropped below naked-eye visibility,
and before month's end was again in conjunction with the sun. On July 11, 1812,
the comet was detected as a faint nebulosity with a tail 10 arcmin long, and
by the 31st, Wisniewsky described it as faint, yellowish and 1.5 arcmin in
diameter--no tail was visible. The comet was last observed on August 17, 1812.
W. Herschel painstakingly measured the diameter of the nucleus as 428 miles
during the latter months of 1811.
(Great Comet)

1811 II   Discovered: November 16   Discovery Magnitude: 6.5
     This comet was discovered by Pons while searching for comets in Eridanus.
It was then located only 14 degrees above the horizon and was described as a
small, very faint spot, with a nucleus and a small tail. The comet was hardly

visible for the rest of November, due to moonlight, but by early December, its
northward motion and a waning moon made it an easy object in telescopes (magnitude near 7). W. Herschel observed it several times during January, and found
a tail nearly 10 arcmin long and a starlike nucleus. The comet was last seen
on February 16, when the magnitude was near 9.5*. Perihelion had occurred on
November 11 (r= 1.58 AU). The inclination was 31 degrees and a period near
755 years is likely.
(Pons)

1813 I   Discovered: February 4   Discovery Magnitude: 6.5
    Pons discovered this comet in Lacerta and described it as a small, faint
spot without a tail. During February, the comet moved through Andromeda and
Pisces. Other observers included only Bouvard and von Zach. At discovery, the
comet had been only 0.34 AU from Earth, but as this distance increased, a noticeable fading occurred and the comet was last seen on March 11, at a magnitude
near 9*. Perihelion occurred on March 5 (r= 0.70 AU).
(Pons)

1813 II   Discovered: April 3   Discovery Magnitude: 5.5
    This comet was first detected by Pons in Ophiuchus, but was independently
found the next night by Karl Ludwig Harding (Göttingen). It was described as
small, with a bright nucleus and no tail. Pons said he could barely see the
comet with the unaided eye. By late April, during the time of its closest
approach to Earth (0.27 AU on April 30), the comet reached a brightness of
3rd magnitude*. Being visible only from the Southern Hemisphere in May, few
observations were made; however, Don José Joaquim de Ferrer (Cuba) consistently
studied the comet between April 29 and May 18, noting a tail 8 arcmin long in
early May. After the 18th, the comet was lost in the sun's glare. Perihelion
came on May 20 (r= 1.21 AU).
(Pons)

1816   Discovered: January 22   Discovery Magnitude: 7.5
    This small, faint comet received little attention during its brief visit
near the sun. Discovered by Pons (Marseilles), it was only seen by him on one
other day in late January. An observer in Paris did detect the comet on February 1, but thereafter, observations were impossible due to the sun's glare.
Orbital calculations revealed that the comet was between 0.40 and 0.48 AU from
Earth when seen and perihelion occurred on March 1 (r= 0.05 AU).
(Pons)

1818 II   Discovered: December 26, 1817 Discovery Magnitude: 7.0
    This comet was discovered by Pons in the evening sky as a faint nebulous
object without a tail or nucleus. By January 18, the comet had brightened to
magnitude 6*with the first traces of tail being visible and during February,
observers noted a clearly discernible nucleus. Perihelion occurred on February 26 (r= 1.20 AU), and during March, observations were marred due to the
comet's closeness to the sun. In April, it was reobserved at a magnitude near
8.5*and Olbers, as well as other observers, reported notable fluctuations in
brightness. Olbers made the final observation of this comet on May 2.
(Pons)

1818 III   Discovered: November 28   Discovery Magnitude: 7.0
    Pons' discovery of this comet was made while sweeping through the region
near Beta Hydrae in the morning sky. He described it as small, round and easily
observed. The comet brightened rapidly and on December 14, it passed 0.17 AU

from Earth. In the next days, the comet nearly reached 4th magnitude*and remained visible to the naked eye until December 28. Bessel independently discovered the comet at Königsberg on the 22nd, and described it as near magnitude 6, with no tail or nucleus. The final observation was made by Harding
on January 30, when the comet was of magnitude 9.5*. Orbital calculations gave
a perihelion date of December 5 (r= 0.86 AU).
(Pons)

1819 II   Discovered: July 1   Discovery Magnitude: 1.0
     This comet was first detected by Tralles (Berlin) very near to the set-
ting sun. Two days later, Olbers estimated a nuclear magnitude of nearly 1 and
found the tail about 8 degrees long. The comet was widely observed with the
naked eye throughout July, and, also during this time, observers in Palermo
described the tail as split. By early August, the comet was near magnitude
6*and by mid-September, it was about 9.5*. Olbers observed the comet on October 20, and described it as very faint and 2 arcmin across. The comet was
last observed on the 25th. Orbital calculations gave the perihelion date as
June 28 (r= 0.34 AU). The comet also seems to have crossed the face of the
sun on June 26, and, at that time, the Earth may have been involved in its
tail.
(Great Comet)

1821   Discovered: January 21   Discovery Magnitude: 6.5
     Discovered nearly simultaneously by Jean Nicolas Nicollet (Paris) and
Pons (Marseilles) in Pegasus, independent discoveries were later made by
Blanpain (Marseilles) on January 25, and Olbers on the 30th. At discovery,
the comet was small with a tail 30 arcmin long, but when observed by Olbers on
the 30th, it was described as 4 arcmin in diameter with a tail 4 degrees long.
Observing in mid-February, John Herschel saw the comet with the naked eye
(magnitude 4) and reported a tail 5 degrees long which tapered to the north.
By March 1, the tail was 7 degrees long and the total magnitude was near 3*.
The comet became lost in the sun's rays after March 10, and on the 22nd, it
passed perihelion (r= 0.09 AU). On April 1, the comet was recovered in the
evening sky at a magnitude near 2*and with a tail over 8 degrees long. On
April 17, the comet was visible to the naked eye during a full moon, but
fading was rapid thereafter--the comet being last seen on May 4.
(Nicollet-Pons)

1822 I   Discovered: May 12   Discovery Magnitude: 4.5
     Independent discoveries of this comet were made by three men, beginning
with Jean Felix Adolphe Gambart (Marseilles) and followed by Pons (Marliya,
Italy) on the 14th, and Wilhelm von Biela (Prague) on the 16th. Gambart described the comet as possessing a tail and a bright nucleus at discovery. In
late May, the comet was visible to the naked eye in moonlight, but quickly
faded thereafter--becoming a difficult object by mid-June. It was last seen
on June 22, at magnitude 9*, with a final search on the 25th, ending in failure.
Perihelion occurred on May 6 (r= 0.50 AU).
(Gambart)

1822 III   Discovered: May 31   Discovery Magnitude: 5.5
     While searching near the ecliptic for periodic comet Encke, Pons found a
diffuse object without tail or nucleus in Pisces. He described it as condensed
towards the center and round. Movement was to the south and by June 12, it
was in Aquarius. On June 17, the comet passed only 0.13 AU from Earth and the

next evening, it was situated near Canopus and appeared as a bright, circular object near magnitude 3[*]. The comet was last seen on June 25--moonlight and clouds interfering thereafter. Perihelion came on July 16 (r= 0.85 AU).
(Pons)

1822 IV   Discovered: July 13   Discovery Magnitude: 6.5

This comet was discovered by Pons as a faint object, without a tail, moving through Cassiopeia. Independent discoveries were made by Gambart (July 16) and Bouvard (July 20). On July 26, Gambart noted the sudden brightening of a star-like condensation within the coma, which by early August, had disappeared. On August 8, the total magnitude equalled M13 (magnitude 5.8) and by August 29, Olbers noted a tail and considered the comet to be brighter (magnitude 5) than M13 and larger (10 arcmin) than nearby M92. On September 1, the comet was still near 4.5[*], but around the 8th, observers described it as a "blurred star" with a magnitude near 3.5 and a tail 2 degrees long. Though the tail grew to 4 degrees by September 20, the coma rapidly shrank during September, until at mid-month it was only 2.5 arcmin in diameter. By October 5, the coma was only 1 arcmin across. On September 20 and 21, Olbers noted an increase in brightness similar to Gambart's event of late July; however, the decrease in brightness occurred much slower. In early October, the magnitude faded to near 6[*] and the comet was last seen on November 11. Perihelion occurred on October 24 (r= 1.15 AU) and a period near 5,000 years seems likely.
(Pons)

1823   Discovered: December 24   Discovery Magnitude: 2.0

This comet suddenly appeared to observers in Europe after having passed perihelion on December 9 (r= 0.23 AU). Pons estimated a tail length of nearly 4 degrees on the 29th, and not less than 5 degrees in early January. Also, in early January, Olbers estimated the total magnitude as 3, while Harding said the nucleus equalled a 6th-magnitude star. The comet remained visible to the naked eye throughout January and early February, with estimates of 3 (brighter than M31) on January 23, and 5 (brighter than M13) on February 6. According to Wisniewsky (St. Petersburg) the comet dropped below naked-eye visibility in mid-February. The main tail steadily shrank during January; however, between the 22nd and the 31st, several observers described an anomalous tail pointing towards the sun that was nearly as bright as the main tail. Pons saw the comet for the last time on April 1, 1824.
(Great Comet)

1824 I   Discovered: July 14   Discovery Magnitude: 5.0

This comet was discovered by Carl L. Rumker (Parramatta, Australia) while sweeping through Sextans. Rumker was using a telescope no larger than 2 inches in diameter and kept the comet under observation until quite faint in his instruments (near magnitude 8) on August 6. The only other observer was Brisbank (Parramatta), who last detected the comet on August 11. Orbital calculations revealed a perihelion on July 12 (r= 0.59 AU). After this date, the comet's retrograde orbit took it rapidly away from both the sun and Earth.
(Rumker)

1824 II   Discovered: July 23   Discovery Magnitude: 7.0

This comet was first seen by Scheithauer (Chemnitz, Germany) and later independently discovered by Pons (July 24), Gambart (July 27) and Harding (August 2). During late July, the comet was in Hercules and possessed neither a

tail nor a nucleus. In August, the comet increased very slightly in brightness,
although minor short-term fluctuations were reported by several observers, and
by the end of the month, a short tail had become visible. The comet reached
a maximum magnitude of 6*during late September and on the 29th, it arrived at
perihelion (r= 1.05 AU). With the comet-Earth distance changing by less than
0.1 AU between October and late December, the brightness faded very slowly.
By December 10, Argelander found the comet slightly fainter than magnitude 8*
and after December 25, the comet could no longer be seen.
(Scheithauer)

1825 I   Discovered: May 19   Discovery Magnitude: 6.5
     Gambart discovered this comet low over the horizon in Cassiopeia and
described it as round, 2 arcmin in diameter, with a brightness "very pronounced towards the center." On the 28th, the coma was 5 arcmin across and
faint traces of a tail were noticed. Rumker (Australia) independently discovered this comet on June 5. During June, the tail increased from 40 arcmin
to 1.5 degrees, while the coma reached a maximum diameter of 7 arcmin on June
10--shortly before nearest approach to Earth (0.79 AU). During July, Olbers
compared the comet to M81 and M82 and found the comet brighter (magnitude 6),
but as the comet neared conjunction with the sun, it rapidly became a more
difficult object to observe and was lost after July 15. Perihelion came on
May 31 (r= 0.89 AU) and an elliptical orbit with a period near 4,000 years.
seems likely.
(Gambart)

1825 II   Discovered: August 9   Discovery Magnitude: 6.0
     This comet was discovered in moonlight by Pons (Florence, Italy) and was
described as round, bright, but without a tail. The comet was then moving
through Auriga. It was independently discovered by Harding on August 24,
when near Gamma Geminorum. The comet's rapid southern movement ended observations after August 27. Perihelion occurred on August 19 (r= 0.88 AU).
(Pons)

1825 IV   Discovered: July 15   Discovery Magnitude: 6.5
     Discovered by Pons (Marliya) while sweeping for comets in Taurus, this
comet was described as a diffuse object with a nucleus and traces of a tail.
Independent discoveries were made by Biela (Josephstadt) on July 19, and
Dunlop (Parramatta) on July 21. The comet's apparent motion was small during
these discoveries and later orbital calculations showed it was then 2.46 AU
and 3.06 AU from the sun and Earth, respectively. Perihelion was expected on
December 11 (r= 1.24 AU), and the comet steadily brightened after discovery.
In August, Harding noted the tail extending nearly 2 degrees and on the 27th,
the comet became visible to the naked eye (magnitude near 5.5). At the beginning of September, the magnitude was nearly 4*and the tail extended over
8 degrees, and one month later, the magnitude had increased to 2*, while the
tail was estimated as 12 degrees long. With bad seeing on October 12, the
tail was still found to be 14 degrees in length and drawings by Dunlop between
October 5 and November 8, indicated the tail was split into five branches. The
comet was lost in the sun's glare after December 24, but reappeared in April
1826. The final observation came on July 8, when the magnitude was between
9 and 10. The comet's orbit is decidedly elliptical, with a period near
4,500 years.
(Pons)

1826 II   Discovered: November 7, 1825   Discovery Magnitude: 8.5
     First detected in Eridanus by Pons (Florence) as a small, round object
with a possible nucleus. Over the next several days, the comet changed position slowly and later orbital calculations revealed distances of 2.80 AU and
1.93 AU from the sun and Earth, respectively, at discovery. The inclination
of 39 degrees kept the comet near the ecliptic during the several months it
was observed. The comet never became brighter than 8th magnitude and was last
seen on April 10, 1826. Being involved in the sun's glare thereafter, the
comet passed perihelion on April 22 (r= 2.01 AU), and remained hidden from
view until too faint for observation.
(Pons)

1826 IV   Discovered: August 7   Discovery Magnitude: 7.5
     This comet was discovered in Eridanus by Pons and was described as round,
diffuse, with a central condensation and tail. Independent discoveries were
made by Gambart (August 14) and Rumker (September 4). At the beginning of
September, the coma diameter was estimated as 4 arcmin and a tail pointed to
the north. During the last days of September, the comet passed within 0.5 AU
from Earth on its way to an October 9 perihelion (r= 0.85 AU). The total magnitude then equalled 5.5*. By mid-October, the comet was still easily seen in
a small telescope, despite moonlight, hence near magnitude 6*. The tail was no
longer detected after November 6, and the comet was last seen on December 11
(magnitude 9.5). The comet's orbit was inclined only 26 degrees to the
ecliptic and the period was near 6,000 years.
(Pons)

1826 V   Discovered: October 22   Discovery Magnitude: 6.5
     Discovered by Pons in Boötes with a faint tail. Independent discoveries
were made by T. Clausen (Hamburg) on October 26 and Gambart on October 28.
Movement was to the south and the comet rapidly brightened. It became visible
to the naked eye on November 1, and by the 12th, it was reported as very bright
(magnitude between 2 and 3) with a round nucleus and a very long tail. The
comet was lost in the sun's glare shortly thereafter, and Gambart predicted a
transit of the sun's disk during perihelion on November 18 (r= 0.03 AU), but
nothing was detected by observers. The comet reappeared in late November, and
was described by Argelander as possessing two tails on the 28th and 29th. During the first days of December, the tail extended nearly 8 degrees and the
total magnitude was near 4.5*. Professor G. Santini (Padua, Italy) noted a
sharp change in the appearance of the comet in December. Between the 1st and
the 5th, he found the nucleus to be sharp and bright with a small, faint tail;
however, afterwards the nucleus became blurred and faint, while the tail became
rather bright. The comet was followed until January 6, 1827, when Argelander
found it to be a faint object with a tail slightly over 15 arcmin in length.
(Pons)

1827 I   Discovered: December 26, 1826   Discovery Magnitude: 6.5
     Detected by Pons in Hercules and independently by Gambart the next morning,
this comet moved eastward and was observed by Pons until December 31, when the
comet became lost in bright moonlight and overcast skies. The comet was not
again seen until January 17, 1827, and afterwards was widely observed. In
late January, the comet was described by Harding as being small, diffuse, with
a short tail and no nucleus. The magnitude was then near 5*. Olbers noted the
comet's considerable brightness between January 18 and 22. The last observation

occurred in twilight on January 26. Perihelion came on February 5 (r= 0.51
AU).
(Pons)

1827 III   Discovered: August 3   Discovery Magnitude: 6.5
     This comet was discovered in Lynx by Pons (Florence) in the morning sky
and was described as small, but bright. It possessed a tenuous tail as well
as a thin jet after August 20, and on the 23rd, the jet extended 15 arcmin.
The brightness was then estimated as near 4.5, and, soon afterward, the coma
was nearly 7 arcmin in diameter. On August 29, the comet was lost in the sun's
rays as the comet neared its September 12 perihelion (r= 0.14 AU). The comet
was recovered at Mannheim on October 16 as a small, diffuse spot near magnitude 8 or 9, but observations ended after the 17th.
(Pons)

1830 I   Discovered: March 16   Discovery Magnitude: 3.0
     Numerous discoveries of this comet occurred in the Southern Hemisphere as
it approached perigee. One of the first discoverers on March 16 was d'Abbadie,
who described the comet as bright with a glow surrounding it. The comet was
then located only 4 degrees from the south celestial pole and possessed a
tail which extended 5 degrees to the east. Further discoveries were made
during the 17th, including those by seamen aboard a ship in Antarctic waters.
They described the tail as nearly 8 degrees long. The comet passed only 0.15
AU from Earth on March 26, and was then at a maximum magnitude of between 1
and 2. By April 21, Gambart indicated a total magnitude still near 3 and
estimated the tail as 2 degrees long, while on May 16, Olbers indicated a
magnitude near 6 and estimated the tail as 1 degree long. The comet was observed continuously until August 17, when Bessel saw it for the final time at
Königsberg. Perihelion occurred on April 9 (r= 0.92 AU).
(Great Comet)

1830 II   Discovered: January 7, 1831   Discovery Magnitude: 2.0
     Among the many independent discoverers of this comet, Herapath (England)
was one of the first to report detailed observations. He described a white
tail 1 to 2 degrees long that protruded nearly perpendicular to the horizon.
The head of the comet was brilliant. On January 15, Biela (Tyrol) observed a
tail 2.5 degrees long with his naked eye before sunrise and on the 23rd, a
3-degree tail was noted by Niccolò Cacciatore (Palermo), who also described
a condensation 20 arcsec in diameter located inside a coma 3 arcmin across.
In mid-February, Friedrich Nicolai (Mannheim) found the coma to be a round,
uniformly-lighted object 3 to 4 arcmin across, and in the first days of March,
when the comet was at perigee (0.5 AU), Olbers described the coma as pale,
diffuse and 20 arcmin in diameter. The comet was last detected on March 19,
at a magnitude near 10*. Perihelion had occurred on December 28 (r= 0.13 AU).
(Great Comet)

1832 II   Discovered: July 19   Discovery Magnitude: 7.5
     Gambart discovered this comet in Hercules and described it as small,
diffuse, without a tail or nucleus. An independent discovery by Harding on
July 29 indicated little change in the comet's appearance, except for the
appearance of a bright, point-like nucleus. During early August, several
observers reported rapid variations in the light of the nucleus. Although
the comet was heading for perihelion, its distance from Earth increased so
rapidly that it was last seen on August 27, at a magnitude near 9*. The comet

arrived at perihelion on September 26 (r= 1.18 AU).
(Gambart)

1833   Discovered: September 30   Discovery Magnitude: 6.0
      Having passed perihelion on September 10 (r= 0.46 AU), this comet was
discovered by Dunlop (Parramatta) in Libra after sunset on September 30. The
comet was then just below naked-eye visibility and possessed a tail extending
2 arcmin. The comet was only seen by a handful of observers in Australia, before the final observation on October 16. On that date the magnitude was near
7*and the moon prevented observations thereafter. Northern observers were not
notified of the comet's appearance until late 1834. The comet's orbit has an
inclination of only 7 degrees and, despite the short time seen, seems to be
decidedly elliptical. The period could be as short as 3.5 years.
(Dunlop)

1834   Discovered: March 8   Discovery Magnitude: 3.5
      While comet hunting in the eastern part of Sagittarius, Gambart found this
comet near the horizon where no stars were visible to the naked eye. On the
10th, the comet was seen near 4 Capricorni--despite fog--as a circular, faint
object less than 5 arcmin across. Cloudy weather and, finally, bad positioning
in the sky caused the comet to be lost to Gambart after the 10th. This comet
was independently discovered in the Southern Hemisphere by Dunlop on March 20.
It was then located in northeastern Capricornus in the morning sky and was
described as small, with a diameter of 1.5 arcmin. Moonlight prevented observations until April 1, and on the 3rd, the comet had arrived at perihelion (r=
0.51 AU). On the latter date, Dunlop estimated the comet's brightness as magnitude 4.5. Dunlop last observed the comet on April 14, when the comet had
moved into Pisces. He then described it as of magnitude 7 with a coma 1 arcmin
in diameter. The inclination of the orbit was only 6 degrees, but the orbital
period is not likely to be less than 1,400 years.
(Gambart)

1835 I   Discovered: April 20   Discovery Magnitude: 7.5
      The Russian astronomer Boguslawski (Breslau) discovered this comet in the
evening sky in Crater and described it as small, very faint, with a diameter
between 3 and 4 arcmin. The comet faded slowly thereafter since perihelion
had occurred on March 20, at the rather large heliocentric distance of 2.04 AU.
By the end of April, the magnitude was 8*and the tail had become broad and
pointed to the east. A near-stellar nucleus was also occasionally seen. The
comet was last observed on May 27, at a magnitude near 9*.
(Boguslawski)

1840 I   Discovered: December 3, 1839   Discovery Magnitude: 5.5
      This comet was discovered near Gamma Virginis by Johann Gottfried Galle
(Berlin) and was described as a nearly homogeneous nebulosity with a stellar
nucleus and a short tail. Taylor (Madras, India) independently discovered the
comet on January 6, 1840. On January 8, an observer in Germany noted the comet's
appearance as similar to that of a nebulous star of magnitude 4.5, with a small
tail. The comet was no fainter than magnitude 9.5*when last seen on February
10. Perihelion had occurred on January 4, 1840 (r= 0.62 AU).
(Galle)

1840 II   Discovered: January 25   Discovery Magnitude: 8.5
      Galle discovered this comet near Rho Draconis as a small, faint, nebulous
object with a noticeable condensation. At the beginning of February, Argelander

indicated a brightness near 7.5 and estimated the coma as no larger than 1
arcmin in diameter. The comet may have undergone an outburst on March 8,
however, moonlight interfered with further observations until the 16th, at
which time the comet was reported as back to normal. The comet reached
a maximum brightness of 6.5* on March 19, but, thereafter, fading was rapid--
primarily due to twilight and low altitude. The comet was last seen on April
3. Perihelion occurred on March 13 (r= 1.22 AU).
(Galle)

1840 III   Discovered: March 7   Discovery Magnitude: 6.5
     Galle's third comet discovery in three months was located near the
Pegasus-Cygnus border on March 7, and was described as possessing a bright
tail several degrees long. The comet remained visible for an additional 21
days before entering the sun's glare, and when last seen had brightened to a
magnitude near 5.5*. Perihelion came on April 2 (r= 0.75 AU). The orbit bears
a close resemblance to that of the comet of 1097.
(Galle)

1840 IV   Discovered: October 26   Discovery Magnitude: 9.0
     This comet was discovered near Omicron Draconis by Dr. Carl Bremiker
(Berlin) using a comet-seeker and was described as a "faint nebula." Observations were continuous during November, as the comet slowly brightened and
the nucleus was detected on several occasions resembling a star of magnitude
10 or 11. Perihelion occurred on November 14 (r= 1.48 AU). During the first
days of December, the comet was closest to Earth (0.86 AU) and seemed as
bright as a 7th-magnitude star. By December 22, the coma had reached a maximum size of 1.3 arcmin, but, as with previous observations, no tail was
visible. By late January 1841, the comet was near magnitude 10*, and when last
observed by Bremiker on February 16, it was near 11*. Orbital calculations
indicate the orbit is elliptical with a period near 286 years.
(Bremiker)

1842 II   Discovered: October 28   Discovery Magnitude: 8.5
     Laugier (Paris) discovered this comet north of Zeta Draconis in the even-
ing sky. He described it as a very faint object without a tail. During November, the comet became brighter as it passed within 0.5 AU of Earth; however,
the inclination of 106 degrees--nearly perpendicular to the ecliptic--caused
observing conditions to deteriorate as the comet moved rapidly southward. On
November 5, Argelander noticed a trace of a tail and by the next evening, he
reported it as extending between 12 and 15 arcmin. A central condensation was
observed on several occasions during November, which later elongated in the
direction of the tail. Moonlight hampered observations near mid-month, but
good conditions after November 20 allowed a few other observations before
the comet entered the sun's glare. On the 21st, the tail was again observed
and the total magnitude was near 6*. The comet was last seen on the 27th.
Perihelion occurred on December 16 (r= 0.50 AU).
(Laugier)

1843 I   Discovered: February 5   Discovery Magnitude: 3.5
     Although several observations of this comet were made in the Southern
Hemisphere preceding perihelion, none were well documented in terms of physical descriptions--the discovery magnitude even being a rough estimate based
on later observations. This comet's presence became widely known on February
27--the day of perihelion-- when the comet passed 0.006 AU from the sun and

was observed in broad daylight slightly over 1 degree from the sun's limb.
The general appearance was that of an elongated white cloud possessing a tail
nearly 1 degree long and a brilliant nucleus. The total magnitude must have
then been near -7. Passengers aboard the Owen Glendower, at sea off the Cape
of Good Hope, had one of the best views as the "short, dagger-like object"
closely followed the sun towards the western horizon as the day progressed.
The comet remained visible during daylight hours into the first days of March.
The comet was still involved in twilight on March 1, but observers still estimated the tail as 30 degrees long. With a clear view of the comet on the
4th, estimates of the tail length varied from 69 to 90 degrees. On March
6, the nucleus was described as very bright with a diameter of 12 arcsec and
was centrally located inside a coma 45 arcsec across. When the comet was
seen on the 8th, one observer found the tail to be 35 degrees long and
described the nucleus as being as bright as Jupiter (magnitude -1). Up until
this time observations were restricted to observers in the Southern Hemisphere,
but, as March progressed, northern observers finally contributed to the large
number of observations being amassed around the world. One of the first
northern observers was Edward Cooper (Nice, France) who first perceived the
tail by accident on the 12th of March--believing it to be a cloud formation.
The next evening another accidental sighting of the tail led him to study the
"cloud's" movement with the stars--thus making its origin non-atmospheric.
Cooper and other northern observers caught sight of the whole comet beginning
on the 17th, when the tail was reported as near 40 degrees long and the total
magnitude was no fainter than 0.* Precise measures of the tail length by
Julius Schmidt (Hamburg) from March 21 to March 30 indicated a decrease of
from 64 degrees to 38 degrees. On the latter date, the magnitude had already
dropped below naked-eye visibility. The brightness decrease continued to be
rapid during April and on the 15th, Encke and Argelander described the comet
as only a "bright spot" near the ephemeris position (magnitude 9.5). The comet
was last seen on April 19, by observers in America. This comet was a member
of the Kreutz sungrazing family and appears to travel in an elliptical orbit
with a period near 513 years.
(Great March Comet)

1843 II   Discovered: May 3   Discovery Magnitude: 7.5
     Felix Victor Mauvais (Paris) discovered this comet in the evening sky
near the Cygnus-Pegasus border. He described the comet as a faint, oval object
3 arcmin in diameter with a bright nucleus. The comet passed perihelion on
May 6 (r= 1.62 AU). During June, Argelander commented that the bright nucleus
allowed observations even in moonlight and Keller estimated the coma to be
3 arcmin in diameter, but with no tail. At full moon on the 12th, the magnitude was probably near 6.5,* and by August 1, the brightness had dipped to
8.5.* Fading thereafter limited observations to only a few observers and the
comet was last seen on October 2, when the magnitude was near 10.*
(Mauvais)

1844 II   Discovered: July 8   Discovery Magnitude: 6.5
     This comet was discovered by Mauvais in Hercules on July 8, and independ-
ently by d'Arrest (Berlin) on July 10. It was then described as bright, with
a coma 3 arcmin in diameter and a small bright nucleus. After discovery, the
comet steadily moved away from Earth. On August 17, observers in Washington
estimated the total magnitude as 9 and the tail was determined as 6 arcmin long.

In late September, the comet was in conjunction with the sun. After passing perihelion on October 17 (r= 0.86 AU), the comet reappeared on October 27, and by November 10 was of naked-eye visibility. The tail then extended 10 arcmin. The comet was followed until March 10, 1845, when the magnitude had decreased to near 9.[*]
(Mauvais)

1844 III   Discovered: December 17   Discovery Magnitude: 0.0
     Having passed perihelion on December 14 (r= 0.25 AU), this comet was first discovered on the evening of December 17 in Guinea when only 11 degrees from the sun. Within the next 4 days, at least 5 independent discoveries were made by observers in the Southern Hemisphere as the comet moved away from the sun. When first detected, the comet's tail extended nearly 4 degrees, but by December 29, it had increased to 8 degrees and was gently curved to the north. During the first days of January, 1845, the tail reached a maximum length of 10 degrees and observers noted the nucleus as near magnitude 2.5. Thereafter, the comet faded rapidly. By the end of the month, it was no brighter than 5th magnitude.[*] Curiously, as the main tail shrank during January, an anomalous tail appeared pointing over 1 degree toward the sun for a short time near mid-month. During the first week of February, the comet passed close to comet 1844 II in the sky and was then independently discovered by 3 observers between the 5th and 7th. By February 8, the comet had dropped below naked-eye visibility and, when last seen by observers at the Cape of Good Hope on March 12, the magnitude was near 10.[*]
(Great Comet)

1845 I   Discovered: December 28, 1844   Discovery Magnitude: 6.5
     Using a comet-seeker, d'Arrest (Berlin) discovered this comet near 15 Cygni during a search for comets. Perihelion came on January 8 (r= 0.91 AU), and the comet's brightness varied little during the next couple of weeks as its increasing distance from the sun was countered by a decreasing distance from the Earth. By the end of January, the comet was described as blurred, without a tail or nucleus and the magnitude was near 7.5;[*] however, during the first weeks of February, the comet brightened and when at perigee on the 17th (0.22 AU), the magnitude was near 6.[*] By early March the comet had already faded to a magnitude near 8.5[*] and when last seen on March 30, it was described as a very faint cloud near magnitude 11.[*]
(d'Arrest)

1845 II   Discovered: February 25   Discovery Magnitude: 7.5
     Independent discoveries of this comet were made by Italian astronomer Francesco de Vico in Ursa Major on February 25, and Hervé Faye (Paris) on March 6. The latter observer described the comet as diffuse with a central condensation, but no tail. In mid-March, C. A. F. Peters (Altona) described the nucleus as 15 arcsec in diameter and indicated a total magnitude near 6.5. During April, the comet's southward motion took it through Cancer and into Puppis. Perihelion occurred on April 21 (r= 1.25 AU) and the comet was followed until May 1, when the magnitude must have been near 9.[*]
(de Vico)

1845 III   Discovered: June 2   Discovery Magnitude: 1.5
     This comet was first detected by Colla (Paris) with the naked eye in the morning sky. Independent discoveries were made by George P. Bond (Harvard Observatory, Massachusetts) and Lieutenant Rice on June 3, and Richter (Berlin)

on June 6. Perihelion occurred on the 6th (r= 0.40 AU), and Wichman (Königsberg) independently found the comet in the evening sky on June 8. When first
seen, the comet's nucleus was very bright and a conical tail extended 1 degree.
On June 6, the nucleus was estimated as near magnitude 3 and on the 7th, observers found a tail 2.5 degrees long and a total magnitude near 0*. By June
10, a tail nearly 5 degrees long was noted and estimates of the diameter of
the nucleus ranged from 10 to 36 arcsec. Argelander reported variations in
both the brightness and tail structure between the 10th and the 16th, and on
the final date the total magnitude had faded to 3*. Rapid fading continued
throughout the latter half of June, and the comet was last seen on July 2.
(Great June Comet)

1846 I   Discovered: January 24   Discovery Magnitude: 7.0
     De Vico (Rome) found this comet in Eridanus and described it as bright
and diffuse, with a diameter of 3 arcmin. Having passed perihelion on January 22 (r= 1.48 AU), and with the distance from Earth slowly increasing, the
comet gradually faded. It was observed during full moon in March, and must
then have been between magnitude 7 and 8. The comet was last seen on May 2,
when Argelander estimated the brightness to be between 9 and 10 in his 13-cm
refractor. During the comet's time of visibility, a faint nucleus was sometimes seen and a short tail was detected in February and March. The orbit is
decidedly elliptical with a period near 2,700 years.
(de Vico)

1846 V   Discovered: July 29   Discovery Magnitude: 9.0
     Although this comet had passed perihelion on May 28 (r= 1.38 AU), it was
independently discovered on the same evening two months later, by de Vico
(Rome) and Hind (London). The comet was then described as a faint nebulosity,
invisible in a comet-seeker, with a star-like point near the coma's center.
On August 12, Argelander indicated a total magnitude near 9.5 and apparently
made the final observation of the central bright point within the coma. By
late August, the comet was near magnitude 10* and when last seen on October
18, it was described as a hardly noticeable nebulosity (magnitude 11).
(de Vico-Hind)

1846 VII   Discovered: May 1   Discovery Magnitude: 7.5
     T. Brorsen discovered this comet near the Pegasus-Vulpecula border and
described it as a large, blurred, but round nebulosity without a nucleus. The
comet had passed Earth within 0.35 AU a few days earlier and was rapidly moving
north when first seen. Independent discoveries were made on May 2, by Wichman
(Königsberg) and de Vico (Rome). The comet became brighter during May as it
neared the sun. On May 8, Schmidt (Bonn) observed the comet in his comet-
seeker despite a full moon (magnitude 6.5), but even under high magnification
he failed to see the nucleus. In mid-May, the comet passed within 20 degrees
of the north celestial pole and observers then reported it barely visible with
the naked eye (magnitude 5.5). Schmidt reported traces of a tail a few days
later and by the 24th, it extended 20 arcmin. Perihelion came on June 5 (r=
0.63 AU). The comet was last seen on June 15, with later observations being
affected by twilight. After conjunction with the sun, the comet was too faint
for further observations.
(Brorsen)

1846 VIII   Discovered: September 23   Discovery Magnitude: 9.0
     This comet was discovered by de Vico (Rome) as a relatively faint object

near the Ursa Major-Camelopardalis border. By mid-October, observations could
be made in moonlight (magnitude near 8.5) and on the 18th, Hind (London) independently discovered the comet in Coma Berenices. The comet was last seen on
October 26, at a magnitude near 9*. Perihelion came on October 30 (r= 0.83 AU).
(de Vico)

1847 I   Discovered: February 6   Discovery Magnitude: 7.5
     While searching for comets in the evening sky, Hind (London) discovered
a faint nebulosity about 1 degree north of Beta Cephei. During the next hour
and a half, the comet moved 8 arcmin to the east-southeast. By the 11th, Hind
commented that the comet was "even fainter now than when I first detected it."
Despite this apparent fading at a time when the comet was nearing both the sun
and Earth, other observers after the 13th indicated a rapid increase in
brightness. This latter pattern was to be expected according to later orbits.
On February 19, the comet had become visible to the naked eye in very dark
skies (magnitude 6). The comet was independently discovered by Bond (Cambridge,
Massachusetts) on March 4, and was described as near magnitude 5 with a short
tail. By the 16th, Hind remarked that the brightness was near 4 and estimates
of tail length by other observers gave values as high as 4 degrees. As the
comet neared perihelion, it became a more difficult object in twilight. On
the 22nd, no tail could be perceived and by the 24th, Bond made the final
observation of the comet when it was located only 3 degrees above the horizon.
     Perihelion came on March 30 (r= 0.04 AU), and at noon, Hind observed the
comet using an 18-cm refractor stopped down to 8 cm. The comet was then described as a small, faint coma with a sharp 8-arcsec-diameter nucleus. The
tail was represented by two rays 40 arcsec long. After perihelion the comet
remained lost in the sun's glare until April 22, at which time it had faded
significantly with a greatly increased distance from Earth (1.7 AU). The
comet was last seen two days later.
(Hind)

1847 II   Discovered: May 7   Discovery Magnitude: 9.5
     This telescopic comet was discovered by Professor Colla (Parma) in the
evening sky near 21 Leo Minoris as a very faint object moving to the northwest.
The comet was then located at distances of 2.14 and 1.79 AU from the sun and
Earth, respectively. The comet changed little in appearance during the next
two months. Perihelion came on June 5 (r= 2.12 AU), and, thereafter, the comet
was moving away from both the sun and Earth. However, in early July, the
magnitude seems to have brightened to nearly 8th magnitude, even though the
comet was then farther from both the sun and Earth than when discovered. By
early September, the brightness had faded to 9 and when observed on November
2, it was close to magnitude 10*. The comet was last seen on December 30.
(Colla)

1847 III   Discovered: July 4   Discovery Magnitude: 6.5
     Mauvais (Paris) discovered this comet when it was located only 11 degrees
from the north celestial pole. The comet was then described as a round nebulosity with a tail nearly 5 arcmin long. The motion was determined as southerly.
On July 15, Bond (Cambridge, Massachusetts) independently discovered the comet,
with Schmidt (Bonn) then describing it as 5 arcmin in diameter with a fan-
shaped tail 8 arcmin long. Fading was more rapid than expected during the
remainder of July and into early August, when at the latter time Schmidt indicated a magnitude no brighter than 8. Perihelion came on August 9 (r= 1.77 AU),

and afterwards, fading was naturally rapid due to increasing distances from
the sun and Earth. The tail, which remained visible during most of August,
finally vanished after September 12, and at the end of October, the comet was
lost in the sun's glare.

After conjunction with the sun, the comet reappeared in December, with a
brightness of magnitude 10.5* being indicated by mid-month. By early February
1848, observers at Vienna Observatory found the comet at the limit of visibility of a 15-cm refractor (magnitude 11.5), and at the end of the month,
Bond found the comet near NGC3599, to which it was equal in brightness (magnitude 12), and noted that it had a star-like nucleus. The comet was last observed on April 22, 1848, by Bond using a 38-cm refractor (magnitude 13).
The orbit is elliptical with a period near 44,000 years.
(Mauvais)

1847 IV   Discovered: August 30   Discovery Magnitude: 8.0
Schweizer (Moscow) discovered this small, faint comet near Epsilon Cassiopeiae and described its movement as towards Psi Cassiopeiae. Schmidt (Bonn)
observed the comet on September 18 and 19, and described it as 3 arcmin in
diameter and very faint (magnitude near 9). The comet was last seen on November 28 (magnitude 11). Perihelion had occurred on August 9 (r= 1.48 AU).
(Schweizer)

1847 VI   Discovered: October 2   Discovery Magnitude: 6.0
Maria Mitchell (Nantucket, Massachusetts) had searched long hours for
comets on nearly every clear evening she had, but on the evening of October 2
(Universal Time), her searching lasted only a few minutes before this bright
comet appeared in her telescope. Three other observers independently found
this comet during the next 10 days: de Vico (Rome) found it as it passed
within 5 degrees of the north celestial pole on the 4th; W. R. Dawes (Cran-
brook, England) discovered it in Draco on the 7th, and described it as resembling "a hazy star of 5th magnitude" to the naked eye; and Madame Rumker (wife
of the director of the Hamburg Observatory) found it on the 11th, near Eta
Herculis and described it as 30 arcmin in diameter and near magnitude 4. The
comet passed perigee on October 12 (0.19 AU), and was described as 30 arcmin
across with a magnitude of 3.5*. A tail seemed nonexistent during this time,
although reports of a narrow 2 degree long tail were made on two occasions. The
comet was lost in twilight after November 14, and passed perihelion on the same
day (r= 0.33 AU). Observations after conjunction with the sun were resumed
on December 10, and continued until January 4, 1848, when the magnitude would
have been near 11*.
(Mitchell)

1848 I   Discovered: August 8   Discovery Magnitude: 7.0
Dr. Adolf Cornelius Petersen (Altona) discovered this comet in the evening
sky less than 4 degrees east of Beta Aurigae. He described the comet as small,
but bright. On August 15, Schmidt (Bonn) indicated a magnitude near 6.5 and
by the 25th, the comet was seen in morning twilight at a magnitude near 6.
The comet was last seen on August 27. Perihelion came on September 8 (r= 0.32
AU). The only estimate of the coma diameter was made by Hind (London) on
August 19, when he determined it as 1.5 arcmin.
(Petersen)

1849 I   Discovered: October 26, 1848   Discovery Magnitude: 8.0
This telescopic comet was discovered 8 degrees south of Phi Draconis by

Petersen (Altona) in the evening sky. It then possessed a distinct nucleus and was moving southward. Schmidt (Bonn) observed the comet on November 1, in moonlight (magnitude 7.5), and estimated the coma as 3 arcmin in diameter. On the 15th, he saw the first traces of a fan-shaped tail extending 5 arcmin. Bond (Cambridge, Massachusetts) independently discovered the comet on November 25, and then reported a sharp nucleus and a tail nearly 20 arcmin long. On December 18, observations indicated a magnitude near 5.5 and a tail extending about 10 arcmin. On the same evening, Bond gave the tail length as nearly 2 degrees and also reported a trace of a secondary tail. On December 22, the tail was 30 arcmin long, but, thereafter, it rapidly shrank until, by month's end, it extended only 8 arcmin. In January 1849, the comet became a more difficult object as it neared perihelion. On the 15th, it was located only 10 degrees above the horizon and was described as a small nebulous object with a bright nucleus, but no tail. A couple of days later, the magnitude was found to be near 4.5. Perihelion came on January 19 (r= 0.96 AU), and in the next few days, observers struggled to catch the final glimpses of the comet before it passed behind the sun. When last seen very near the horizon on the 26th, the comet was near magnitude 4.5*.
(Petersen)

1849 II   Discovered: April 15   Discovery Magnitude: 6.5
     Goujon (Paris) discovered this comet in Hydra and described it as a large nebulosity with a bright nucleus. The comet moved northward and in late April, it had brightened to magnitude 5.5* and was located in Leo. By mid-May, the comet was near magnitude 6* and possessed a short, wide tail and a coma 6 arcmin in diameter. Perihelion came on May 26 (r= 1.16 AU), and thereafter, the comet slowly faded as it moved away from both the sun and Earth. In July, observers indicated a magnitude slightly brighter than 10 and when last seen on September 22, it was described as a faint, blurred object (magnitude 10.5).
(Goujon)

1849 III   Discovered: April 11   Discovery Magnitude: 7.5
     G. Schweizer (Moscow) discovered this comet in the evening sky near Beta Coronae Borealis. A few hours later, the comet was independently discovered in Cambridge, Massachusetts, and on the 14th, it was accidentally found by A. Graham (Markree, Ireland). The latter observer described the comet as "easily seen with an ordinary telescope. The nucleus tolerably bright; but badly defined. The coma much diffused." The comet moved southwestward and at the end of April, when at perigee (0.21 AU), it was located in Virgo at a magnitude near 5*. No tail was then present. In early May, the comet became a more difficult object due to its low position in the sky and was last observed on the 9th. Perihelion came on June 8 (r= 0.89 AU), and the comet remained in conjunction with the sun until late August. It then reappeared as a faint object and remained in reach of large telescopes only until the 27th, when the magnitude was near 10.5*. The orbit was elliptical with a period near 13,500 years.
(Schweizer)

1850 I   Discovered: May 1   Discovery Magnitude: 9.0
     This comet was discovered nearly 2 degrees north of Sigma Draconis by Petersen (Altona), who described it as a faint telescopic object moving northward. Around May 8, the comet was described as a very faint nebulosity (magnitude 8.5) with a diameter of 3 arcmin. Petersen then described the central condensation as possessing a granular appearance similar to a cluster of faint

stars. By May 30, the first traces of a tail were noticed and the nucleus
was estimated by Schmidt (Bonn) to be near magnitude 9.5. On June 10, Schmidt
reported no tail, but measured the coma as 4.5 arcmin in diameter and estimated
the nucleus as still near magnitude 9.5. By June 28, the tail extended 28
arcmin and the coma had increased to 6.5 arcmin. In the next few days, the
tail and coma continued to grow as the comet neared perigee (0.5 AU), and on
July 14, the tail was 1.5 degrees long, the coma was 10 arcmin in diameter, the
total magnitude was near 4.5*and the nuclear magnitude equalled 8. During the
remainder of July, the comet became steadily smaller and fainter as it headed
away from the Earth. Perihelion came on July 24 (r= 1.08 AU) and the comet
was lost in twilight after the 28th. It reappeared in the southern sky after
conjunction with the sun on September 5, and was then estimated as 5 arcmin in
diameter. As the comet grew fainter, observers reported it to increase in
size and the central condensation gradually disappeared. When the comet was
last seen on October 16, the brightness had faded to near 10th magnitude.*
(Petersen)

1850 II   Discovered: August 30   Discovery Magnitude: 8.5
    G. P. Bond discovered this comet in the evening sky in Camelopardalis,
about 10 degrees due north of Alpha Persei. Using a 38-cm refractor at Harvard
Observatory, he described the comet as faint and "it presents a very feeble
concentration of light towards the center." The comet was independently discovered by four other observers between September 5 and 14: Brorsen (Senftenberg Observatory) found it on the 5th, as a bright object (magnitude 6.5) with
diffuse edges; Mauvais (Paris) discovered it on the 9th, as a whitish nebulosity nearly 3 arcmin in diameter; Robertson (Markree, Ireland) also found it
on the 9th, and said that under small magnification the comet was very large
and resembled a cluster; and on the 13th, Clausen (Dorpat, Estonia) found the
comet in Lynx. Around mid-September, the comet was widely observed. It was
described as bright (magnitude 6) and contained a sharp central nucleus. An
occasional tail was noticed by some observers. On September 18, the comet was
at perigee (0.40 AU). Afterwards, the increasing distance from the Earth was
countered by a decreasing distance from the sun, thus causing the brightness to
remain near 6th magnitude*well into October. In early October, bright twilight
made observations difficult and the comet was lost after the 13th. Perihelion
came on October 19 (r= 0.57 AU), and the comet was recovered 5 degrees above the
horizon on the 28th, when Bond indicated a magnitude near 5.5. By November 7,
the comet had faded to near magnitude 7*and was last observed on the 14th.
(Bond)

1851 III   Discovered: August 1   Discovery Magnitude: 7.5*
    This comet was discovered at Senftenberg Observatory in the evening sky by
Brorsen. It was then nearly 3 degrees northeast of the globular cluster M3, in
Canes Venatici, and was described as small, but quite bright. Full moon on the
12th, made observations difficult, but by August 23, the comet seems to have
brightened to magnitude 6.5. Perihelion occurred on August 26 (r= 0.98 AU),
and, afterwards, the comet seems to have undergone a sudden change in appearance.
G. Rumker (Hamburg Observatory) had established the fact that a "very distinct
and fine" nucleus existed between August 4 and 29, but observers in September
found no nucleus and indicated an unusually rapid fading. On September 22, the
comet was near magnitude 10.5,* and when last seen on the 30th, it was near 11.
Curiously, this fading and subsequent disappearance took place as the comet was

approaching Earth. It was at perigee on October 12 (0.51 AU), and Brorsen
could then find no trace of the comet using a 10-cm refractor (magnitude
fainter than 10.5) at a time when the predicted brightness, based on pre-perihelion observations, should have been near 7.
(Brorsen)

1851 IV   Discovered: October 22   Discovery Magnitude: 4.0
    This comet was 21 days passed perihelion (r= 0.14 AU) when it was discov-
ered by Brorsen at Senftenberg Observatory. Brorsen then described the comet
as bright, with a shining nucleus and a bright tail more than 1 degree long.
After receiving Brorsen's telegram, Petersen observed the comet the next evening and found it to possess two tails, with the smaller one turned towards the
sun. The comet was last seen with the naked eye on October 24, when Littrow
(Vienna) described it as bright with a 6th-magnitude nucleus which "glittered
like a star." He also described the main tail as very bright and 30 arcmin
long, while the sunward tail was small, faint and 8 arcmin long. As the comet
rapidly moved away from both the sun and Earth, it became a difficult object
in November. On the 14th, Argelander (Bonn) indicated a magnitude near 9, and
added that the tail was clearly visible. On the 17th, the magnitude was near
10.5[*], and when last observed on the 21st, Littrow reported the tail as still
being 32 arcmin long.
(Brorsen)

1852 II   Discovered: May 16   Discovery Magnitude: 9.5
    Chacornac (Marseilles) discovered this comet a short distance from Iota
Cephei and described it as a small, faint and diffuse comet without a tail or
nucleus. Independent discoveries were made by Petersen (Altona) on May 17, and
Bond (Cambridge) on the 18th. The latter observer described the comet as rather
faint, round and 2 arcmin in diameter. On May 24, the comet passed only 1 degree from the north celestial pole. By the 26th, the magnitude was estimated
as near 10, and when last seen on June 15, Bond indicated a magnitude near 11.
Perihelion had occurred on April 20 (r= 0.91 AU).
(Chacornac)

1853 I   Discovered: March 6   Discovery Magnitude: 6.5
    Professor A. Secchi (Rome) discovered this comet near the Eridanus-Lepus
border at a declination of nearly -16 degrees. He described it as a bright
object 5 arcmin in diameter with several bright points in the denser part of
the coma. The comet was independently found on March 8, by Charles W. Tuttle
(Cambridge Observatory) and Schweizer (Moscow) and also on March 10, by Dr.
Hartwig (Leipzig). The comet had already passed perihelion on February 24
(r= 1.09 AU), and when first observed was near perigee (0.5 AU). On March 13,
observers in Bonn also observed bright points of light in the comet's coma.
By the end of the month, the comet had faded to magnitude 8.5[*] and possessed
a coma 2 arcmin in diameter. Fluctuations in brightness were reported at the
beginning of April, but twilight ended observations after the 18th, when the
magnitude was near 10.5[*].
(Secchi)

1853 II   Discovered: April 5   Discovery Magnitude: 7.5[*]
    This comet was found in the dawn sky by Schweizer (Moscow). It was then
in Aquila and was described as 3 arcmin in diameter, with a faint nucleus, but
no tail. By mid-April, the magnitude had brightened to 6.5[*] and the nucleus was
noticeably brighter than when discovered. Observations in the Northern Hemi-

sphere ended after April 24, and as the comet passed within 0.08 AU of Earth
on the 29th, the spectacle was visible to only southern observers. At that time,
the comet had a magnitude near 0* with a tail 5 degrees long and a coma 10
arcmin in diameter. During the next three days, the comet changed its appearance quickly as the nucleus went from magnitude 1 to 7 and the tail decreased
from 4 degrees to 1 degree. Perihelion occurred on May 10 (r= 0.91 AU), and
the comet was last seen on June 11. The orbit was found to be decidedly
elliptical with a period near 781 years.
(Schweizer)

1853 III   Discovered: June 11   Discovery Magnitude: 7.5
     Wilhelm Klinkerfues (Göttingen Observatory) discovered this comet about
10 degrees south of Theta Ursae Majoris. It was then described as bright, with
a tail 3 to 4 arcmin long. In late June, the nucleus was estimated as near
magnitude 10.5 and a short fan-shaped tail extended 3 arcmin. At the end of
July, observers indicated a total magnitude near 7 and a coma diameter of
about 1.5 arcmin. By August 11, the total magnitude was near 5.5* and in the
next few days the rapid brightening allowed the comet to remain visible to
the naked eye as it roamed deeper into evening twilight. The tail reached
a maximum length of over 12 degrees on August 28, and by the 30th, the tail
could not be observed since the comet set with the sun; however, the comet
remained visible in broad daylight until September 4. Perihelion came on
September 2 (r= 0.31 AU), and the next day Hartnup (Liverpool) studied the
comet with a 22-cm refractor. He described the nucleus as sharp-edged and
round with a diameter of 9 arcsec. The total magnitude has been estimated
as brighter than magnitude -1. The closest distance the comet came to the
sun was 8 degrees. The comet remained lost in the sun's glare between the
4th and the 13th, but on the latter date, observers reported a tail 5 degrees
long. Fading was rapid thereafter, and the comet was last observed on January 10, 1854, at a magnitude near 10*
(Klinkerfues)

1853 IV   Discovered: September 12   Discovery Magnitude: 7.5
     This comet was discovered by C. Bruhns (Berlin) in Lynx and was described
as a large, faint nebulous object resembling a star cluster. By September
21, the comet was near magnitude 6.5* but a rapid brightening thereafter
brought the magnitude up to 4.5 on the 30th. On the latter date, observers
estimated a tail at least 15 arcmin long and the nucleus was reported as
becoming prominent at a magnitude between 4 and 5. On October 3, the brightness was equal to Theta Leonis (magnitude 3.3), while the tail was nearly
1 degree in length. After October 4, the comet was lost in the sun's glare
and passed perihelion on the 17th (r= 0.17 AU). After conjunction with the
sun, the comet reappeared on November 27, and was described as a very faint,
diffuse nebulosity about 3 arcmin in diameter. The comet was last observed
on December 12, at a magnitude near 9*
(Bruhns)

1854 I   Discovered: November 25, 1853   Discovery Magnitude: 8.5
     While observing in the constellation Cassiopeia, Robert van Arsdale
(Newark, New Jersey) discovered a "small, round and bright" comet. An independent discovery by Klinkerfues (Göttingen Observatory) was made on December
2, and the brightness was estimated on both dates as being faint. Bright
moonlight soon after discovery made the comet a very difficult object to ob-

serve, but the comet continued to brighten and was reported as about magnitude 7.5 on December 25, 1853. The comet passed perihelion on January 4,
1854 (r= 2.05 AU), and although it was then expected to become slightly
brighter, the comet remained nearly invisible in telescopes due to moonlight.
Traces of a tail had been reported in late December, and observers continued
to observe the tail throughout January, with the length decreasing from 4
arcmin on the 1st to only a trace by the end of the month. The comet slowly
faded after early January and was last seen at Bonn on March 1, 1854, at a
magnitude near 10.5*.
(van Arsdale)

1854 II   Discovered: March 23   Discovery Magnitude: 2.0*
     This comet was discovered in Cassiopeia shortly before sunrise by de
Menciaux (Overton, England), who described the tail as situated perpendicular to
the horizon. Perihelion came on the 24th (r= 0.28 AU), and the comet was lost
in twilight until March 28, when detected in the evening sky. In the following days, the comet was observed throughout the world as a bright object
of magnitude 0*with a tail nearly 5 degrees long and a nucleus near magnitude
1.5. Fading was rapid and by April 8, the total magnitude had dimmed to 5.5*,
while the nucleus had faded to 6. By mid-April, the tail had decreased to
less than 1 degree long and the nucleus was described as 3 to 4 arcmin in
diameter with a magnitude near 8. The comet was last seen on April 28, when
observers in India described it as a small white cloud. The orbit bears a
close resemblance to that of the comet of 1677.
(Great Comet)

1854 III   Discovered: June 5   Discovery Magnitude: 6.0
     Klinkerfues discovered this comet in Triangulum and, shortly thereafter,
Argelander described it as very bright with a substantial condensation towards
the center. By mid-June, observers gave magnitude estimates near 5 and also
reported a coma 4 arcmin in diameter, with no nucleus, and a narrow tail 1
degree long. Perihelion came on June 22 (r= 0.65 AU), and on June 24, van
Arsdale (Newark) independently discovered the comet (magnitude near 4). At the
end of June, the comet magnitude had faded to 5*and the tail extended nearly
1 degree. On July 11, Schmidt estimated the coma diameter and tail length as
1.3 arcmin and 7 arcmin, respectively, and added that the total magnitude
was near 6.5. By July 14, observers reported a distinct decrease in brightness
with only a trace of tail still visible. The comet was last seen on July 30,
when it was described as very faint. The orbit is similar to that of the comet
of 961.
(Klinkerfues)

1854 IV   Discovered: September 11   Discovery Magnitude: 8.0*
     This comet was first seen by Klinkerfues and was thereafter independently
discovered by Bruhns on September 12, van Arsdale on the 13th, Giovanni Battista
Donati (Florence) and Mitchell (Nantucket) on the 18th, and Gussew (Vilna, Russia)
on the 21st. The comet was generally described as faint, diffuse and large.
The comet was then located in the north polar region and, when discovered by
Mitchell, was nearly joined to the galaxy M82 in Ursa Major. Increasing distance from the Earth was countered by a decreasing distance from the sun so
that the comet remained near magnitude 7.5*right up to the day of perihelion on
October 28 (r= 0.80 AU). During September the coma had been estimated as about
5 arcmin in diameter. With increasing distances from both the Earth and sun

the comet faded rapidly and was last observed by Colla (Parma) on December 3, using a 9.6-cm refractor. Colla had reportedly observed this comet between December 22 and 24, but later orbital calculations revealed that he had instead made prediscovery observations of comet 1854 V instead.
(Klinkerfues)

1854 V   Discovered: December 22   Discovery Magnitude: 8.0
     This comet was discovered on January 14, 1855, by F. A. Winnecke (Berlin) and Dien (Paris), with the former observer describing the comet as a faint nebulosity near Gamma Scorpii. Sometime later, however, Colla (Parma) was found to have actually observed this comet on December 22, 1854, but at the time thought he was continuing his observations of comet 1854 IV. Having passed perihelion on December 16, 1854 (r= 1.36 AU), the comet slowly faded after discovery. At the end of January, the brightness had decreased to near 8.5* and observers measured the coma as nearly 5 arcmin in diameter. By mid-February, the brightness was near 9.5*, while the circular coma had shrank to about 2 arcmin. The comet was last seen on April 23, as an exceedingly faint spot of magnitude 11.5*.
(Winnecke-Dien)

1855 I   Discovered: April 11   Discovery Magnitude: 8.5
     Discovered near Delta Corvi in the evening sky by Schweizer (Moscow), this comet was a faint object with a starlike nucleus when viewed through an 8-cm refractor. It had passed perihelion on February 5 (r= 2.19 AU), and after discovery, fading was very slow. In early May, Winnecke (Berlin) reported the comet slightly brighter (magnitude near 8) at a time when it should have been slightly fainter than at discovery. He also reported a coma diameter of 40 arcsec. By May 18, the comet had faded to about magnitude 9.5* and was last observed on June 5. The orbit is elliptical with a period near 500 years.
(Schweizer)

1855 II   Discovered: June 3   Discovery Magnitude: 6.0
     Donati (Florence) discovered this comet 4 days after it had passed perihelion (r= 0.57 AU) and described it as slightly fainter than the globular cluster M13 (hence, a magnitude near 6.0). Independent discoveries were made by Dien (Paris) and Klinkerfues (Göttingen Observatory) on the 4th. Bruhns (Berlin) observed the comet on June 5, and described it as a nebulous mass slightly less than 1 arcmin in diameter, without any tail and comparable in brightness to the globular cluster M2 in Aquarius (magnitude 6.3). The comet remained low over the horizon after discovery and on June 14, it had reached an elevation of only 19 degrees. The comet was then near magnitude 8*. Moon-light during the next several days made observations impossible and the comet was seen one last time on June 30, in Cancer, at a magnitude near 10*. Despite the short duration of visibility, an orbital calculation by George van Bies-broeck (1916) gave an elliptical orbit with a period of 252 years, though this period is considered very uncertain.
(Donati)

1855 IV   Discovered: November 13   Discovery Magnitude: 7.5
     This comet was discovered nearly 2 degrees north of Alpha Sextantis by C. Bruhns (Berlin) and was described as a feeble nebula. It moved rapidly westward thereafter, but remained little changed in brightness through peri-helion on November 25 (r= 1.23 AU). At the beginning of December, the mag-nitude brightened rapidly as the comet neared perigee, and when closest to

Earth on the 6th (0.25 AU), it was described as visible to the naked eye (magnitude 5) with a coma 15 arcmin in diameter. By the end of December, the comet had already faded to a magnitude near 9.5*and was last observed on January 3, 1856 (magnitude near 10.5).
(Bruhns)

1857 I   Discovered: February 23   Discovery Magnitude: 7.5
     Professor d'Arrest (Leipzig) discovered this comet near the Vulpecula-Pegasus border and described it as rather bright with a diameter of 1.5 arcmin. On February 26, an independent discovery was made by van Arsdale, who found the comet near Kappa Pegasi. During the first days of March, the comet had brightened to about magnitude 6.5*and possessed a nucleus inside a 3 arcmin diameter coma. No tail was noticeable then, although Winnecke (Bonn) said the coma was extending towards the sun. The extension was also noticed by others, and on March 17, Schmidt (Olmutz) measured it as 2 arcmin long. The comet passed perihelion on March 21 (r= 0.77 AU), and the magnitude was then near 6*. A faint trace of tail nearly 5 arcmin long appeared on the 26th, and remained visible through April 11. On this latter date, the comet was seen with the naked eye by Schmidt (magnitude near 5). The comet steadily faded thereafter, and was last seen on May 2.
(d'Arrest)

1857 III   Discovered: June 23   Discovery Magnitude: 8.0
     Klinkerfues (Göttingen Observatory) described this comet as of moderate brilliancy in his telescope when he discovered it between Beta and Epsilon Persei. Independent discoveries were made by Dien (Imperial Observatory, Paris) on June 24, and Habitch (Gotha) on the 25th. A few days later, observers estimated the coma as 3 arcmin in diameter with traces of a tail. The comet brightened during July, as it headed for perihelion, with estimates of magnitude 7 on the 2nd, and 5.5 on the 13th. During the same time the coma shrank to 0.9 arcmin by the 10th, and the tail steadily grew to 10 arcmin long by the 12th. Observations thereafter were difficult due to the comet's movement into morning twilight. Perihelion came on July 18 (r= 0.37 AU), and the comet was last seen at low altitude the next morning.
(Klinkerfues)

1857 IV   Discovered: July 26   Discovery Magnitude: 8.5
     C. H. F. Peters (Dudley Observatory, New York) discovered this comet nearly 10 degrees north of Alpha Persei in Camelopardalis. Peters then described it as "very faint, showing no sort of nucleus and an ill-defined outline." Independent discoveries were made on July 28 by Dien (Imperial Observatory, Paris) and on July 30 by Habitch (Gotha) and Donati (Florence). On July 31, the coma was estimated as 3 arcmin in diameter and irregularly shaped, with a slight tail. The comet's brightness changed very little during August, as the decreasing solar distance countered the effects of an increasing distance from Earth. On August 20, the coma diameter was estimated as 1.5 arcmin and a nucleus near magnitude 9 was seen. Perihelion came on the 24th (r= 0.75 AU), and fading was rapid thereafter. The magnitude on September 25 was near 9.5*, and when last seen on October 22, it was near 10.5*. Orbital calculations revealed an elliptical orbit with a period of 235 years. Although comets are named after their discoverers, Peters proposed to call this comet the Olcott Comet, "after the very beloved and esteemed name of the distinguished citizen who is identified with the history of the erection of this observatory." Despite this

proposed tribute, the comet is still known today as comet Peters.
(Peters)

1857 V   Discovered: August 20   Discovery Magnitude: 8.0*
     This comet was discovered by Klinkerfues (Göttingen Observatory) in Camel-
opardalis when 13 degrees from the north celestial pole. On August 23, the
comet was located only 9 degrees from the pole, but thereafter began a more
southerly movement. Observers at this time indicated a coma diameter of
5 arcmin, and by the 27th, the comet had reached naked-eye visibility. By
mid-September, the magnitude was near 4*and the tail extended 2.5 degrees from
a coma 4 arcmin in diameter. By the end of the month, the tail had increased
to 4 degrees and the total magnitude was considered to be near 3. Perihelion
came on October 1 (r= 0.56 AU), and the comet remained visible only until the
3rd, when it was seen in very bright twilight at a time when no other stars
were visible. The orbit seems to be elliptical with a period near 2,500 years.
(Klinkerfues)

1857 VI   Discovered: November 10   Discovery Magnitude: 9.0
     Donati (Florence) and van Arsdale (Newark) discovered this comet only 5
hours apart just south of Iota Draconis. The comet was then described as
small and faint with a nebulous appearance. Soon afterward, the coma was measured as 3 arcmin in diameter and possessed no visible condensation. This appearance and the magnitude of 9*remained unchanged through the remainder of
November. During December, fading was rapid and the comet was last seen on
the 19th, at a magnitude near 10.5*. Perihelion occurred on November 19 (r=
1.01 AU).
(Donati-van Ardsale)

1858 IV   Discovered: May 22   Discovery Magnitude: 7.0
     C. Bruhns was comet hunting in the morning sky when he discovered this
comet near Tau Andromedae. He described it as easily observed with a coma
diameter of 3 or 4 arcmin. The comet passed perihelion on June 5 (r= 0.54 AU),
and was then at its maximum brightness of 6.5*with a tail 30 arcmin long. By
June 10, it had faded to 7.5, according to d'Arrest, and was last seen on July
15 (magnitude 10).
(Bruhns)

1858 VI   Discovered: June 2   Discovery Magnitude: 7.5*
     Donati (Florence) discovered this comet near Lambda Leonis and described
it as a feeble round nebulosity about 3 arcmin in diameter. He announced his
discovery with some reserve as this object was located in the same part of the
sky as comet 1858 III (periodic comet Tuttle-Giacobini-Kresak), which had been
discovered one month before; however, Donati's comet proved to be new. Movement
was very slow after discovery and by June 19, the comet had advanced only 106
arcmin to the north-northeast. Brightening was also slow and when independently
discovered in rapid succession by Horace P. Tuttle (Harvard College Observatory),
Henry M. Parkhurst (Perth Amboy, New Jersey) and Maria Mitchell (Nantucket, Massachusetts) between June 28 and July 1, the comet had brightened by less than
half a magnitude. A more rapid brightening began in August, with a magnitude
of 6.5*being estimated on the 5th, and 5.5*being estimated by the 20th. Around
the time of the latter observation, mathematicians had produced calculations
which indicated the comet would become very bright at the end of September. In
late August, the comet occupied such a position in the heavens as to allow it to
rise before the sun and set after it. This circumstance led some observers to

believe that two different comets had appeared in the skies.

The comet developed into one of the most impressive comets of the century during September, beginning with the appearance of a distinctive curved tail on the 6th. Among the many observatories conducting extensive programs of observation was Harvard College Observatory, which on September 12 noted a sudden increase in both the brightness and the length of the tail. The total magnitude was then estimated as near 3, while the 10-arcsec diameter condensation resembled a star of magnitude 5. The tail then extended 6 degrees. By September 16, the nucleus was said to outshine the brightest stars of the Big Dipper (magnitude 1.5), and Winnecke (Pulkovo Observatory) caught the first glimpse of one of two narrow, straight rays which later formed a tangent to the strong curve of the main tail. On September 25, the main tail extended 11 degrees and by the 28th, it was 19 degrees long.

Perihelion came on September 30 (r= 0.58 AU), and afterwards the comet became even more spectacular as it headed for perigee (0.5 AU) near October 9. On October 2, the total magnitude outshone Arcturus (hence, a magnitude near -1) and the tail extended 25 degrees. By the 5th, the magnitude was said to be comparable to Arcturus (magnitude 0.5), while the tail had grown to 35 degrees. The comet was described as being most spectacular on October 10-- when nearest Earth. The tail stretched 60 degrees across the sky in "a magnificent scimitar-like curve," and the nucleus was described as being as prominent as a 1st-magnitude star. Afterwards, as the distances from the sun and Earth increased, the comet rapidly faded. On the 15th, the magnitude was still near 1[*] and the tail extended 15 degrees, but by October 31, the brightness was fainter than Beta Aquilae (magnitude near 4). The comet was last seen with the naked eye on November 8 (magnitude 5.5) and observations continued until March 4, 1859 (magnitude near 10).

Numerous estimates of the size of the nucleus were made during September and October, 1858. Though some of these indicated a diameter of nearly 2000 miles, estimates made during the close approach to Earth established a diameter nearer to, and probably smaller than, 500 miles. The comet appears to move in an elliptical orbit with a period near 2000 years.
(Donati)

1858 VII   Discovered: September 6   Discovery Magnitude: 7.5

Tuttle (Harvard College Observatory) discovered this comet in the evening sky in Perseus. The comet was widely observed in October, as it neared both perihelion and perigee. The latter occurred near October 3 (0.5 AU), and the comet was described as a bright object of substantial size. In the next couple of days, observations could be made with the naked eye (magnitude 5.5) and by the 8th, the coma was estimated as 5 arcmin across without a tail. Perihelion came on the 13th (r= 1.43 AU), and fading was rapid thereafter. The comet was last seen on November 10, as a very faint object (magnitude near 10.5).
(Tuttle)

1859   Discovered: April 2   Discovery Magnitude: 7.5

William Tempel (Venice) found this comet near Gamma Ursae Minoris. On April 18, Peters (Altona) described the comet as 4 arcmin in diameter with a weak condensation at the center. The brightness had changed little since the discovery. American observers were not informed of the new comet until a rash of independent discoveries occurred in late April. On April 23, James C. Watson (Ann Arbor, Michigan) found the comet near 17 Lyncis and described it

as bright (magnitude 6.5) and elongated in a direction opposite the sun. Other
discoveries were made by Tuttle at Harvard Observatory (April 27) and James
Ferguson at Washington (April 28). In the middle of May, the comet was visible
in twilight (magnitude near 5) and possessed a coma nearly 5 arcmin in diameter with a tail 15 arcmin long. Shortly thereafter, it was lost in the sun's
glare. Perihelion came on May 29 (r= 0.20 AU), and when the comet had exited
the sun's glare at the end of June, it was descibed as faint and diffuse with
a magnitude near 10. The final observation was made on July 1, 1859.
(Tempel)

1860 I   Discovered: February 27   Discovery Magnitude: 6.5

While observing the star Mu Doradus, the French astronomer E. Liais (Director of the Brazilian Coast Survey, Olinda, Brazil) discovered a very unusual
comet which consisted of two separate components. The larger comet was of
magnitude 6 or 7 and possessed a 9th-magnitude nucleus. The coma was nearly
a half arcmin in diameter with a noticeable inner coma 7 or 8 arcsec across.
The secondary comet was 4 arcsec in diameter with a central condensation. The
larger and brighter component preceded the fainter comet across the sky and,
judging by the location of the nucleus inside the coma, Liais remarked on the
noticeable elongation of the comet away from the sun. His drawings showed a
short, broad tail of undetermined length and two luminous jets directed towards
the sun on his first night of observation. At the same time he drew the nucleus
as divided into three condensations. The smaller comet was located 73 arcsec
from the main body and was described as perfectly round. The luminous jets
persisted until March 11, and the next evening the secondary component was
nowhere to be seen. The final observation of the primary component came on
March 14, 1860. The comet was observed by Liais only during the 17 days of
observation and parabolic orbits were calculated for both bodies. Calculations
by various mathematicians indicate a split date in December 1859--nearly 2
months before the February 17 perihelion (r= 1.20 AU). The orbit bears some
resemblance to comet 1942 IV.
(Liais)

1860 II   Discovered: April 17   Discovery Magnitude: 7.5

Having passed perihelion on March 6 (r= 1.31 AU), this comet was discovered
by George Rumker (Hamburg Observatory) near Theta Persei when 2 AU from Earth.
Fading was quite rapid as the comet moved away from both the sun and Earth and
the brightness dropped 3 magnitudes in the next month. Estimates of the coma
diameter were consistently near 2 arcmin during May and no tail was ever detected. The comet was last seen on June 12.
(Rumker)

1860 III   Discovered: June 18   Discovery Magnitude: 3.5

This comet passed perihelion on June 16 (r= 0.29 AU), and was first seen
two days later in Italy. During the next three days, independent discoveries
were made by at least 100 people around the world as the comet appeared in the
evening sky visible to the naked eye and possessing a tail several degrees long.
The comet began being viewed in dark skies around June 24, and observers then
saw a tail 20 degrees long. Afterwards, as the comet neared Earth, the magnitude increased and the tail decreased and by the 28th, magnitude estimates
were as high as 1.5, while the tail was estimated as 15 degrees long. The
magnitude remained near 2*until July 6, when the comet was nearest Earth (0.5
AU), and thereafter rapidly declined. On July 11, the magnitude was near 3*,

with a tail 1 degree long, but by the 18th, the brightness was down to 4.5*.
Further independent discoveries were made during the latter half of July, as
the comet became visible to Southern Hemisphere observers. By September 12,
the magnitude was estimated as near 8.5* and the final observation came on
October 18, when observers at the Cape of Good Hope indicated a magnitude
near 11.
(Great Comet)

1861 IV   Discovered: October 24   Discovery Magnitude: 7.5
     Tempel (Marseilles) discovered this small comet in Leo. It moved north-
ward and was again seen in Marseilles on the 25th, and in Paris on the 26th.
On the latter date, the coma was described as nearly 4 arcmin across. Bad
weather and bright moonlight prevented further observations. Orbital calculations indicated a perihelion on either September 21 or 22 (r= 0.65 to 0.68
AU). There has been some conjecture that an object seen by Tuttle (Harvard
Observatory) near the north celestial pole on November 14 might be this same
comet. This object was then described as extremely faint and was not seen on
any other date by Tuttle.
(Tempel)

1861 I   Discovered: April 5   Discovery Magnitude: 7.5
     A. E. Thatcher (New York) discovered this comet in Draco and described it
as a tailless nebulosity 2 arcmin in diameter with a central condensation. The
comet brightened slowly during April as it drew closer to the sun and Earth and
was near magnitude 5.5* on the 28th. An independent discovery was then made by
Baeker (Nauen, Germany) with the naked eye. On May 4, observers indicated a
magnitude near 3.5. The coma was estimated as 8 arcmin in diameter and the tail
was at least 1 degree long. Between May 9 and 10, the comet passed less than
0.3 AU from Earth and was described as 20 arcmin across with a tail 1 degree
long and a magnitude near 2.5*. Afterwards, the comet's decreasing distance from
the sun countered the slowly increasing distance from Earth and between May 15
and 28, the brightness declined from a magnitude of 3.5 to 4*. Afterwards, the
comet was lost in twilight and perihelion came on June 3 (r= 0.92 AU). The comet
was recovered in the Southern Hemisphere on July 30, as a nebula without any
visible nucleus and was followed until September 7, when observers indicated
a magnitude near 10. The orbit was firmly established as elliptical with a
period near 415 years.
(Thatcher)

1861 II   Discovered: May 13   Discovery Magnitude: 4.5*
     Jerome L. Tebbutt (Windsor, New South Wales) discovered this comet with
the naked eye. The comet steadily brightened as it approached the sun and
Earth and in the first days of June, observers indicated a magnitude near 2
and a tail about 5 degrees long. The comet was followed in the Southern Hemisphere until June 11, when the brightness was near 1* and the tail extended 40
degrees; subsequently it was lost in the sun's glare. Perihelion came on
June 12 (r= 0.82 AU), and the comet suddenly appeared to observers in the
Northern Hemisphere on June 29.
     On June 30, the comet was located only 0.13 AU from Earth and was described
as being as bright as Saturn (magnitude 0) with a coma 30 arcmin in diameter
and a tail 100 degrees long. During the same day, Earth apparently passed
through the tail, according to some predictions, but despite some unconfirmed
accounts of unusual daytime darkness and a yellowish sky, most reports indicated

no perceptible effects. The tail length and brightness rapidly decreased
during July. From values of near 100 degrees in the first days of July,
the tail decreased to 40 degrees on the 8th and 30 degrees on the 12th. The
magnitude was compared to various stars during the month with the following
estimates being made: 0.0 on the 1st, 2.4 on the 7th, 2.7 on the 12th, 3.5
on the 17th and 4.5 on the 24th.

Naked-eye visibility ended in mid-August and the comet continued to be
observed through the end of the year, when a description on the 21st of December indicated a magnitude near 8.5. During 1862, the comet became a difficult
object with the brightness being near 10*in early February, and near 11*when
last seen on May 1. Calculations revealed a decidedly elliptical orbit with a
period near 409 years.
(Great Comet)

1861 III   Discovered: December 29   Discovery Magnitude: 7.5
This comet was discovered 22 days after it passed perihelion (r= 0.84 AU)
by Tuttle (Harvard College Observatory) in Virgo. On January 8, 1862, Winnecke
(Pulkovo Observatory) independently discovered the comet and described it as
nearly 4 arcmin in diameter with an intense condensation. Movement was very
rapid as the comet approached perigee (0.3 AU on January 14) and by the 22nd,
it was located in Cepheus only 9 degrees from the north celestial pole. Fading
was unusually fast during late January, as the comet changed from 8.5*on the
19th to near 10.5*when last detected on February 2. The predicted magnitude
for the latter date, based on estimates in December and early January, should
have been near 9.5.
(Tuttle)

1862 II   Discovered: July 2   Discovery Magnitude: 4.5
Schmidt (Athens) discovered this comet 10 days after it had passed perihelion (r= 0.98 AU). It was then situated near Beta Cassiopeiae and was described as 22 arcmin in diameter. Independent discoveries were made by Tempel
(Marseilles) a few hours after Schmidt, and by Bond (Cambridge, Massachusetts)
and Simons (Albany, New York) on July 3. The comet passed only 0.10 AU from
Earth on July 4, and attained a magnitude of 4.5*with a coma 34 arcmin in
diameter and a tail 30 arcmin long. Schmidt's accurate measures of the coma
diameter continued throughout July as the comet moved rapidly away from the sun
and Earth. On the 14th he found it to be 9.7 arcmin in diameter and this decreased to 6.0 arcmin on the 19th and 2.0 arcmin on the 29th. During the
same time period, the magnitude faded from 5.5*on July 7 to nearly 10*by the
30th. The comet was last observed on the 31st of July.
(Schmidt)

1862 IV   Discovered: November 28   Discovery Magnitude: 6.5
This comet was discovered by Respighi (Bologna) in the morning sky. It
was then located in Virgo and was described as 3 arcmin in diameter with a
nucleus. An independent discovery was made by Bruhns (Leipzig) on December 1,
at which time the comet had advanced into morning twilight. On December 16,
Bruhns described the comet as faint due to its low position in the sky, but
added his opinion that it could have been visible to the naked eye if more
favorably placed in the sky (magnitude 5.5). The comet was lost in twilight
soon afterward, and passed perihelion on December 28 (r= 0.80 AU). It remained
hidden in the sun's glare until mid-February, when it was recovered on the 18th
(magnitude 9.5), and was followed for only 2 more days before becoming too

faint for further observations.
(Respighi)

1863 I   Discovered: December 1, 1862   Discovery Magnitude: 9.0
     Bruhns (Leipzig) discovered this comet in Sextans and described it as 2
arcmin in diameter and very faint. Still two months from perihelion, the
comet slowly brightened to about magnitude 8.5*on December 15, and 7.5*by
January 24, 1863. On the latter date, the coma was estimated as 1 arcmin in
diameter with a bright nucleus. Perihelion came on February 3 (r= 0.79 AU),
and by the 9th, observers indicated a magnitude near 7. The coma was then
1 arcmin across and the nucleus was called "very bright." The comet was
considered brightest in mid-February, and traces of a tail were reported by
several observers between the 13th and 18th. Afterwards, the comet rapidly
faded as it moved away from the sun and Earth. On March 1, it was near magnitude 8.5*and when last seen on March 13, it was hardly recognizable at a
magnitude near 10*.
(Bruhns)

1863 II   Discovered: April 12   Discovery Magnitude: 5.5
     Having passed perihelion on April 5 (r= 1.07 AU), this comet was discov-
ered in Aquarius by Klinkerfues (Göttingen Observatory). He described it as
bright with traces of a tail. On April 14, the comet was independently discovered by Donati (Florence) as it was moving through Delphinus. The comet
slowly faded as it moved away from the sun and Earth, and by the end of May,
the magnitude was near 7*. Also during May, the comet passed 10 degrees from
the north celestial pole. By mid-August, the comet was described as circular
with a diameter of 30 arcsec. At the same time, observers indicated a total
magnitude slightly fainter than 9, with a nucleus near magnitude 11.5. In
mid-September, the brightness had declined to 9.5*and on October 1, it was
near 10. The comet was followed until November 15, when observers at Pulkovo
Observatory indicated a magnitude near 12.5. The comet's orbit is elliptical
with a period near 45,000 years.
(Klinkerfues)

1863 III   Discovered: April 13   Discovery Magnitude: 5.5
     On the evening after full moon, Respighi (Bologna) was searching for
comets when he spotted a bright one in Pegasus. His description indicated a
nucleus as bright as a star of 6th magnitude and a tail was estimated as 40
arcmin long. The comet moved to the northeast and was independently discovered
on the 14th by Baeker (Nauen) and on the 16th by Winnecke (Pulkovo Observatory).
The latter observer described the tail as 3 degrees long and indicated a magnitude near 4.5. Perihelion came on April 21 (r= 0.63 AU), and observers indicated a magnitude near 4 with a fan-shaped tail 1 degree long. Although considered slightly brighter on the 22nd, a rapid fading set in thereafter and by
May 9, it was near 6th magnitude*. Observations continued until June 2, when
the magnitude must have been fainter than 9*. The orbit is elliptical with a
period near 17,800 years.
(Respighi)

1863 IV   Discovered: November 5   Discovery Magnitude: 4.0
     This comet was discovered by Tempel (Marseilles) about 2 degrees north of
Delta Crateris and was described as bright with a tail 2 degrees long. Perihelion came on the 6th (r= 0.71 AU), and the comet was independently discovered 8 days later by Schmidt (Athens), who described it as visible to the naked

eye (magnitude 4.0) with a tail 3.5 degrees long. The magnitude reached a
maximum of 3*in mid-November when the comet was at perigee (0.7 AU) and
observers then described the tail as nearly 10 degrees long. Fading was
relatively slow thereafter as the magnitude dropped to 4.5*on November 22, and
5*by December 3. When observed in late January, no tail was visible and a
magnitude near 8 was indicated. The comet was last seen on February 10, 1864
(magnitude 9). The elliptical orbit has a period near 15,800 years.
(Tempel)

1863 V   Discovered: December 28   Discovery Magnitude: 6.5
        This comet was first discovered on the day of perihelion (r= 0.77 AU) by
Respighi (Bologna) near the Hercules-Lyra border. The brightness of the comet
brought about independent discoveries by Baeker (Nauen) on January 1, 1864,
and Karlinski (Krakow, Poland) and Watson (Ann Arbor, Michigan) on January 9.
The early observations simply indicated a circular coma and a small tail, but
observations on January 9 indicated an 8th-magnitude nucleus with a coma 4
arcmin in diameter and a tail 30 arcmin long. The comet moved rapidly towards
the north and on the 29th, it was located in Cassiopeia. On January 31, it
arrived at perigee (0.18 AU) and was estimated as near magnitude 5.5, with an
elongated nucleus and a jet extending in the direction of the tail. The comet
continued its rapid motion into February, as it moved through Perseus, Taurus
and into Orion by the 10th. The next evening, the comet was described as very
faint (magnitude near 9). The comet was followed until March 1, when the magnitude must have been near 10*. An attempt by Schmidt to find the comet on the
4th ended in failure.
(Respighi)

1863 VI   Discovered: October 10   Discovery Magnitude: 7.5
        Baeker (Nauen) discovered this comet in Leo, with Tempel (Marseilles)
making an independent discovery 5 days later. The latter observer described
the comet as a diffuse nebula with a nucleus situated on the sunward side of
the geometric center of the coma. The comet was then located 1.7 AU and 2.1
AU from the sun and Earth, respectively. Although approaching the sun and
Earth, the comet's appearance changed little during November as observers consistently estimated magnitudes of 7 and 11 for the coma and nucleus, respectively. The coma was then nearly 4 arcmin in diameter, while the tail extended
over 10 arcmin. During December, the total magnitude increased to nearly 6*,
while the nuclear magnitude reached 9.5. Perihelion came on December 29 (r=
1.31 AU), and at the turn of the year, the comet passed only 0.14 AU from
comet 1863 IV--both being in the same field of view of a wide-angle comet-
seeker. Thereafter, the comet was followed until April 14, when the total magnitude was near 11*.
(Baeker)

1864 I   Discovered: September 10   Discovery Magnitude: 8.5
        Having passed perihelion on July 28 (r= 0.63 AU), this comet was discov-
ered by Donati (Florence) as a very faint object nearly 1½ months later. Although the comet was moving rapidly away from the sun, its decreasing distance
from the Earth caused it to fade slowly during the short time it was visible.
On September 15, the magnitude was near 9*and an estimate of the coma diameter
gave 1.5 arcmin. By month's end, the comet had faded to magnitude 9.5*and
observations officially continued only until October 10, when the comet was
near magnitude 10*. There is a possibility that further observations were made

on October 19 and 20, by Reslhuber (Kremsmunster), though he said he could
not see the comet with full certainty.
(Donati)

1864 II   Discovered: July 5   Discovery Magnitude: 6.0
        Discoveries of this comet were made by Tempel (July 5), Respighi (July 6)
and Kowalczyk at Krakow (July 13). Descriptions at that time indicated the
comet was round, without a tail, and nearly 4 arcmin in diameter. An ill-defined neculeus had a magnitude near 9. By July 30, the comet had brightened to
magnitude 4.5*, while the coma had a diameter of 6 arcmin with a slight elongation away from the sun. At the beginning of August, observers began to view
the results of an approaching perigee. On the 3rd, the coma diameter was 10
arcmin, with no tail present. On the 5th, the coma had swelled to 15 arcmin
and observers reported a very narrow tail 3 degrees long. On August 6, Schmidt
(Athens) gave the coma diameter as 32 arcmin and estimated the magnitude as
between 2 and 3. He also reported a "gossamer-like tail" which was nearly
30 degrees long. Perigee came on August 8 (0.10 AU), and three days later the
comet was independently discovered in the Southern Hemisphere by Qnaife (Australia) and Moesta (Santiago) with the naked eye. The latter observer reported
the head as being as bright as a star of magnitude 2, with a coma nearly 1
degree in diameter. Although the tail was then invisible to the naked eye,
telescopes revealed it as 40 arcmin in length. Perihelion came on the 16th
(r= 0.91 AU), and afterwards, the magnitude dropped quickly. The comet dropped
below naked-eye visibility at the end of August, and by September 8, no tail
was visible--even in telescopes. In mid-September, the comet was observed in
moonlight (magnitude near 7.5) and observations continued until October 5
(magnitude near 9). This object holds the distinction of being the first comet
to have its spectrum observed. This feat was accomplished on August 5 and 6,
by Donati (Florence). The comet's orbit is elliptical with a period near
3,900 years.
(Tempel)

1864 III   Discovered: July 23   Discovery Magnitude: 9.5
        Donati and Toussaint (Florence) discovered this comet in Coma Berenices
and described it as very faint with a coma 2 arcmin across. The comet was not
quite as difficult to observe during August, as it brightened to about magnitude
8.5*by mid-month. Donati spotted a 15-arcmin-long tail on the 3rd, and a stellar
nucleus was reported by other observers. The comet was lost to Northern Hemisphere observers after mid-August as it moved southwestward; however, Moesta
(Santiago) began observations on August 30, when he detected a tail 20 arcmin
long and a magnitude near 8*. The comet was lost in evening twilight after mid-
September. Perihelion came on October 11 (r= 0.93 AU), and Moesta began observing the comet again on November 2. At the end of the month, the comet's rapid
southern motion had brought it to within 8.5 degrees of the south celestial
pole and during December, it moved steadily northward. Northern Hemisphere
observers finally spotted the comet again in mid-January, 1865. It was then
described as near magnitude 9*, with an 11th-magnitude nucleus. The coma was
then 1 arcmin in diameter and the tail extended 7 arcmin. On January 21, the
coma was 2 arcmin in diameter and by the 27th, the nuclear brightness had declined to magnitude 12, with the tail still extending 5.5 arcmin. The comet
was followed until February 25, 1865, when the total magnitude had faded to
near magnitude 11*. The orbit is elliptical with a period near 55,000 years.
(Donati-Toussaint)

1864 IV   Discovered: December 15   Discovery Magnitude: 6.5
     This comet was discovered by Baeker (Nauen) just north of Scutum and was
described as a very bright telescopic comet. The comet moved eastward and
passed perihelion on December 22 (r= 0.77 AU). By the 23rd, it had developed
traces of a tail, but observers could see no nucleus. On the 29th, Engelman
(Leipzig) estimated the coma diameter to be 1 arcmin, with a small tail pointing away from the sun. The comet's magnitude was estimated as being as bright
as a star of magnitude 6 and a nucleus 15 arcsec in diameter could be seen..
The next evening, Engelman said the comet seemed brighter (magnitude 5.5) and
this aspect continued almost to the end of January 1865, despite the fact that
the comet was moving away from both the sun and Earth. On January 25, the
comet was described as large and very bright, with a prominent central condensation. During the first week of February, observers found the comet to still
be very bright in moonlight with a strong condensation. A rapid fading began
thereafter, and observations on February 20 and 25, indicated brightnesses of
7.5 and 8.5, respectively. Low altitude prevented further observations of the
comet after the latter date.
(Baeker)

1864 V   Discovered: December 31   Discovery Magnitude: 9.0
     This comet passed perihelion on December 28 (r= 1.11 AU), and was discov-
ered two days later by Bruhns (Leipzig). At discovery, the comet was described
as a diffuse nebula 2 arcmin in diameter. Although the comet was moving away
from the sun during January, it was also moving rapidly towards the Earth. The
outcome of this condition should have made the comet a bright telescopic object
by January 24, but instead, the comet faded during the month. On January 21,
it was described as exceedingly faint and difficult to measure (magnitude near
9.5) and when last seen on January 29, the comet could hardly be seen by observers at Leipzig (magnitude near 10.5). The comet was at perigee on the 24th
(0.3 AU). Observations were attempted in February, but no trace of the comet
could be found.
(Bruhns)

1865 I   Discovered: January 17   Discovery Magnitude: 2.0
     After having approached the sun on a course that kept it in twilight, this
bright comet passed perihelion on January 14 (r= 0.03 AU), and then suddenly
appeared to observers in the Southern Hemisphere on the 17th and 18th. Numerous independent discoveries were made, beginning with Abbott (Hobart Town,
Australia) and Moesta (Santiago) who supplied the first detailed observations of
the comet when it was only 9 degrees from the sun's limb at sunset. When first
discovered, the tail was nearly 15 degrees long and this increased to 26 degrees
by January 24th. At this time, the tail was noticeably curved with a sharply
bounded southern edge and a diffuse northern edge. By January 30, the tail had
declined to a length of 12 degrees. During the first half of February, the tail
declined from a length of 5 degrees to 3 degrees and the magnitude fell from 4
to 4.5.* The comet was still barely visible to the naked eye on February 22, but,
thereafter, a rapid fading set in. By mid-March, the magnitude was near 9*and at
the end of the month observers indicated a brightness near 9.5.* The comet was
followed until May 2, when the magnitude must have been near 10.5.*
(Great Southern Comet)

1867 III   Discovered: September 26   Discovery Magnitude: 6.5
     This comet was independently discovered by Baeker (Nauen) and Winnecke

(Honigesstein) on the same night while moving through Ursa Major. Around
October 4, the comet was described as a bright nebula nearly 4 arcmin in
diameter with a strong central condensation and a short, wide tail. The
comet's distance from Earth changed little during October, and the comet's
brightness changed little early in the month. By mid-October, the comet had
entered Canes Venatici and, although the brightness had not changed, the tail
had grown to a length of 10 arcmin and a nucleus 35 arcsec across had become
visible. As the comet drew closer to the sun it brightened to magnitude 6*
by the 23rd, with estimates of the tail length ranging between 15 and 20 arcmin.
Thereafter, observations became more difficult as each day passed as the comet
edged closer to the horizon and deeper into twilight. On October 27, the tail
could only be traced for 5 arcmin and the sharp nucleus, when compared to 8
stars of known brightness, was found to be near 8th magnitude. When last seen
on October 31, the comet was near magnitude 4.5*, with no visible tail. Perihelion came on November 7 (r= 0.33 AU).
(Baeker-Winnecke)

1868 II   Discovered: June 13   Discovery Magnitude: 6.5
     Winnecke (Karlsruhe) discovered this comet in Perseus, with an independ-
ent discovery coming from Marseilles Observatory on the same night. The next
evening astronomers to whom the discovery had been announced observed the
comet and described it as a bright telescopic object with a tail and a nucleus
24 arcsec in diameter. By June 17, the comet had brightened to 6th magnitude*
and a tail 1.5 degrees long was visible in a comet-seeker. On the 20th, naked-
eye observations were begun as the comet reached magnitude 5*. The tail was
then described as 3 degrees long. Perihelion came on June 26 (r= 0.58 AU),
and observers then reported the magnitude as near 4.5. The comet was then
situated near the horizon and possessed a coma 5 arcmin across with a tail 1
degree long. On June 30, the comet was at perigee (0.60 AU). On July 9, the
magnitude was still near 5*and the tail was still about 2 degrees long and by
the 15th, these values had declined to only 5.5 and 1 degree, respectively.
Due to the closeness to the horizon, the comet was only viewed until July 17.
(Winnecke)

1869 II   Discovered: October 12   Discovery Magnitude: 7.5
     This comet was discovered two days after it had passed perihelion (r=
1.23 AU) by Tempel (Marseilles). The comet was then described as relatively
faint in a low-power telescope and was moving slowly southeasterly through
Sextans in the morning sky. As October progressed, the comet's magnitude
changed little as its distance from Earth decreased at a rate sufficient to
counter the effects of an increasing solar distance. During this time, the
comet was consistently described as circular with a diameter of 1.5 arcmin and
no tail. Although the magnitude was still near 8*on November 13, the comet's
decreasing distance above the horizon made it impossible to observe after that
date.
(Tempel)

1870 I   Discovered: May 30   Discovery Magnitude: 7.0
     Tempel (Marseilles) and Winnecke (Karlsruhe) independently discovered this
comet nearly simultaneously near the Andromeda-Pisces border. It was then
described as bright with a diameter of 2.5 arcmin. The first traces of a tail
appeared on June 5, and by the 8th, it extended 20 arcmin. Also, on the latter
date, the coma was 3 arcmin across and the magnitude was near 6.5*. Despite

decreasing distances from the sun and Earth, the comet changed little during
June and into July. Observers indicated a magnitude near 7 in mid-June, with
a diameter of 2.5 arcmin and only traces of a tail. During the latter half of
June, observers indicated a magnitude near 7 and a short, wide tail extending
nearly 5 arcmin. The comet was last seen on July 10, at a magnitude near 6.5*
and with a tail 20 arcmin long. Moonlight prevented observations over the
next few days and when searches were again resumed in dark skies, the comet
had moved too far south for observation. Perihelion came on July 14 (r= 1.01
AU). *

(Winnecke-Tempel)

1870 II   Discovered: August 28   Discovery Magnitude: 7.0

This comet was discovered in Cetus by Jerome Eugene Coggia (Marseilles)
and was described as a bright, circular object 2 arcmin across with a central
condensation. Perihelion came on September 2 (r= 1.82 AU), and in the next
few days observers indicated a coma diameter between 3 and 5 arcmin. The
total magnitude was still near 7*and the condensation possessed a magnitude
near 10.5. During the final week of September, the comet took on an oval
shape, but with no definite tail being visible. The total magnitude then
equalled 7*and the star-like nucleus still possessed a magnitude near 10.5.
Between September 30 and October 17, observers no longer saw a prominent
nucleus (last seen near magnitude 12), but instead reported many luminous
points within the coma. During November, the magnitude remained near 7.5;
however, fading became more rapid in December. Observers also reported the
comet becoming larger and more diffuse near mid-month, and when last seen on
the 25th, the magnitude must have been near 10.

(Coggia)

1870 IV   Discovered: November 24   Discovery Magnitude: 7.5*

While searching for comets in the morning sky, Winnecke (Karlsruhe) came
across a diffuse object near Gamma Virginis. He described it as a bright
telescopic object with a diameter near 2.5 arcmin and a strong condensation.
The comet became a more difficult object in the next few days as it moved
deeper into morning twilight and was last seen on December 1. With the comet
approaching both perigee and perihelion, observers searched the evening sky
after the first week of December, but with no results. Schmidt (Athens) made
continuous searches between the 9th and 23rd, and concluded that the comet
must have faded quickly, since anything brighter than magnitude 11 should
have been found. Perihelion came on December 20 (r= 0.39 AU).

(Winnecke)

1871 I   Discovered: April 7   Discovery Magnitude: 8.0

This comet was discovered by Winnecke near the double cluster in Perseus
and was described as small and pale. Independent discoveries were later made
by Borrelly (Marseilles) on April 13, and by Lewis Swift (Marathon, New York) on
April 15. Observations began in full force in Europe on April 10, when observers reported a tail 4 arcmin long with a nucleus. By the end of April, the
nucleus had brightened to a magnitude of 8, while the total magnitude had
increased to near 6.5* The comet continued to brighten during May, although
a decreasing altitude began to interfere with observations shortly before mid-
month. On the 6th, the tail extended 8 arcmin and the coma was estimated by
Schmidt (Athens) as 2.2 arcmin across. Though the tail was near 10 arcmin long
on the 9th, it had become invisible after the 12th, while the coma decreased

by only 0.1 arcmin. When observed on the 16th, the comet was only 4 degrees above the horizon. The magnitude was near 5.5*, while the nucleus was still being reported as near magnitude 8. Afterwards, observations were impossible. Perihelion came on June 11 (0.65 AU), and the comet remained a difficult object until August 5, when observed at the Cape of Good Hope, as a relatively bright telescopic object. Due to low altitude, no further observations were made. (Winnecke)

1871 II   Discovered: June 14   Discovery Magnitude: 8.0
     While searching for comets, Tempel (Milan) discovered this comet near Beta Ursae Majoris and described it as diffuse with a diameter near 4 arcmin. Although the comet was approaching the sun, the increasing distance from Earth caused it to slowly fade during the next month. At the end of June, the comet was near magnitude 8.5* and possessed a tailless coma with a nucleus. During July, the brightness seemed to hold near 9* as the comet neared its July 27 perihelion date (r= 1.08 AU), although the nucleus noticeably faded from 11.5 to 12.5, according to Schmidt (Athens). After perihelion, the comet began to move towards Earth. The result of changes in both total and nuclear brightnesses was somewhat reversed over July, with the total magnitude fading rapidly from magnitude 9 to 10.5* by September 10, and the nuclear magnitude dimming by only 0.5 magnitude during the same time. A tail was observed on only one occasion during this comet's apparition and this was simply as an extension of the coma on August 9. The coma remained at a diameter of nearly 3 arcmin between June and early September. The comet was last observed on September 20, as a faint object near magnitude 11*.
(Tempel)

1871 IV   Discovered: November 3   Discovery Magnitude: 8.5
     Tempel (Milan) discovered this comet in Scutum and described it as a faint object with no nucleus or tail. Within the next few days, the comet moved almost directly south and was described as about 2 arcmin in diameter with a slight elongation to the coma. The brightness changed little during the first two weeks of visibility as the increasing distance from Earth countered the effects of an approaching perihelion. However, a rapid increase in brightness began at the end of November, as the comet dropped below the horizon for Northern Hemisphere observers. Perihelion came on December 20 (r= 0.69 AU), and the comet was seen as a 7th-magnitude object by observers in the Southern Hemisphere during January. It passed within 5 degrees from the south celestial pole on February 2, and, thereafter, headed rapidly northward. The final observation came on February 21, when the comet was detected as a very faint object at the Cape of Good Hope (magnitude near 9). This comet's orbit is elliptical with a period near 2,000 years.
(Tempel)

1873 IV   Discovered: August 20   Discovery Magnitude: 6.5
     Borrelly (Marseilles) found this comet near the Auriga-Lynx border in the morning sky. He described it as bright with a coma diameter of 3 arcmin and a strong condensation. By the end of the month, observers indicated a magnitude near 6, and added that a 9th-magnitude nucleus had become visible. The first traces of a tail were detected on September 2, but they vanished a few days later as the comet's altitude decreased. Perihelion came on September 11 (r= 0.79 AU), and the comet was last detected on the 21st. Despite the short span of observation, the orbit was found to be elliptical with a period

near 3,400 years.
(Borrelly)

1873 V   Discovered: August 23   Discovery Magnitude: 6.5
     This comet was discovered as a bright telescopic object by Henry (Paris).
It was then located near the Lynx-Camelopardalis border in the morning sky. At
the end of the month, observers reported a sharply bounded coma nearly 10 arcmin
in diameter with a star-like nucleus and a narrow tail nearly 1 degree long. The
tail grew slightly during the first couple of days in September, but soon became difficult to detect as the comet entered twilight on its way to conjunction with the sun. The comet was last seen on September 10*, at a magnitude
near 4*, when only 3 degrees above the horizon. Perihelion came on October 2
(r= 0.38 AU), and the comet remained hidden from view until mid-November, when
Johann Palisa (Pola, Austria) recovered it. He continued observations until the
28th of November, when the magnitude must have been near 10*. The orbit is
elliptical with a period near 54,000 years.
(Henry)

1874 I   Discovered: February 21   Discovery Magnitude: 8.5
     Shortly before dawn, Winnecke (Strasbourg) discovered this comet in Vul-
pecula and described it as 2 arcmin in diameter. Over the next three days,
observers reported an 11th-magnitude star-like nucleus and traces of a tail.
On the 25th, the comet had become larger with a distinctly non-stellar nucleus,
and 24 hours later, the condensation was hardly detectable inside a coma over
3 arcmin in diameter. Although the brightness should then have been slightly
brighter than at discovery, it was reported to have become fainter than magnitude 9. The comet was not seen again. A search on March 10, revealed nothing at a time when the magnitude should have been near -2. The comet was at
perihelion on that date (r= 0.04 AU), and remained lost in the sun's glare
until late March. Searches by Winnecke on March 23, April 6 and April 7
failed to find the comet near the predicted position.
(Winnecke)

1874 II   Discovered: April 12   Discovery Magnitude: 6.5
     This comet had already passed perihelion on March 14 (r= 0.89 AU) when
discovered by Winnecke (Strasbourg) in the morning sky near Beta Aquarii on
April 12. The comet was then described as a bright telescopic object, 4 arcmin
in diameter, with a strong central condensation. Independent discoveries were
made by Borrelly (Marseilles) on April 15, and by Tempel (Milan) on the 18th.
A rapidly decreasing distance from Earth caused the comet to brighten up to
the time of perigee during the first week of May (0.5 AU). The circular coma
then reached a diameter of 10 arcmin and possessed a stellar nucleus near magnitude 11. Afterwards, the rapidly increasing distance from Earth caused the
comet to fade to magnitude 7.5* by June 4, and when last seen on the 17th, it
was hardly brighter than magnitude 10.5*.
(Winnecke)

1874 III   Discovered: April 17   Discovery Magnitude: 7.5
     This comet was discovered by Coggia in Camelopardalis and was described
as a bright telescopic comet. Within a few days of discovery, observers were
reporting a coma nearly 4 arcmin in diameter and a very faint nucleus. At the
beginning of May, the first traces of tail appeared and the nucleus had become
brighter and more star-like, and by month's end, with a total magnitude near
6.5, the tail was extending over 20 arcmin and the nucleus was near magnitude

8. Also, during May, the comet's positional change of hardly 2 degrees southward indicated a close approach to Earth during July when it was due to cross
the ecliptic plane. The comet brightened rapidly during June, though its position in northern skies only changed by another 2 degrees by month's end. On
June 8, the tail extended 1 degree and, a few days later, the comet was barely
visible to the naked eye. On June 19, observers indicated a total magnitude
near 3.5 and a nuclear magnitude near 5. The tail was estimated as 3 degrees
long and very narrow, with a width of less than 8 arcmin at its end.

At the beginning of July, the comet entered Lynx and, soon afterward, began to move more rapidly southward. On the 2nd, the total magnitude was near
1.5*, the nuclear magnitude was near 2.5 and the tail extended 5 degrees. Perihelion came on July 9 (r= 0.68 AU), and the comet was then described as about
magnitude 1.5*, with a nucleus of 2nd magnitude and a tail 13 degrees long. On
July 13, the brightness was estimated as greater than Capella (hence, near magnitude 0). The tail extended nearly 20 degrees and contained a dark line emanating from the nucleus, according to Bruhns (Leipzig) and Etienne L. Trouvelot
(Harvard Observatory). The latter observer also indicated that the nucleus was
surrounded by a series of hoods. These hoods were seen by other observers in
the next few days and seemed to originate on the sunward side of the nucleus
only to sweep back into the tail. On July 18, the comet was near perigee (about
0.25 AU) but, surprisingly, the nucleus had become invisible. The tail then
extended 48 degrees. The tail was longest on July 21, when Schmidt (Athens)
measured the length as 65.8 degrees. A few days later, the comet had dropped
below the horizon for Northern Hemisphere observers.

Observers in the Southern Hemisphere began observations on July 27. They
indicated a rapid fading at first, followed by a very slow fading as the comet
neared the south celestial pole. During October, the comet had moved to within
15 degrees of the pole, but observations ceased on the 19th, when Thome (Cordoba)
indicated a magnitude near 11. The comet's orbit is elliptical with a period
near 13,700 years.
(Coggia)

1874 IV   Discovered: August 19   Discovery Magnitude: 8.5
Coggia discovered this comet in the morning sky in Taurus and described
it as a faint circular object without a nucleus. The comet was then one month
past its perihelion passage (r=1.69 AU on July 18) and was just exiting the
sun's glare when discovered. On August 22, the comet was described as nearly
3 arcmin in diameter with a faint nucleus near magnitude 11.5 and at the end of
the month, the brightness had faded to 9.5*. During September, the brightness
faded by another magnitude and at the beginning of October, the comet had moved
into Orion. On October 10, the total magnitude was near 10.5* and the nucleus
could still be faintly seen at a magnitude near 12. Observations continued
until November 15, when the comet had faded to near magnitude 11*. The comet's
orbit is elliptical with a period of 306 years.
(Coggia)

1874 V   Discovered: July 26   Discovery Magnitude: 7.5
Borrelly (Marseilles) discovered this comet in Draco while searching for
comets in the morning sky. It was then moving at a rate of 1 degree per day
due to a somewhat small distance from Earth (0.59 AU). Observations by other
observers on the 30th, indicated a diameter of 4 to 5 arcmin with an eccentric
condensation. From the time of discovery until late August, the comet approached

the sun while receding from Earth and did not become brighter than magnitude
7.* Curiously, observers reported an apparent brightness fluctuation on August
10, when the comet was noticeably fainter and smaller than it had been on the
3rd. Perihelion came on August 27 (r= 0.98 AU), and, afterwards, the comet
faded rapidly. On September 3, it was described as circular with a sharp nucleus and a total magnitude near 8.5 and by October 9, it was described as
large and diffuse, with a magnitude near 9.5. The comet was last detected on
October 21, when near magnitude 11.* The orbit is elliptical with a period
near 24,400 years.
(Borrelly)

1874 VI   Discovered: December 7   Discovery Magnitude: 8.5
     Having passed perihelion on October 19 (r= 0.51 AU), this comet was dis-
covered by Borrelly (Marseilles) when near opposition. Three days later,
observers reported a magnitude near 9 with a coma diameter greater than 4
arcmin. The comet continued to fade rapidly, and by January 3, 1875, Bruhns
(Leipzig) indicated a magnitude near 10, when he described the comet as faint
and diffuse with a diameter of 2 arcmin. The comet was last seen on January 7.
(Borrelly)

1877 I   Discovered: February 9   Discovery Magnitude: 6.5
     Having passed perihelion on January 19 (r= 0.81 AU), this comet was dis-
covered shortly after perigee (0.45 AU) by Borrelly (Marseilles) on February
9, and Pechule (Copenhagen) on the 10th. The former observer then described the
comet as a bright, circular object. Thereafter, although the comet moved away
from the sun and Earth, it was considered brightest on the 18th, when slightly
fainter than the Andromeda Galaxy (hence, near magnitude 5.5) and nearly 20
arcmin in diameter. On the 20th, the diameter had decreased to 19.5 arcmin,
according to Schmidt (Athens), who used a comet-seeker. Other observers using
refractors at that time found the coma only 9 arcmin across. At the beginning
of March, only a slight condensation was visible and on the 9th, observers
described the comet as ill-defined and diffuse. The magnitude had faded to
9.5* by March 17, when the coma was only 1.5 arcmin across, and by April 1, the
comet could hardly be detected at a magnitude near 10.5.* It was last seen on
the 3rd.
(Borrelly)

1877 II   Discovered: April 6   Discovery Magnitude: 7.0*
     This comet was discovered by Winnecke (Strasbourg) in Pegasus and was
described as possessing a faint nucleus near magnitude 10.5 and a faint double
tail up to 1 degree long. In the next three days, observers reported a coma
diameter up to 4 arcmin across and a strong condensation. Block (Odessa) independently discovered the comet on April 11. By April 15, Winnecke described
the comet as near magnitude 5.5, with a main tail 2 degrees in length and a
fan-shaped secondary tail.
     Perihelion came on April 18 (r= 0.95 AU), and on the 19th, a total magnitude
near 6.5 was estimated by Tupman (Greenwich). The coma was then estimated as
3.5 arcmin in diameter. At the beginning of May, observers reported the comet
as visible to the naked eye (magnitude 5.5) with a strong condensation and a
nucleus. On the 6th and 7th, the magnitude was already estimated as 6.4, with
a tail 12 arcmin long, and by the 15th, the magnitude had dropped to 7.0,* with
a tail 54 arcmin in length. Around this same time, the comet was within 10
degrees of the north celestial pole.

By mid-June, the total magnitude was near 8.5*while the nuclear magnitude
was estimated as near 10. The comet was last seen on July 13, at a magnitude
near 10*and a nuclear magnitude near 11.5. The comet's orbit is elliptical
with a period near 19,800 years.
(Winnecke)

1877 III   Discovered: April 10   Discovery Magnitude: 8.0
     This comet was first detected by Block (Odessa), who then thought he had
found a hitherto unknown nebula in Cassiopeia. He described it as faint, but
saw no urgency in making a report. L. Swift (Rochester, New York) independently
found the comet on the 11th, and on the 14th, Borrelly (Marseilles) also made
a discovery. Returning to the region on April 16, Block realized the object
was a comet. During the next few days, observers described the comet as 1.5
arcmin across with a strong condensation and a slightly oval shape. By April
26, the magnitude seemed near 7.5 and the next day, the comet arrived at perihelion (r= 1.01 AU). Fading was rapid thereafter, and the comet's magnitude
was near 10*on May 11. During the remainder of the month, the condensation
faded to invisibility and the comet was last seen on June 5. The orbit is
elliptical with a period near 10,700 years.
(Swift-Borrelly-Block)

1877 V   Discovered: October 2   Discovery Magnitude: 7.5
     This comet actually passed perihelion on June 27 (r= 1.07 AU), and was
not discovered until three months later. Tempel(Archetri, France) then found
the comet in the evening sky near the Aquarius-Cetus border. The comet was
then located 1.82 AU and 0.90 AU from the sun and Earth, respectively. On
October 4, the magnitude was still near 7.5*and a tail extended nearly 5 arcmin.
A nucleus near magnitude 10 was detected on the 5th, and the next evening,
observers estimated a coma diameter of 1.5 arcmin with a tail 5 arcmin long.
By October 11, the comet had faded to 9th magnitude* Observers also reported
a nucleus of magnitude 11.5 and a tail 3 arcmin long. Fading continued to be
rapid and the comet was near magnitude 11.5*when last seen on October 31.
Curiously, a study by E. M. Pittich in 1969 indicated the comet could have
been discovered 198 days earlier since it was well placed in the northern sky
and of sufficient brightness. This has led some astronomers to believe this
comet was discovered shortly after an outburst in brightness.
(Tempel)

1877 VI   Discovered: September 14   Discovery Magnitude: 9.5*
     Coggia (Marseilles) discovered this comet 3 days after it had passed
perihelion (r= 1.58 AU). Although the comet was then moving away from the
sun, it was rapidly approaching Earth and, thus, became brighter in October as
it rapidly headed south. Traces of a nucleus were noted soon after discovery
as well as a coma 1.5 arcmin in diameter. By early October, the nucleus had
become noticeably brighter and on the 15th was estimated as near magnitude
12. On the latter date, the comet was near magnitude 8.5*with a coma still
1.5 arcmin across. The brightness changed little over the next month, but
after mid-November, fading was rapid. The comet was last detected on December 10, at a magnitude near 10.5*
(Coggia)

1878 I   Discovered: July 8   Discovery Magnitude: 7.5
     This comet was discovered by L. Swift (Rochester, New York) in the evening
sky near Alpha Ophiuchi. It was then 0.50 AU from Earth and near opposition.

The next evening, Peters (Clinton, New York) remarked that he could not detect a nucleus in an 18-cm refractor. On July 19, the comet had faded to
magnitude 8.5*, due to an increasing distance from Earth, and perihelion came
on the 21st (r= 1.39 AU). The comet was last detected at low altitude on
July 24, when the magnitude was near 9*.
(Swift)

1879 II   Discovered: June 17   Discovery Magnitude: 6.5
     L. Swift (Rochester) discovered this comet near the double cluster in
Perseus. On the 21st, he described it as bright with a short tail. The
comet had passed perihelion on April 27 (r= 0.90 AU) and, according to
Pittich's study in 1969, the comet could have been discovered 139 days earlier,
although it was in northern skies for only half that period. This has fostered
the belief that an outburst may have occurred shortly before discovery. European observations began on June 21, when the comet was described as 3 arcmin in
diameter and visible in moonlight (magnitude 6). The comet had faded to a
magnitude near 8*on July 9, in moonlight, and by the 18th, detailed observations gave a coma diameter of 1 arcmin with a 10.5-magnitude condensation 15
arcsec across. The comet was last seen on August 24, when the central condensation was near magnitude 12.5 and the coma measured 0.7 arcmin across. The
comet's total magnitude was then near 11.5*.
(Swift)

1879 IV   Discovered: August 24   Discovery Magnitude: 10.5
     E. Hartwig (Strasbourg) was searching for comets using a 16-cm seeker when
he found this comet slightly above the Big Dipper with a rapid southward motion.
The comet was then described as very faint with a diameter of 1 arcmin. Perihelion came on the 29th (r= 0.99 AU), and the comet slowly faded thereafter as
it moved away from both the sun and Earth. Under very clear skies at Leipzig
on September 13, Bruhns reported a diameter of 4 arcmin and indicated a magnitude near 11.5. The comet was last seen on the 18th.
(Hartwig)

1879 V   Discovered: August 21   Discovery Magnitude: 7.5
     Like the previous comet, this object was found in Ursa Major and was moving
rapidly southward; however, the similarity ends there as this comet was 3
magnitudes brighter when discovered by Palisa (Pola, Austria) and was still
over a month from perihelion. The first descriptions after discovery indicated this comet was a small, but very bright telescopic object. The nucleus was
"well-shaped" and possessed a magnitude near 9. During the first days of
September, observers indicated a coma diameter of 2 arcmin and a nuclear magnitude near 8. By September 13, the coma had grown to 5 arcmin, with the comet
and nucleus being near magnitudes of 6.5*and 8, respectively. The first traces
of a tail were also noticed at that time. By October 4, the comet was near 6th
magnitude*with a diameter of 5.5 arcmin and a condensation. Perihelion came
on October 5 (r= 0.99 AU), and, although fading was slow, the comet was last
seen on October 22, at a magnitude near 7*. Observations thereafter were affected by twilight.
(Palisa)

1880 I   Discovered: February 1   Discovery Magnitude: 3.0
     This comet passed perihelion on January 28 (r= 0.005 AU), and was first
detected at Cordoba, Argentina, as only a tail extending above the horizon in
the evening sky. Although then invisible below the horizon, the comet's head

was located only 7 degrees from the sun. The next evening, independent discoveries were made in South America, South Africa and Australia, and the comet was then described as a luminous streak extending nearly 30 degrees from the southwest horizon towards the south celestial pole. The comet's head was first detected on February 3, when Dr. B. A. Gould (Director of the National Observatory of the Argentine Republic at Cordoba) caught a glimpse of it in twilight and haze a short distance above the horizon. He described it as nearly 3 arcmin across with no visible nucleus. The tail was then described as 40 degrees long. On February 4, Mr. Eddie (Grahamstown, South Africa) made the first detection of the nucleus, which he described as faint and straw-colored. He also estimated the total magnitude to equal 47 Tucanae (magnitude near 3.5) and found the tail to be 50 degrees long and not as curved as on previous days. The comet faded rapidly and was below naked-eye visibility by the 12th. On the 17th, the nucleus was a very difficult object to observe and the total magnitude must have been near 7.5[*]. The comet was last seen on February 20, when barely visible with the Cordoba equatorial. The comet never became visible in the Northern Hemisphere and is a member of the sungrazing family of comets.
(Great Southern Comet)

1880 II   Discovered: April 7   Discovery Magnitude: 7.5
     J. M. Schaeberle (Ann Arbor, Michigan) discovered this comet in Camelopardalis when only 4 degrees from the north celestial pole. A tail 3 arcmin long was then visible. By mid-April, observations indicated the appearance of a star-like nucleus near 10th magnitude and a tail 3 arcmin long. At the beginning of May, the total magnitude was still near 7.5[*], but the tail had become fan-shaped and 6 arcmin long. By month's end, the brightness had dropped to 8[*] and the nucleus had suddenly faded to near 14th magnitude. The tail was then only 1 arcmin long. The comet was lost in twilight after June 8, and finally reached perihelion on July 2 (r= 1.81 AU). Afterwards, the comet remained in twilight until September, and was then observed only until the 12th, when near magnitude 8[*]. The brightness had varied little during the duration of visibility. Pre-perihelion observations were unchanged from day to day since the comet was moving away from Earth as it approached the sun. Post-perihelion observations were basically unchanged since the comet approached Earth as it moved away from the sun. Orbital calculations revealed the comet travels in a hyperbolic path.
(Schaeberle)

1880 III   Discovered: September 29   Discovery Magnitude: 5.5
     This comet was discovered by Hartwig (Strasbourg) 22 days after it had passed perihelion (r= 0.35 AU). It was then described as a bright object with a tail 50 arcmin long and had only recently exited from the evening twilight. An independent discovery was made the next evening by Harrington (Ann Arbor, Michigan). During the first days of October, observations indicated the following: total magnitude near 5.5; nuclear magnitude near 7; coma diameter 7 arcmin; and a tail length near 2 degrees. By October 3, the tail was estimated as nearly 4 degrees long and, thereafter, the increasing distances from the sun and Earth caused the comet's physical parameters to decline. On October 9, the magnitudes of the coma and nucleus were 6.5[*] and 9, respectively, while the coma diameter was 17 arcmin and the tail length was 1.2 degrees. By the end of October, the magnitudes of the coma and nucleus were 9.5[*] and 12, respectively, while the coma diameter was 8 arcmin and the tail length was

near 20 arcmin. The comet was last seen on November 30, as a faint object
(magnitude near 11.5) with an ill-shaped nucleus.
(Hartwig)

1880 V   Discovered: December 16   Discovery Magnitude: 7.5
     While observing during a lunar eclipse, Pechule (Copenhagen) discovered
this comet in Aquila and described it as 1 arcmin in diameter and not particularly bright. Having passed perihelion on November 9 (r= 0.66 AU), the
comet was becoming fainter as its distance from both the sun and Earth increased. At the end of December, the comet was near 8th magnitude*and a
nucleus of magnitude 9 was visible. Observers were then reporting a coma
diameter of slightly over 1.5 arcmin; however, Schmidt (Athens) detected a
very faint outer coma which he measured as nearly 9 arcmin in diameter. By
January 23, observers reported a total magnitude of 9.5 and a nuclear magnitude of 11. Schmidt's estimate for the diameter of the faint outer coma
at that time was nearly 10 arcmin. This estimate came at a time when the
comet was located 1.9 AU from the Earth. Observations continued until March
31, 1881, when the comet was described as very faint (magnitude near 11.5).
(Pechule)

1881 II   Discovered: May 1   Discovery Magnitude: 7.0
     L. Swift (Rochester) discovered this comet near Theta Andromedae in the
morning sky. The comet was described as bright and heading toward the sun.
On May 4, the comet was described as circular with a diameter of 2 arcmin
(magnitude near 6.5). The comet was observed only until May 12, when Peters
(Leipzig) and Borrelly (Marseilles) considered it too difficult to observe due
to twilight and moonlight. Perihelion came on May 20 (r= 0.59 AU).
(Swift)

1881 III   Discovered: May 22   Discovery Magnitude: 3.0
     This comet was first seen by Tebbutt (Windsor, New South Wales), who, dur-
ing a brief scan of the western sky, spotted a strange hazy-looking object.
Upon examination with a marine telescope, the object was resolved into two
stars and a 3rd-magnitude comet. Independent discoveries were made by Ellery
(Melbourne) on May 23, and Gould and Davis (Cordoba) on the 25th. The former
observer estimated the nucleus as near magnitude 5. During the first few days
of June, the comet had brightened to magnitude 2*and the tail had grown to
nearly 6 degrees. Estimates of the coma then ranged from 10 to 20 arcmin.
The comet was lost in evening twilight after June 11.
     Perihelion came on June 16 (r= 0.73 AU), and the comet appeared in north-
ern skies on June 22, when it possessed a total magnitude of 1*and a noticeable tail. The tail grew from a length of 10 degrees on the 24th, to 20 degrees
on the 25th, and, thereafter, began to slowly shrink. At the end of the month
it had declined to 12 degrees. The total magnitude and the nuclear magnitude
remained virtually unchanged through the end of June, as they were consistently estimated as near 1*and 2, respectively. During the days between June
27 and 30, three jet-like plumes were detected emanating from the nucleus and
pointing towards the sun. As the comet's distances from both the sun and Earth
increased, these plumes rapidly vanished. During July, the comet faded from
magnitude 2 to 5*and on the 23rd, it was an easy object to observe when only
8 degrees from the north celestial pole. At the beginning of August, the
total magnitude was near 6*and by the beginning of September, it was near 8*.
By December 20, observers described the comet as less than 3 arcmin in diameter

and the total magnitude near 11.5.* The comet was then impossible to measure.
Observations continued until February 15, 1882, when it had faded to nearly
magnitude 12.5.* The comet's orbit is elliptical with a period near 2,400
years.
(Great Comet)

1881 IV   Discovered: July 14   Discovery Magnitude: 6.0
     Schaeberle (Ann Arbor, Michigan) discovered this comet in the evening
sky in Auriga. He then described it as a bright telescopic object, barely
visible to the naked eye. During the remainder of the month, the comet brightened to magnitude 5.5* as it neared both the sun and Earth. Observers also
described a tail 1 degree long and a nucleus near magnitude 10.
     During August, the comet rapidly brightened from a magnitude of 5.5* on
the 2nd to about 3* on the 25th. The tail reached a maximum length of 10 degrees between the 21st and the 25th, and the nucleus obtained a maximum brightness of 5th magnitude on the 25th. Perihelion came on August 22 (r= 0.63 AU),
and perigee occurred near the 25th, when the comet was nearly 0.55 AU from
Earth. Observers also reported a faint, curving tail to the east of the main
straight one. During mid-August, the comet was located in the north circumpolar region and provided the uncommon spectacle of two bright comets circling
the pole at one time, although its companion, 1881 III, easily outshone it for
most of that time. As September began, the comet moved rapidly southward. On
the 5th, the total magnitude was near 5*, while the nuclear magnitude was estimated as 9. By the 13th, when at an altitude of less than 5 degrees, the
total magnitude was estimated as near 6*. Observations in the Northern Hemisphere ceased after this date.
     Observations in the Southern Hemisphere began on September 10, and by the
17th, the comet had rapidly faded to a magnitude near 7.5*. On the latter date,
Tebbutt (Windsor, New South Wales) described the comet as a small nebulous
object with a central condensation, but no tail. The comet was observed until
October 21, when it had reached a magnitude near 12*. Although mathematicians
have commonly provided only a parabolic orbit for this comet, there is a possibility that the orbit is hyperbolic.
(Schaeberle)

1881 VI   Discovered: September 18   Discovery Magnitude: 6.5
     Edward Emerson Barnard (Nashville, Tennessee) discovered this comet in the
evening sky 4 days after it had passed perihelion (r= 0.45 AU). The comet was
then located in Virgo and was described as a bright telescopic object with a
round coma 2 arcmin across. A faint tail was detected by Barnard the next evening and on the 28th, the length was estimated as about 7 arcmin. The comet's
distances from both the sun and Earth increased after discovery, causing the
comet to fade rapidly. On October 3, it was near magnitude 7.5*, but was still
estimated as 2 arcmin in diameter. By the 8th, the comet had faded to magnitude
8.5* and on the 11th, it must have been near 9*. The comet was last detected on
October 28, at a magnitude near 11*.
(Barnard)

1881 VIII   Discovered: November 17   Discovery Magnitude: 8.5
     L. Swift (Rochester) discovered this comet in Cassiopeia. It was then
located about 15 degrees from the north celestial pole and was described as a
faint, slightly condensed nebula. Perihelion came on November 20 (r= 1.92 AU),
and afterward, the comet was moving away from both the sun and Earth. The

comet was not observed in Europe until November 25, due to a telegraphic error.
On that date, Winnecke (Strasbourg) could easily see the comet in an 11.5-cm
seeker. He described it as nearly 4 arcmin across and indicated a magnitude
near 8. Magnitude estimates seemed to vary during December, with observers
indicating magnitudes between 7.5 and 9.5--depending on the size of telescope
used. The coma was consistently reported as near 3 arcmin. By month's end
the comet was commonly described as very faint and observations ended on January 12, 1882. The comet's orbit is elliptical with a period near 2,740 years.
(Swift)

1882 I   Discovered: March 18   Discovery Magnitude: 7.0
    C. S. Wells (Dudley Observatory, New York) discovered this comet in Her-
cules as a 7th-magnitude object with a tail 30 arcmin long. The story goes
that Dudley Observatory had just been reopened after being rebuilt and re-
equipped and several prominent citizens were visiting its director, Lewis Boss.
One citizen remarked on the numerous comets being discovered at other observatories in the United States and, with that, Boss turned to his assistant and
jokingly said: "You see, Mr. Wells, you must discover a comet." Wells did so
only one week later.

    At discovery, the comet was located 2.04 AU and 1.74 AU from the sun and
Earth, respectively. Its small coma diameter of 0.5 arcmin and the long tail
of nearly 8 arcmin gave it the appearance of a large comet in miniature, according to Boss on March 19. By early April, the comet possessed a sharp
nucleus of magnitude 10, a coma diameter of 1 arcmin and a tail 20 arcmin long.
Despite decreasing distances from both the sun and Earth, the comet changed
little in appearance until the middle of May, when observers estimated a total
magnitude near 6. The tail was then reported as nearly 1 degree long. By
June 2, the total magnitude had brightened to 2*, while the nuclear magnitude
was near 3. Observers also reported a coma diameter of 2 arcmin and a tail
length of 5 degrees. On June 6, the total magnitude was near 0.5*and, with
strong twilight, estimates of the coma and tail were few and far between. On
the 7th, the comet could only be observed for 10 minutes after sunset, but,
beginning on the 9th, the comet had become bright enough to allow observations
even at noon. On June 10, Schmidt (Athens) detected the comet when it was
located only 2.8 degrees from the sun's limb. He described it as a white,
very dense point with diffuse edges. He estimated it as nearly 6 arcsec in
diameter and found no trace of a tail (magnitude near -5). Perihelion came on
June 11.03 (r= 0.061 AU), and 16 hours later, observers at Dudley Observatory
detected the true nucleus at a diameter of 0.75 arcsec. The coma was then
described as faint and 10 arcsec across.

    After a further daytime observation on June 12, the comet remained hidden
in twilight until June 16. Three days later, observers at Rio de Janeiro,
Brazil, described the nucleus as very bright and the tail as 45 degrees long.
The total magnitude must then have been near 1*. The comet faded rapidly thereafter, as it headed away from both the sun and Earth. At the end of June, it
was near 3rd magnitude* and by mid-July, it had faded to magnitude 5.5*. Observers on August 2 described the comet as 3.5 arcmin in diameter with a small
condensation and a magnitude near 7.5 and four days later, it had already faded
to 8.5*. The comet was last detected on the 16th. The orbit seems to be elliptical with a period near 1 million years.
(Wells)

1882 II   Discovered: September 1   Discovery Magnitude: 0.0
The "Great September Comet" of 1882 was first discovered in the morning
sky by a group of Italian sailors in the Southern Hemisphere. Numerous independent discoveries were made over the next few days in Australia, New Zealand,
South Africa and South America. Though observations are scarce over these
first few days, Gould (Cordoba) managed to collect the reports of several
local people which indicated the comet was as bright as Venus on the 5th, with
a brilliant tail. Gould's own observations were delayed due to bad weather
until the morning of the 14th. Another professional astronomer, Cruls (Rio de
Janeiro Observatory) was first notified of the comet on the 10th, and he is
credited with being the first to inform European observatories of the comet
by sending a telegram shortly after he confirmed the comet's existence on the
12th. September 10 marked the day that numerous experienced observers began
to study the comet and compiled descriptions indicate a total magnitude near
-2 and a nuclear magnitude near 3. The tail was still very bright, but due to
strong twilight, it only seemed to extend for 3 or 4 degrees. On the 14th, the
comet had moved to within 12 degrees of the sun and had attained such a brightness so as to allow it to be visible in broad daylight (magnitude near -4).

Although observers were awed by the sight of the comet on the 14th, there
was still much to come. The comet was followed right through perihelion on
the 17th (r= 0.008 AU), and, incidentally, was independently discovered at
that time by Dr. Common (Ealing, England), who had been searching the sun's
vicinity ever since he had heard the report of a comet discovery during the
total solar eclipse of May 17, 1882. Observers reported a striking contrast
between the comet's silvery radiance and the sun's reddish-yellow glare as
the comet neared the sun on the 17th. Dr. Elkin (Cape of Good Hope) described
the brilliancy of the nucleus as scarcely fainter than the sun's limb and estimated that the nuclear brightness had to then be greater than -10. (Laboratory
studies of today indicate the comet to have possessed a total magnitude between
-15 and -20!) Many observers followed the comet continuously "right into the
boiling of the limb"--a circumstance unheard of in the history of comets. As
the comet touched the sun's limb "it vanished as if annihilated" and several
observers thought the comet had passed behind the sun; however, the truth of
the matter is that the comet transited the sun and left no traces of a dark
nucleus on the solar disk. The next morning, Sir David Gill (Cape of Good
Hope) described the comet as "an astonishing brilliancy as it rose behind the
mountains on the east of Table Bay, and seemed in no way diminished in brightness when the sun rose a few minutes afterward. It was only necessary to shade
the eye from direct sunlight with the hand at arm's length, to see the comet,
with its brilliant white nucleus and dense white, sharply bordered tail of
quite half a degree in length." Throughout the day on the 18th, observers in
Cordoba referred to the comet as the "blazing star near the sun." The brightness persisted through the end of September, with estimates of the nuclear magnitude still being near 0 on the 24th. On the 27th, it was reported that 10
degrees of tail were still visible in bright twilight after stars of magnitude
1 had vanished.

Interest shifted to the comet's nucleus beginning on September 30, when it
was observed to be double, and at the beginning of October, the number of condensations within the coma equalled 4 or 5. Curiously, without knowledge of
this apparent splitting, some observers began to report comet-like objects

a few degrees from the comet's head: the first was observed by Schmidt (Athens)
on October 9, 4 degrees to the southwest of the main comet; Barnard (Nashville)
discovered 6 or 8 objects 6 degrees southwest of the main comet on the 14th--an
observation confirmed in Europe; and, finally, William R. Brooks (New York)
found an object to the northeast of the main comet around the 21st. Only
Schmidt's object was observed for more than one day and the exact relation of
these bodies to the main comet has never been completely unraveled. They were
not physically connected to 1882 II at the time of their observations, but they
did travel in the same direction and at the same rate. Concerning the condensations within the main comet's coma, observations continued for four of them
until the end of February 1883, and individual orbits could be calculated for
each. Astronomical investigators have pinpointed the date of splitting as
September 17--the day of perihelion.

Aside from the multiple nuclei, other features were also well observed as
the comet faded. The tail remained longer than 20 degrees throughout October,
and, for a few days after the 16th, was even accompanied by an anomalous sunward tail which was visible to the naked eye. The total magnitude was near 0*
during the first week of October, and by the end of the month it had faded to
near 3*. The comet remained a spectacular object in Southern Hemisphere skies
through the end of the year and in mid-January, Gould still reported it as
visible to the naked eye (magnitude near 5) with a tail nearly 15 degrees long.
Naked-eye visibility ended in mid-February (magnitude near 6) and the comet
became more and more difficult to detect until it was last seen on June 1, 1883,
by Thome (Cordoba) at a magnitude near 10*. Orbital calculations gave elliptical
orbits for all four nuclei with the periods being 672 years for nucleus A, 761
years for nucleus B (the main comet), 871 years for nucleus C, and 953 years
for nucleus D. This comet is a member of the Kreutz sungrazing family and the
orbit bears a very close resemblance to 1965 VIII.
(Great September Comet)

1882 III   Discovered: September 14   Discovery Magnitude: 10.0
     Barnard (Nashville) discovered this comet in the morning sky as a result
of regular searches for comets. It was then in Gemini and was described as
circular with a diameter of 2 arcmin and some central condensation. The comet
moved southward and brightened as it approached the sun and Earth. By the end
of September, observers indicated a magnitude near 8.5, despite moonlight, and
by mid-October, it was near 7.5*. The comet remained tailless until October 10,
when Schmidt (Athens) reported traces of a faint extension nearly 9 arcmin long.
The central condensation became stronger and brighter during October, and, by
the 17th, was near magnitude 8. The comet was too far south for observation in
the Northern Hemisphere after the 17th. Using an approximate orbit by Hind,
Gill (Cape of Good Hope) was the first observer in the Southern Hemisphere to
find the comet on November 11. Perihelion came on November 13 (r= 0.96 AU),
and the comet was observed daily by Tebbutt (Windsor) between November 30 and
December 8, when the magnitude had faded to near 9*. The comet was too faint
for observation thereafter.
(Barnard)

1883 I   Discovered: February 24   Discovery Magnitude: 6.5
     This comet was independently discovered by Brooks (New York) and L. Swift
(Rochester) on the same evening. It was then described as a bright telescopic
object with a coma nearly 4 arcmin across and a faint tail. The comet had

passed perihelion on February 19 (r= 0.76 AU), and, upon discovery, was heading away from both the sun and Earth. At the end of the month, observers were
reporting a tail at least 1 degree long and a coma nearly 4 arcmin across. In
addition to the main tail, Giovanni Schiaparelli (Italy) also detected a faint
secondary tail. Although he continued to observe this secondary tail until
March 3, no other observer reported such a sighting. By March 1, the main tail
had decreased to a length of 20 arcmin and observers were reporting the appearance of a star-like nucleus near magnitude 10. The comet's total magnitude was
then between 6.5 and 7[*]. The comet faded as expected throughout March, and by
the end of the month was near 8th magnitude[*]; however, according to Schmidt's
accurate measures of the comet's coma, the comet grew during the month. He
estimated a diameter of 4.2 arcmin on the 8th, and 7.4 arcmin on the 29th. By
April 4, Schmidt found the comet to be 10.5 arcmin across. The comet continued
to fade and was followed until April 24, when Schmidt indicated a magnitude
near 10.
(Brooks-Swift)

1883 II   Discovered: January 7, 1884   Discovery Magnitude: 5.0
     Having passed perihelion on December 25, 1883 (r= 0.31 AU), this comet
was moving away from both the sun and Earth when discovered by Ross (Elstern-
wick, Australia) on the evening of January 7. The comet was then located in
Grus and was described during the next couple of days as being barely visible
to the naked eye (magnitude near 5.5). On January 12, the coma was estimated
as 2 arcmin in diameter with a faint condensation and a stubby tail. The
comet faded rapidly and by February 1, was near magnitude 9[*]. The comet was
last seen on February 19, in Melbourne, Australia. An orbit by Moravi in 1910,
indicated the comet travelled in an elliptical orbit with a period of 64.6
years; however, in 1972, Brian Marsden and Zdenek Sekanina found the comet's
short duration of visibility allowed only the calculation of a parabolic orbit.
They added that the 64.6-year period "is completely fictitious."
(Ross)

1885 II   Discovered: July 8   Discovery Magnitude: 11.0
     Barnard discovered this comet at Vanderbilt University Observatory (Nash-
ville) as a very faint object about 1 arcmin in diameter and with a small condensation. Although he estimated a total magnitude near 11, there is a strong
indication, based on observations over the following few weeks, that Barnard
had actually estimated the magnitude of the condensation. Observations during
the remainder of July indicated a total magnitude near 9.5 and a coma diameter
of about 1.5 arcmin. After mid-month, the central condensation was replaced
by a star-like nucleus of magnitude 11. Perihelion came on August 6, and the
perihelion distance of 2.51 AU was greater than any other comet observed up
to that time, except that of 1729. Since discovery, the Earth had been rapidly
moving away from the comet and by mid-August, the total magnitude had declined
to near 10.5[*], with a coma 1 arcmin across. Also, during August, a tail was
occasionally reported extending up to 4 arcmin from the coma. By the end of
the month, the magnitude was near 11[*] and the comet was last observed on September 4.
(Barnard)

1885 III   Discovered: September 1   Discovery Magnitude: 8.5
     Brooks (Phelps, New York) discovered this comet in Canes Venatici and des-
cribed it as circular with no tail and a diameter of 2 arcmin. Strongly sus-

pecting the object as a comet before observing any movement, Brooks telegraphed
word of his discovery to L. Swift, director of Warner Observatory; however,
Swift delayed transferring the news to Kiel, Germany (the European center for
astronomical telegrams) until the discovery was confirmed, since numerous
nebulae were situated in this region of the sky. Confirmation was made on
September 3, but before the telegraph reached Europe, an independent discovery
was made by Common (Ealing, England). Having passed perihelion on August 10
(r= 0.75 AU), the comet was slowly fading as its distance from the sun increased and the distance from Earth remained basically stationary. By mid-
September, observers reported a coma diameter of 3 arcmin, but with little
trace of the condensation observed at discovery. The total magnitude had by
then faded to 9.5.[*] As the month progressed, the diameter remained unchanged
though the comet's circular outline became more irregular and the total magnitude continued to steadily decline. The comet was last observed on October
5, as a very faint object 3 arcmin across and near magnitude 11.5. During the
comet's duration of visibility it faded about 3 magnitudes; however, the predicted change should have only been about 1.5 magnitudes. Some researchers
believe the comet could have undergone an outburst in brightness about the
time of discovery. The orbit is elliptical with a period of 275 years.
(Brooks)

1885 V   Discovered: December 26   Discovery Magnitude: 7.5

     Having passed perihelion on November 26 (r= 1.08 AU), this comet was dis-
covered one month later in Aquila by Brooks (Phelps, New York). The next evening, Barnard (Nashville) made an independent discovery and described the comet
as circular, with a condensation, but no tail. The comet had just entered the
evening sky after exiting the solar glare and was heading away from both the
sun and Earth. By the end of December, the comet had already faded to near
magnitude 8.5.[*] A tail was then described as faint and short, and a nucleus
was estimated as near magnitude 9.5. Fading was slower thereafter, but, due
to a low altitude, the comet was not observed frequently after mid-January. It
was, however, followed until March 1, 1886, when it had faded to a magnitude
near 11.5.[*] Some researchers believe the comet may have undergone an outburst
in brightness shortly before discovery.
(Brooks)

1886 I   Discovered: December 1, 1885   Discovery Magnitude: 8.5

     This comet was discovered in Pisces in the evening sky by Fabry (Paris)
and was described as faint, with a nucleus near magnitude 11. During the next
few days, observers confirmed Fabry's initial description, but added a coma
diameter of 2 arcmin and a westward motion. The comet was found 4 months prior
to perihelion passage and slowly brightened as it neared the sun and Earth. It
reached a magnitude of 7.5[*] in January 1886, and a brightness of 6.5[*] during
February. Also, in February, observers described the coma diameter as still
near 2 arcmin and reported the nucleus as near magnitude 8.

     The comet attained naked-eye visibility during the first week of March,
and at the beginning of April, it had brightened to a magnitude near 4.[*] On
April 2, jet-like formations were noted near the nucleus; however, these were
no longer visible on the 4th. The tail was described as straight and 1 degree
long on April 3, and after the comet passed perihelion on the 6th (r= 0.64 AU),
the tail continued to grow as the comet neared the Earth. On April 9, the tail
extended 3 degrees and by the 25th, it was very narrow and 10 degrees long to

the naked eye. Also, on the latter date, the total magnitude had brightened
to nearly 0*. The comet was last seen by Northern Hemisphere observers on
April 26, when it was located only 3.5 degrees above the horizon and only 14
degrees from the sun.

Observations in the Southern Hemisphere began on May 1, when the comet
was at perigee (0.20 AU). It was then near magnitude 2*, with a coma 15 arcmin
across and a straight, sharply outlined tail 9 degrees long. The nucleus was
then described as well-condensed and possessed sunward jets curving back into
the tail. These jets were reported off and on, depending on the size and
power of the telescope used, until May 20. The tail rapidly decreased in size
during the first week in May and by the 7th was only slightly longer than 2
degrees. The comet was last detected with the naked eye on May 12 (magnitude
near 5.5) with moonlight interfering on the following days. On May 14, the
nucleus was described as star-like with a magnitude of 9. The comet was followed until July 30, when the total magnitude was near 10.5*. Mathematicians
have established that the orbit was hyperbolic with an eccentricity near
1.0003.
(Fabry)

1886 II   Discovered: December 4, 1885   Discovery Magnitude: 9.5
This comet was discovered in Taurus during the evening hours by Barnard
(Nashville) and was described as faint, with only a slight condensation. Observations during the next few days revealed a small coma diameter of 0.5 to 1.0
arcmin, a short tail and an 11th-magnitude nucleus. By the end of December,
the total magnitude was near 8.5, while the nuclear magnitude had increased
to 9. A faint, fan-shaped tail was also present. As the comet approached the
sun and Earth it brightened to near magnitude 8*by mid-January; however, by
late January, the comet-Earth distance was increasing as Earth moved toward the
opposite side of the sun. Subsequently, the nucleus remained near magnitude 9,
and by early March, the total magnitude had increased to only 7. Thereafter,
the comet-Earth distance rapidly decreased and by the end of March, observers
reported a total magnitude near 6.5, a tail over 10 arcmin long and a coma 3
arcmin in diameter.

During April, the comet became visible to the naked eye as it brightened
another magnitude by the 30th. The tail also increased to about 1 degree and
the nucleus brightened to nearly 6th magnitude. Perihelion came on May 3 (r=
0.48 AU), and observers reported the star-like nucleus had vanished, while the
tail had grown longer. The total magnitude was then between 4 and 4.5*and the
tail reached a maximum length of 3 degrees 3 days later. Observations thereafter became more difficult as the comet's rapid southward movement caused it
to approach the horizon. The comet was last seen by Northern Hemisphere observers on May 16, when the full moon and a low altitude caused it to be difficult
to observe despite an apparent magnitude near 3.5*. Observations began in the
Southern Hemisphere on May 26, when the comet was at perigee (0.24 AU). Fading was rapid during June as the distances from both the sun and Earth increased
and by July 5, the comet was described as faint and diffuse. It was last seen
on July 26, when it appeared as a faint, ill-shaped, luminous spot near magnitude 11*.

Both 1886 II and 1886 I were visible in the evening sky and arrived at
perihelion one month apart. These facts by themselves mean very little, but
when coupled with the comets' paths across the sky and the orbital elements, it

becomes apparent that these comets bore a striking resemblance to one another.
Both comets travelled paths across the sky that were basically parallel--with
1886 II lagging behind by one month. Both comets moved northwestward prior
to perihelion and rapidly southward after perihelion. Also after perihelion,
the comets passed the Earth at distances near 0.2 AU. In terms of the orbital elements, the comets had inclinations which differed by only 2 degrees,
arguments of perihelion which differed by only 7 degrees and eccentricities
which equalled 1.0002. The only large differences in orbital elements were
32 degrees in the ascending nodes and 0.17 AU in the perihelion distances.
(Barnard)

1886 III   Discovered: May 1   Discovery Magnitude: 4.5
     Brooks (Phelps, New York) discovered this comet in the morning sky in
Pegasus. He described it as bright with a tail more than 1 degree in length
and a very faint nucleus. On May 2, observers described the nucleus as yellowish with a magnitude near 7.5 and estimated the tail as extending nearly 12
arcmin. Perihelion came on May 4 (r= 0.84 AU), and, thereafter, the comet
rapidly faded as it headed away from both the sun and Earth. On May 6, the
tail was described as split at a distance of 6 arcmin from the nucleus and on
the next morning, observers reported the split as occurring at a distance of
8 arcmin from the nucleus. By May 20, the comet had faded to near magnitude
7.5*and possessed a tail 12 arcmin long. On the 21st, the nucleus was no
longer visible and the tail was less than 10 arcmin in length. The comet was
followed until June 3, when it was described as a hardly visible shapeless
object.
     Between May 4 and 12, Pechule (Copenhagen) described a nebulous streak
emanating from the nucleus and ending in a second, somewhat blurred nucleus
from which extended the comet's tail. A secondary nucleus was also reported
by W. Luther on May 7, but a comparison with Pechule's object indicates that
Luther's secondary nucleus was on the other side of the main nucleus. It is
interesting to note that also on May 7, V. Knorre commented that the tail
seemed to emanate from a sharp tip, but reported no nucleus at this point.
This tip was in the precise location of Pechule's object and may explain
Pechule's mistake while using a lower power telescope. Luther's object has
been called a possible knot in the comet's tail by some researchers, though
this conclusion is not definite.
(Brooks)

1886 V   Discovered: April 28   Discovery Magnitude: 8.0
     While comet hunting in the evening sky, Brooks (Phelps, New York) discover-
ed a fairly bright, circular object 1 arcmin in diameter near Beta Cassiopeiae.
At the beginning of May, observers reported the coma as 2 arcmin across with an
11th-magnitude nucleus near its center and by mid-May, the comet had brightened to a magnitude near 6.5*with a nucleus of the 8th magnitude. The comet
became a more difficult object during the next week as it entered the evening
twilight on its way to conjunction with the sun. On the 25th, it was near magnitude 5.5*with a coma diameter of 2.3 arcmin and a diffuse nucleus of magnitude
6.5. The comet was lost after the 30th. Perihelion came on June 7 (r= 0.27 AU),
and the comet was recovered in Southern Hemisphere skies on July 3, as an 8th-
magnitude*object 1.5 arcmin in diameter. On the 5th, the central condensation
was described as "insignificant." The comet was followed until July 30, when
it had faded to a magnitude near 10.5* Throughout the comet's duration of

visibility, the coma remained circular with no trace of a tail. The orbit is
elliptical with a period of 768 years.
(Brooks)

1886 VIII   Discovered: January 24, 1887   Discovery Magnitude: 10.0[*]
     This comet passed perihelion on November 28, 1886 (r= 1.48 AU), but was
not discovered until nearly two months later, when Barnard located it near the
Lyra-Hercules border in the morning sky. He then described it as circular with
a central condensation, but no tail. The coma was estimated as 1 arcmin in
diameter. The comet slowly faded as it moved away from the sun and Earth,
with observers indicating a magnitude near 10.5 during February and March. On
April 27, the total magnitude was near 11[*], and Auguste H. P. Charlois (Nice)
described the coma as small, circular, with a 13th-magnitude nucleus. The
comet was last seen on May 23, as a small, very faint object near magnitude
11.5[*].
(Barnard)

1886 IX   Discovered: October 5   Discovery Magnitude: 8.0
     This comet was discovered by Barnard (Nashville) in the morning sky as it
moved through Sextans. He described it as bright and circular, with traces of
a tail. The next morning, independent discoveries were made by Hartwig (Bamberg) and Pechule (Copenhagen). On October 8, observers estimated a coma diameter of 3 arcmin with a trace of tail extending to the west. The total magnitude was then near 7.5[*]. By the end of October, the coma had grown to 5 arcmin
and a faint, but distinct tail extended westward for 8 arcmin. The total magnitude was then slightly brighter than 6[*], since the comet was barely visible to
the naked eye in dark skies, and the sharp nucleus possessed a magnitude near
8.
     On November 9, observers estimated a magnitude near 5.5 and a tail length
of nearly one-half degree. The coma diameter was estimated as 10 arcmin, which
is unusual for a comet located at distances of 1 AU from the sun and 1.4 AU
from Earth. Shortly thereafter, the tail became split with the southern portion being shortest and sharply outlined and the northern portion being longest
and more diffuse. Towards the end of the month, the northern tail developed
radial rays close to the comet's head. Naked-eye estimates of the comet's
tail lengths on November 24, gave values of 20 and 40 arcmin for the southern
and northern tail, respectively. On the 29th, photographs revealed the southern tail to have shrunk to 15 arcmin, while the northern tail became longer and
more pronounced. The total magnitude on the latter date was near 3.5[*].
     At the beginning of December, estimates of the comet's total magnitude
were near 3.3 and the larger tail was nearly 7 degrees long. The secondary
tail was still visible as it curved southward for 20 arcmin. The comet was
nearest Earth on the 6th (0.96 AU), and from that date until perihelion on the
16th (r= 0.66 AU), the brightness changed very little. The tail rapidly declined after perihelion as the distances from both the sun and Earth increased.
On the 20th, the main tail extended 7 degrees, but, four days later, it had
shrunk to 5 degrees, with the short tail still visible and 20 arcmin long. On
December 26, the total magnitude was about 4.5 and the nucleus was visible as
a small bluish disk. At the beginning of January 1887, the comet began to
rapidly approach the horizon as it neared conjunction and it was last seen in
bright twilight on January 13.
     After conjunction, Finlay (Cape of Good Hope) became the first Southern

Hemisphere observer of this comet when he accidentally rediscovered it on
April 29, at a magnitude near 10. Observations were few over the next couple
of months, due to the comet's faint magnitude, and it was last detected on
June 17, at a magnitude near 11.* The comet's orbit is hyperbolic with an
eccentricity of 1.0004.
(Barnard-Hartwig)

1887 I   Discovered: January 18   Discovery Magnitude: 1.5
      This comet was first discovered shortly after sunset by a farmer at
Blauwberg, South Africa. A few hours later, an independent discovery was made
by Thome (Cordoba), who described the comet as a narrow ribbon of light which
contracted near the sun. Thome added that he could discern no coma or condensation of any type so that his estimate of the comet's position was only
rough. The comet had passed perihelion on the 11th (r= 0.005 AU), and when
discovered was located 14 degrees from the sun. On January 20, Todd (Adelaide,
Australia) estimated the comet's magnitude as comparable to the Magellanic
Clouds (magnitude near 2) and the next evening, in bright twilight, Thome
estimated the tail as 25 degrees long. Thome estimated the comet's head as
then being about 5 degrees above the horizon, though no nucleus or condensation
could be detected. On the evening of January 22, observers estimated a tail
length of nearly 40 degrees and Finlay (Cape of Good Hope) described the tail
as resembling a faint, narrow, luminous ribbon with no trace of a head or condensation. Observers on the 24th indicated a total magnitude near 3. The
comet rapidly became more diffuse over the next few days and by the 27th was
described by Todd as bearing a head which resembled a nebulous mass cut by a
wide crevasse or discontinuity. He added that another discontinuity existed
above the head in the tail. At the same time, Cruls (Rio de Janeiro) estimated
the dimensions of the tail as 52 degrees long and one-half degree wide. The
next evening, with the comet having climbed further above the horizon, Tebbutt
(Windsor, New South Wales) reported the tail as extending to near Alpha Eri-
dani (over 30 degrees long), but, like other observers before him, he could
not discern any trace of a head or condensation. The comet was last observed
on January 30, at a magnitude near 4.5.* The tail was estimated to extend nearly 15 degrees and as before possessed no visible head or nucleus. Observations
thereafter were impossible due to moonlight. Due to the comet's apparent lack
of a coma or nuclear condensation, the few positions obtained were very discordant. Consequently, all previously published orbits for this comet are
considered hypothetical, based on the assumption that this comet was a member
of the Kreutz sungrazing group.
(Great Southern Comet)

1887 II   Discovered: January 23   Discovery Magnitude: 8.5
      Discovered near Phi Draconis by Brooks (Phelps) in the evening sky, this
comet was described as a highly condensed nebula moving slowly towards Cepheus.
As the comet approached the sun, it passed perigee on February 12 (1.28 AU), and
was then described as 2 arcmin across with a star-like nucleus near magnitude
10.5. No tail was then visible and the total magnitude was still near 8.5.*
About the same time, the comet passed within 10 degrees of the north celestial
pole. Afterwards, the comet's increasing distance from Earth countered the
decreasing distance from the sun and when perihelion came on March 17 (r= 1.63
AU), the total magnitude was still estimated as near 8.5.* The coma was then
described as about 1.5 arcmin in diameter with a star-like nucleus near its

center. The comet faded rapidly during April. On the 12th, the total magnitude was near 9*, while the nucleus was near magnitude 10.5, and, when last seen
on the 23rd, the total magnitude was near 9.5*. Observations thereafter were
impossible due to the comet's low position in the western sky. The orbit is
elliptical with a period near 1,000 years.
(Brooks)

1887 III   Discovered: February 17   Discovery Magnitude: 10.5
    While searching for comets in the evening sky using a 15-cm refractor,
Barnard (Nashville) discovered this object in Puppis. He described it as very
faint and moving rapidly northwestward. The comet was then near perigee (0.24
AU) and, thereafter, moved rapidly away from Earth as it approached the sun.
On the 24th, observers described the comet as brighter (magnitude near 9.5)
with a diameter between 2 and 4 arcmin. The coma was slightly condensed and,
although no tail was visible, it was elliptical in shape. Between February 18
and March 14, the central condensation steadily faded as the comet receded
from Earth--with the magnitude changing from 10.5 to 12.5*. Observers on March
19, indicated the total magnitude was near 10 and the coma diameter was near
2.5 arcmin. Perihelion came on March 28 (r= 1.01 AU), and the comet was found
to be very faint (magnitude near 10.5) due to its low altitude. It was followed until April 10, when the total magnitude had declined to nearly 11.5*.
(Barnard)

1887 IV   Discovered: May 13   Discovery Magnitude: 9.5
    Barnard (Nashville) discovered this comet in the evening sky near the
borders of Hydra, Scorpius and Centaurus. It was then described as faint and,
the next evening, Barnard reported the presence of a faint tail. Within the
next few days, observers in the United States and Europe described the comet as
nearly 2 arcmin in diameter with a near-stellar nucleus of about magnitude 11.5.
The total magnitude was still near 9.5* and the faint tail was still visible.
The comet's orbit was inclined only 17 degrees to the ecliptic and the result
was a slowly decreasing distance from the Earth, as that planet, which was
located closer to the sun, slowly caught up with the comet. Perigee came after
the first week of June (0.38 AU), and observers then described the comet as
near magnitude 9 with a coma diameter of 1.5 arcmin. The nucleus was estimated
as about 10th magnitude and the tail was measured as 3.6 arcmin long and 1.4
arcmin wide. Perihelion came on June 17 (r= 1.39 AU), and, thereafter, Earth
slowly pulled farther away from the comet. By mid-July, the total magnitude had
faded to nearly 10*. At the same time, the coma diameter was estimated as about
1 arcmin and a small, fan-shaped tail was visible. The comet was followed
until August 12, when it had faded to a magnitude of 11.5*. The orbit is
elliptical with a period near 6,700 years.
(Barnard)

1888 I   Discovered: February 19   Discovery Magnitude: 3.0
    Mr. Sawerthal (Royal Observatory, Cape of Good Hope) discovered this comet
with the naked eye in Telescopium and, on the next morning, an independent discovery was made by Nolau (Brankholm, Australia). On the 25th, Eddie (Cape of
Good Hope) reported that the comet possessed a highly condensed nucleus near
magnitude 4 and a curving tail 8 degrees long. Although the next several days
were hampered by moonlight, Eddie made an observation on March 4, and described
the comet as of total magnitude 3.5 with a tail 3 degrees long. Through a
telescope, he could see a double jet formation which curved from the apex to

the tail. Perihelion came on March 17 (r= 0.70 AU), and naked-eye estimates
then gave a total magnitude of 4, a nuclear magnitude slightly fainter than
18 Aquarii (hence, near 5.7) and a tail length of 6 degrees. The comet's
distances from the sun and Earth slowly increased after perihelion and the
comet was last detected with the naked eye in mid-April, when the tail was
still nearly 2 degrees long.

About mid-March, the comet had moved far enough to the north to allow
observations in both the Southern and Northern Hemispheres. On March 20 and
22, A. Charlois (Nice Observatory) suspected the comet's nucleus to be split
and on the 30th, L. Cruls (Rio de Janeiro) independently discovered the second
nucleus at a distance of 2.96 arcsec from the main one. The second nucleus
was observed by several other observers during April, as its distance from
the main nucleus increased to 6.3 arcsec by the 16th. During this same
interval, some observers gave fragmentary accounts of a possible third condensation. The last accurate measure of the secondary nucleus was made on May
11, by C. F. Pechule (Copenhagen) when it was 15 arcsec away, but reports of
its existence continued until June 5, and, possibly, even until June 24, although the indicated separation of 7 arcsec is inconsistent with all of the
previously reported observations. Modern researchers have assigned the date
of splitting somewhere between March 2 and 11.

Moonlight interfered with observations in early May, and by mid-month,
the total magnitude was near 7.* On May 18, the comet was faint in moonlight
as observed through a telescope, and, on the 19th, the nucleus was estimated
as near magnitude 9.5. Beginning on May 20, the comet underwent an outburst
in brightness with the total magnitude being 5 and, the next morning, the
nucleus was of magnitude 5.8, with two wide lateral branches emanating from
the head. On May 22, the comet began to fade and on June 1, the lateral
branches had spread backward, like a fountain, to join the tail. On June 2,
the total magnitude was near 5.5 and by the 9th, it had fallen to near 6. One
week later, the comet had finally declined to its expected brightness of 8.0.
By mid-July, observers indicated a total magnitude near 9.5, a tail 10 arcmin
long and a very faint nucleus near magnitude 11.5. By mid-August, the total
magnitude had faded to near 10.5*and the comet was last seen on September 4,
with a tail 5 arcmin long and a total magnitude near 11.5.* A possible observation by L. Swift on September 25 is still open to doubt up to the present
time. The orbit is elliptical with a period near 2,200 years.
(Sawerthal)

1888 III   Discovered: August 8   Discovery Magnitude: 8.5
Having passed perihelion on July 31 (r= 0.90 AU), this comet was discovered by Brooks (Geneva, New York) in Ursa Major as it headed slowly away from
the sun and Earth. At discovery, it was described as 0.5 arcmin in diameter
with a star-like nucleus of 11th magnitude and a faint tail 5 arcmin long.
Brooks' total magnitude estimate of 9 to 10 may, in fact, have referred to a
central condensation containing the nucleus, since observations at the end of
August, strongly indicate a total magnitude near 9. Also, at the end of August, observers reported a tail 3 to 6 arcmin long. By September 23, the comet
was described as near magnitude 9.5 with a very faint star-like nucleus. Observations continued into October, and the comet was seen for the final time
on the 30th, when observers in Vienna indicated a magnitude near 11.5. The
orbit is elliptical with a period near 971,000 years.
(Brooks)

1888 V   Discovered: October 31   Discovery Magnitude: 9.5
This comet passed perihelion on September 13 (r= 1.53 AU), and remained
hidden in the sun's glare until about mid-October. While searching for comets
in the morning sky on the 31st, Barnard (Lick Observatory, California) discovered this comet in Hydra and described it as 1 arcmin across with a diffuse
nucleus near magnitude 11.5 and a tail 3 arcmin long. Although the comet was
then moving away from the sun, its decreasing distance from Earth caused it to
fade very slowly during the next several months. During the first half of
November, observers continually estimated the coma diameter as 2 arcmin and
estimated the nucleus as near magnitude 10.5. Moonlight halted observations
during the last half of November, and by early December, observers found the
total magnitude near 9.5 and the coma diameter near 1 arcmin. The nucleus
had taken on a stellar appearance and was estimated as about magnitude 11.5.
Perigee occurred during early January 1889 (1.4 AU), and observers then
described the comet as 2 to 3 arcmin in diameter, with a nuclear magnitude
near 11 and a total magnitude near 10. With the distances from the sun and
Earth increasing thereafter, the comet began to change more rapidly, although
the total magnitude faded slower than expected. By the end of February, the
comet was estimated as 2 arcmin in diameter with a nucleus near magnitude
12.5 and a total magnitude near 10.5. One month later, the total brightness
had declined another one-half magnitude, while the nucleus was then estimated
to be of magnitude 14 and the coma was found to be 1 arcmin across. At the
beginning of May, the stellar nucleus had disappeared and the comet was described as very small and faint. It was last observed on May 23, when Barnard
indicated a total magnitude near 12.5. The orbit is elliptical with a period
near 2,400 years.
(Barnard)

1889 I   Discovered: September 3, 1888 Discovery Magnitude: 8.5
While searching for comets with a 10-cm seeker, Barnard (Lick Observatory)
discovered this comet near the Gemini-Monoceros border and described it as 1
arcmin across with an 11th-magnitude nucleus, but no tail. The next morning,
Brooks (Geneva, New York) made an independent discovery. The comet was then
located 2.55 and 2.82 AU from the sun and Earth, respectively, with both distances decreasing during the next couple of months. On September 13, B. Engel-
hardt (Dresden) described the comet as a bright elongated object and indicated
a total magnitude of 8.5. With a 30.5-cm refractor he observed the nucleus to
be double, although no other observation was made therefter that confirmed
this sighting. A later researcher calculated the most probable date of splitting as August 30, 1888, but the general consensus today is that the secondary
nucleus was a star involved in the coma.
At the beginning of October, the comet was reported as brighter (magnitude
about 8) with a nucleus near magnitude 10 and a coma 3 arcmin across. The
first traces of a tail were noted on the 27th, by Baeker (Strasbourg). By
early November, the total magnitude had increased to 7.5[*] and the star-like
nucleus was estimated as near magnitude 9. Perigee came near the end of November (1.10 AU), and the comet was then described as 6 arcmin in diameter with
a total magnitude of 6[*]. Although the comet-Earth distance increased rapidly
during December, and should have caused a fading of at least one magnitude by
the end of the month, the comet, instead, developed into an impressive telescopic object. Observers during the first days of January 1889 estimated the

total magnitude as near 5.5 and the nuclear magnitude as near 9. On the 4th, Rudolf Spitaler (Vienna) estimated the coma as 3 arcmin in diameter and the tail as 6 degrees long. Conditions for observing the comet deteriorated during the remainder of January, first due to moonlight and, finally, due to twilight as the comet neared conjunction with the sun. The last observation was made on January 28, and the comet passed perihelion on the 31st (r= 1.81 AU).

The comet was recovered near the Pegasus-Aquarius border on June 25, by Spitaler and Holetschek (Vienna). They described it as about magnitude 8 with a coma 3 arcmin across and a nucleus near 9th magnitude. The next morning, the coma was reported to be more developed on the sunward side and by July 3, Barnard found a tail 1 degree long and 3 arcmin wide directed towards the sun. This anti-tail had vanished by mid-July. By the end of the month, observers described the comet as 2 arcmin across with only a slight trace of tail pointing away from the sun. The total magnitude was then near 9.5[*] and the nuclear magnitude was about 10. During August, the total magnitude and nuclear magnitude faded to values of 10.5[*] and 11.5, respectively. The comet was again approaching conjunction with the sun as October began, and, when last seen on the 24th, it was estimated as near magnitude 11[*].

The comet was again seen near opposition in 1890, when Spitaler recovered it on March 28, using the Vienna Observatory's 69-cm refractor. He then described the comet as circular with a slight condensation. Brightness estimates over the next few months by Spitaler and Barnard indicated a total magnitude between 13 and 14 and a coma diameter between 15 and 30 arcsec. After opposition, the comet quickly faded as its distance from Earth increased and it was last seen on September 8. It was then located over 6 AU from both the sun and Earth. There is a possibility that Spitaler recovered the comet near opposition on May 1, 1891, using the 69-cm refractor; however, since he was using an ephemeris and described the comet as barely seen at the limit of his telescope, the observation is considered doubtful. The duration of visibility equalled 735 days--over 200 days greater than the record set by comet 1811 I. The orbit is hyperbolic with an eccentricity of 1.0012.
(Barnard)

1889 II   Discovered: April 1   Discovery Magnitude: 12.0

Barnard (Lick Observatory) discovered this comet in the evening sky and described it as very faint with a coma 10 arcsec across, a tail 15 arcmin long and a star-like nucleus near magnitude 13. It was then located in Taurus and was moving to the southwest. As April progressed, the comet gradually neared evening twilight as Earth rapidly moved away from it, and when last seen on the 29th, the total magnitude was near 11.5[*], the coma diameter was 1 arcmin and the tail was extending about 8 arcmin. Attempted observations by Spitaler (Vienna) on May 1 and 2, were fruitless due to twilight.

The comet arrived at perihelion on June 11 (r= 2.26 AU), and, after conjunction with the sun, it was recovered in the morning sky toward the end of July. Observers then described it as a faint, diffuse object near magnitude 10.5 with a short tail. The comet was at perigee in early October (1.84 AU), and was then near magnitude 10[*]. With the distances from both the sun and Earth increasing thereafter, the comet began to fade rapidly. By late October, the total magnitude was near 10.5[*] while the nucleus was estimated as about 12 and at the end of November, the total magnitude and nuclear magnitude had de-

clined to values of 11.5[*]and 14, respectively. The coma diameter during that
same time interval remained near 1.5 arcmin. Shortly thereafter, the comet
entered evening twilight on its way to conjunction with the sun. Using the
91-cm refractor at Lick Observatory, Barnard recovered the comet at opposition on August 23, 1890. He then described it as a small, faint nebula about
5 arcsec in diameter and indicated a total magnitude near 13.5. The comet was
last seen on September 8. The orbit is elliptical with a period near 1 million
4 hundred thousand years.
(Barnard)

1889 IV   Discovered: July 19   Discovery Magnitude: 3.5
    This comet was discovered on the day of perihelion (r= 1.04 AU) by
Davidson (Australia) with the naked eye. He described it as bright with a
nucleus near magnitude 5 and a coma 5 arcmin across. No tail was reported.
Then located in Centaurus at a declination of -45 degrees, the comet rapidly
moved northeastward, passing perigee (0.35 AU) along the way, and became visible to Northern Hemisphere observers on July 25. The comet was then described as a 4th-magnitude object with a stellar nucleus of magnitude 6 and a coma
nearly 5 arcmin across. A tail was then widely reported as extending 30
arcmin. As the comet moved away from both the sun and Earth, it faded from
a magnitude near 5[*]on August 1, to near magnitude 7[*]by the 31st. Also, during
August, a secondary nucleus appeared.
    A. Ricco (Palermo, Italy) was the main observer of the double nucleus.
Observing between August 3 and 11, using a 25-cm refractor, he reported that
the separation between the components became more distinct every day until
August 8, with the fainter and larger component following the brighter one.
Between the 8th and 11th, the separation increased to about 16 arcsec, but
the secondary nucleus became larger and fainter and was not seen by Ricco after
the latter date. Independent of each other on August 6, both Ricco and Barnard
(Lick Observatory) saw no sign of a secondary nucleus. Barnard was then unaware of Ricco's discovery and was using a 30-cm equatorial. While observing
with a 38-cm telescope in late August, F. Renz also discovered a secondary
nucleus on August 28, and reobserved it on September 2. A study by Sekanina
in 1979 indicated that the rough estimates of separation and position angle
by Renz conformed surprisingly well to the extrapolated position of Ricco's
object. He then concluded that the secondary nucleus was real and very likely
separated from the main nucleus around July 29.
    By late September, the comet had become smaller and more condensed. The
total magnitude was then near 9.5[*]and the tail was described as short and
straight. One month later, the comet had faded to near magnitude 10.5[*]and
the coma was estimated as 30 arcsec in diameter with only a slight elongation
away from the sun. The comet was last observed at a magnitude near 11.5[*]on
November 23. The orbit is elliptical with a period near 9,000 years.
(Davidson)

1890 I   Discovered: December 12, 1889   Discovery Magnitude: 9.5
    Borrelly (Marseilles) discovered this comet near Gamma Draconis and des-
cribed it as a faint and diffuse object 2 arcmin in diameter. Thereafter, the
comet moved rapidly southward toward perihelion and conjunction with the sun;
however, although the altitude was rapidly decreasing, the distance from Earth
was diminishing and the comet became much brighter. Between December 17 and
19, observers indicated a total magnitude near 8.5 and described the coma as

2 arcmin across. No tail or nucleus was then reported. Observations were
affected by moonlight during late December, but when the comet was again
observed on January 5, 1890, it had increased to a magnitude of 7.* By January 9, the total magnitude had brightened to 6.5*and the comet's altitude had
declined to only 5 degrees. A tail rapidly developed thereafter, and by the
16th, it extended 45 arcmin. On the same date, observers indicated the total
magnitude was near 4.5. The comet was last observed on January 20, very close
to the horizon. The tail was then shortened by the low altitude to only 5
arcmin and the brightness must then have been about 3.5 to 4. Perihelion
came on January 26 (r= 0.27 AU), and the comet was not observed after conjunction.
(Borrelly)

1890 II   Discovered: March 20   Discovery Magnitude: 7.5
     Brooks (Geneva, New York) discovered this comet in Equuleus shortly before
dawn and described it as a bright telescopic object with a star-like nucleus
and a short, wide tail. The comet was then located 2.12 AU and 2.77 AU from
the sun and Earth, respectively; however, although these distances decreased
during the next couple of months, the comet's brightness faded to between magnitudes 8.1 and 8.5 by the middle of May. Later observations showed the May
brightness estimates to have been correct, thus indicating the possibility of
the comet undergoing an outburst in brightness about the time of discovery. A
study by E. M. Pittich in 1969 lends support to the outburst theory as he
showed that if brightness had not undergone a drastic change the comet would
probably have been discovered at least 20 days earlier.
     The comet passed perihelion on June 2 (r= 1.91 AU), and was then estimated
as slightly brighter than magnitude 8.5. Although the comet's distance from the
sun increased thereafter, its distance from Earth decreased and the brightness
remained near 8.5*until nearly mid-August. The only physical feature observed
to change during that interval was the tail, which increased from about 6 arcmin
on June 26, to 15 arcmin on July 15, when the comet was near perigee (2.05 AU).
By late August, the total magnitude had faded to 9*and by mid-November, it
was near 10.5* Observations continued into the next year, and on February 28,
1891, the total magnitude was near 11* At the same time, the nucleus was estimated as near magnitude 12 and the fan-shaped tail extended 1 arcmin. The
comet vanished in evening twilight after May 30, but was recovered near opposition on January 6, 1892, in the morning sky. The comet was seen for the last
time on February 5, when it appeared as an ill-shaped spot 1 arcmin across and
near magnitude 12.5* The orbit is hyperbolic with an eccentricity of 1.0003.
(Brooks)

1890 III   Discovered: July 18   Discovery Magnitude: 8.0
     This comet passed perihelion on July 9 (r= 0.76 AU), and was discovered in
Ursa Major by Coggia (Marseilles) 9 days later. Coggia then described the comet
as a bright, circular object 2 arcmin in diameter. The comet moved rapidly
southeastward towards conjunction with the sun. At the same time, its distances from both the sun and Earth increased and the brightness faded rapidly.
By late July, the comet was seen at low altitude at a magnitude near 9*and with
a faint tail. A non-stellar nucleus was then near magnitude 11. The comet was
last detected, in twilight, on August 14, when observers at Lick Observatory
found it at a magnitude near 10.
(Coggia)

1890 IV   Discovered: November 15   Discovery Magnitude: 8.0

    T. Zona, director of the Palermo Observatory, Sicily, discovered this comet in Auriga and described it as very bright with a coma 1 arcmin across and a central condensation. It had passed perihelion on August 7 (r= 2.05 AU), and at discovery was located 2.36 AU and 1.49 AU from the sun and Earth, respectively--with both distances increasing. Observations on November 16 confirmed the comet's initial brightness and established the existence of a faint nucleus near magnitude 11.5; however, observations on the 18th revealed that the total brightness had declined to magnitude 9.5 and no nucleus could then be discerned. The coma diameter was still near 1 arcmin. By November 21, the comet had reached magnitude 10.5 and remained near that brightness at least until December 5. Curiously, between the day of discovery and the 5th of December, the comet should have faded only 0.5 magnitude and an outburst in brightness shortly before the time of discovery seems a distinct possibility. As December progressed, the comet gradually became fainter and by January 2, 1891, the magnitude was near 11.5. The coma was then about 30 arcsec in diameter with a slight trace of condensation. The comet was last detected on January 13, at a magnitude near 12*. The orbit is elliptical with a period near 11,000 years.
(Zona)

1890 VI   Discovered: July 23   Discovery Magnitude: 9.5

    This comet was discovered by William F. Denning (Bristol, England) using a 25-cm reflector. Then located in Ursa Minor, the comet was described as a faint, circular object 1 arcmin across. Although the comet was then approaching both the sun and Earth, it had barely brightened one-half magnitude when it reached perigee during the latter half of August (1.20 AU). Thereafter, the comet slowly faded and near the date of perihelion on September 25 (r= 1.26 AU), the comet possessed a magnitude near 10.5*, with a star-like nucleus of 14th magnitude. Since discovery, the comet had moved almost due south, and after perihelion it was located only a short distance above the horizon. It was last detected in the Northern Hemisphere on October 6, at a magnitude near 11*. Observations in the Southern Hemisphere began on October 1 and continued until November 8, when Thome (Cordoba) indicated a magnitude slightly fainter than magnitude 11. The orbit is elliptical with a period near 61,000 years.
(Denning)

1891 I   Discovered: March 30   Discovery Magnitude: 8.0

    This comet was first discovered by Barnard (Lick Observatory) and was independently detected by Denning (Bristol) the next morning. It was then described as bright with a coma 1 arcmin across, a tail 30 arcmin long and a nucleus near magnitude 10. Afterwards, the comet moved southeastward towards conjunction with the sun and perihelion. On April 5, the magnitude had brightened to 7.5* and, due to the interference of twilight, the tail extended only 6 or 8 arcmin. The comet was last seen on April 14, at a magnitude near 6.5*. Perihelion came on April 28 (r= 0.40 AU), and afterwards, when recovered on May 19, the comet was strictly a Southern Hemisphere object. On that date, the magnitude was near 7*. During the first half of June, the total magnitude was near 8.5* and the coma was described as 1.5 arcmin in diameter with a brightening at the center. By the end of the month, the magnitude had faded to about 10*, and when last seen on July 9, it had faded to near 11th magnitude*.
(Barnard-Denning)

1891 IV   Discovered: October 3   Discovery Magnitude: 8.0
     Barnard (Lick Observatory) discovered this comet in the morning sky and
described it as bright, with a coma diameter of 1 arcmin and a nuclear condensation near magnitude 12. The comet was then located in Puppis at a declination near -25 degrees and moved southward. Subsequently, Lick Observatory was
the only Northern Hemisphere observatory to view the comet, and by October 9,
even they could no longer see the comet due to its increasing southern declination. Following the telegraphic announcement, observers in Sydney, Australia,
found the comet on October 9, and during the next few days, the comet was described as a faint, circular object with a small condensation. Although the
comet was moving toward the sun, its increasing distance from Earth should
have kept the magnitude the same as at discovery on into November; however,
from the time the comet was first observed in the Southern Hemisphere it was
never brighter than 10th magnitude, and this makes the comet a candidate for
a prediscovery outburst in brightness. Perihelion came on November 14 (r=
0.97 AU), and the comet was followed at Cordoba until December 7, when near
magnitude 10.5.
(Barnard)

1892 I   Discovered: March 7   Discovery Magnitude: 4.0
     While searching for comets in the morning sky, Professor L. Swift (Roch-
ester, New York) discovered this comet in Sagittarius. Although found with a
telescope, it could be seen with the naked eye by knowing exactly where to
look. One of the first photographic records of the comet was obtained on
March 11, by Professor H. C. Russell (Sydney Observatory, Australia). His
photo showed 8 tail rays, two of which extended beyond the plate's edge. On
each side of these two long rays, three other streamers were grouped together.
Other photos during the month continued to reveal the multiple structure in the
tail, although the number and relative luster of the rays underwent rapid and
marked changes--especially near perihelion. Visual observers were not treated
to so spectacular a show of the tail, but the comet itself disappointed no
one. Toward the end of March, the total magnitude was near 3, the coma was
5 arcmin across and some observers reported a tail nearly 15 degrees long. At
the beginning of April, with the total magnitude unchanged, the coma and tail
had increased to 10 arcmin and 20 degrees, respectively.
     Perihelion came on April 7 (r= 1.03 AU), and, thereafter, with the dis-
tances from both the sun and Earth increasing, the comet slowly faded. By
April 23, the total magnitude had declined to 4 and the tail was still about
4 degrees long and one month later, these values had declined to 5.5 and 30
arcmin, respectively. The comet finally dropped below naked-eye visibility
in early June, and by month's end, the total magnitude was near 6.5, while an
estimate of the star-like nucleus gave a magnitude of 7.
     Observations continued through the end of 1892, with the total magnitude
declining to 8[*]in August, 9.5[*]in October and 11[*]in December. It was last detected on February 16, 1893, when the total magnitude was near 13[*]. The orbit
is elliptical with a period near 23,000 years.
(Swift)

1892 II   Discovered: March 18   Discovery Magnitude: 10.0
     Denning (Bristol, England) discovered this comet near Delta Cephei while
using a 25-cm reflector. He described it as a small, faint object with a
central condensation, but no tail. Observations through the end of March

indicated a magnitude of 11.5 for the star-like nucleus and a coma diameter of
1 to 2 arcmin. Observations during April were basically unchanged, primarily
due to the comet's large distance (over 2 AU) from both the sun and Earth.
Perihelion came on May 11 (r= 1.97 AU), and, afterwards, the comet faded very
slowly as a decreasing distance from Earth countered the effects of an increasing distance from the sun. By late September, the total magnitude had
declined to only 11 and a nucleus of magnitude 12 was also visible. Observations continued through the end of 1892, and finally ceased on January 12,
1893, when the comet was located in Eridanus as an exceptionally faint object
(magnitude near 13) with a coma diameter of 30 arcsec and a slight condensation. The orbit is hyperbolic with an eccentricity of 1.0004.
(Denning)

1892 VI   Discovered: August 28   Discovery Magnitude: 9.0[*]
    Brooks (Geneva, New York) discovered this comet in Auriga and described
it as very bright with a definite nucleus and a short, faint tail. The comet
was then located 2.16 AU and 2.39 AU from the sun and Earth, respectively, with
both distances decreasing. By late September, the comet was near magnitude
8.5 and possessed a coma 1 arcmin across and a slightly diffuse nucleus near
magnitude 10. By the end of October, the total magnitude had reached 7.5,
the coma was about 2 arcmin across and a star-like nucleus was estimated as
about magnitude 9.2. The comet became faintly visible to the naked eye shortly
after mid-November, with Holetschek (Vienna) estimating the brightness as near
5.5. Telescopic aid then revealed a tail 2.5 degrees long, while photographs
showed a length closer to 5 degrees. The nucleus was about 7.5. Perigee
occurred in early December (0.87 AU), and observations indicated a total magnitude slightly brighter than 6. Due to a rapid southern motion, observations in
the Northern Hemisphere ended about mid-month. The comet passed perihelion on
December 28 (r= 0.98 AU).
    Southern Hemisphere observations had begun in late November, and continued
on into the new year. After January 1893, all traces of a tail had vanished
and the star-like nucleus had diffused into a faint central condensation. By
mid-February, the total magnitude had declined to 8.5[*] but fading was so slow
that the comet had dropped by only one magnitude by mid-April. Observations
continued until July 19, when the magnitude was near 12.5[*] The orbit is hyperbolic with an eccentricity of 1.0004.
(Brooks)

1893 I   Discovered: November 20, 1892   Discovery Magnitude: 10.0
    Discovered near Epsilon Virginis in the morning sky by Brooks (Geneva),
this comet was described as 1 arcmin across with a prominent eccentric nucleus,
but no tail. After one week, observers were describing the comet as slightly
brighter (magnitude 9.5) with a coma 1 arcmin across and a nucleus near magnitude 11. The comet continued to approach the sun and Earth throughout December, and by month's end the total magnitude was near 8, while the coma had
grown to 2 arcmin in diameter and a short, fan-shaped tail had become visible.
Perihelion came on January 7 (r= 1.20 AU), and the comet was then described as
near magnitude 7.5 with a coma diameter of 3 arcmin and a nucleus of magnitude
11.7. Perigee occurred a couple of days later (0.70 AU) and the comet faded
thereafter. By mid-February, the total magnitude equalled 10[*] and on March 8,
it was near 11[*] On the latter date, the comet was rapidly nearing conjunction
with the sun and was last detected on March 11, when the coma diameter was

estimated as about 15 arcsec. Although a parabolic orbit is generally given
in comet orbit catalogs, the orbit may possibly be hyperbolic.
(Brooks)

1893 II   Discovered: June 20   Discovery Magnitude: 5.0
     This comet was first detected near Rho Arietis by W. E. Sperra (Randolph,
Ohio), who immediately mistook the comet for periodic comet Finlay, which was
then traversing the morning sky in that general region. Without an ephemeris
on comet Finlay's movement, Sperra continued believing he was observing comet
Finlay for the next two weeks as his comet slowly brightened. With the brightening came further discoveries during the first days of July. On the 4th,
Roso de Luna (Logrosan, Portugal) found it and, on the 8th, David E. Hadden
(Alta, Iowa) found it without optical aid as a large, hazy star without any
visible tail. Neither observer immediately suspected a cometary nature. The
first person to report the new comet was Alfred Rordame (Salt Lake City, Utah),
who, after discovering the comet in Lynx in the evening sky, immediately reported it to the proper authorities. The next evening, F. Quenisset (Paris) independently found the comet and made a report.
     Perihelion came on July 7 (r= 0.67 AU), and the comet became widely obser-
ved on July 11, when described as near magnitude 3.5 with a tail 12 degrees
long. The tail rapidly increased in length as it grew to 17 degrees by the
12th, and 22 degrees on the 13th, according to Hadden. Other observers generally reported lengths within a degree or two of Hadden's estimates. After the
13th, the tail declined to 10 degrees by the 15th, and only 3 degrees by the
18th. With a decreasing tail length came a rapid fading as the comet moved
away from both the sun and Earth, and during the first days of August, it had
reached a magnitude of 6 and possessed a diffuse nucleus near magnitude 9.5.
Rapidly approaching evening twilight as August progressed, the comet was last
observed on the 17th. After conjunction, it was recovered on November 3, as
a very faint object which was thereafter kept under observation until December
21, when it had faded to between magnitude 12 and 13.[*] The comet's orbit is
elliptical with a period near 44,000 years.
(Rordame-Quenisset)

1893 IV   Discovered: October 17   Discovery Magnitude: 7.0
     Having passed perihelion on September 19 (r= 0.81 AU), this comet was
moving northward through Virgo when discovered by Brooks (Geneva). It was
then described as a bright telescopic object with a distinct nucleus and a
tail 3 degrees long. During the next few days, observers reported a coma 2
arcmin across and a tail which photographically extended 10 degrees and was
irregularly divided into two slightly divergent branches. Unusual tail activity continued into November, with numerous knots, bends and secondary tails
being photographed. On two occasions the tail even separated from the coma,
leaving a dark space between the beginning of the tail and the coma which
lasted only a day or two. By November 9, the photographic tail had shrunk to
one-half degree long. The comet slowly faded after discovery and by mid-November, it was near magnitude 8.5. At the beginning of December, it had faded
to near 9.0. Perigee came on December 5 (1.3 AU), and the comet rapidly faded
thereafter. By mid-December, it was near 10[*] and by the end of the month, when
located only 14 degrees from the north celestial pole, the total magnitude was
estimated as 11.5.[*] The comet was followed until January 27, 1894, when it was
described as a faint spot near magnitude 13. The orbit is elliptical with a

period near 3,500 years.
(Brooks)

1894 II   Discovered: April 1   Discovery Magnitude: 6.5
     Walter F. Gale (Sydney, Australia) discovered this comet near Eta Horo-
logii using a 7.6-cm telescope. He described it as circular and bright, with
a faint tail 1 degree long. The comet was initially too far south for observation in the Northern Hemisphere, but southern observers reported a rapid
brightening as the comet neared both the sun and Earth. Photographs on April
5 revealed a faint double tail and the comet became visible to the naked eye
2 days later (magnitude 5.5). On April 12, the narrow tail extended 2 degrees
and the coma was described as 4 arcmin across. Perihelion came on April 13
(r= 0.98 AU).
     Soon after mid-month, the comet began heading rapidly northward as it
approached Earth and the first northern observer was Andrew E. Douglass
(Lowell Observatory, Arizona) on April 26. He then described it as near 5th
magnitude with a large round coma. Other observers were soon measuring the
coma as near 10 arcmin in diameter with a tail 2 degrees long. Perigee came
on May 1 (0.35 AU), and the comet was described as 15 arcmin across with a
total magnitude near 4. Although the comet receded from both the sun and
Earth thereafter, the tail was considered longest on May 5. Barnard then
photographed it using a 15-cm Brashear camera and noted faint traces of a tail
extending to the edge of the plate, or a little more than 6 degrees.
     By the end of May, the comet had faded to 7th magnitude and the coma had
shrunk to 3 arcmin in diameter. By the end of June, the brightness had dropped
to 9 and one month later, it was near 11.5. The comet was last observed on
August 21, at a magnitude near 14.5* The orbit is elliptical with a period
near 958 years.
(Gale)

1895 III   Discovered: November 22   Discovery Magnitude: 7.5
     This comet was discovered in the morning sky near Upsilon 1 Hydrae by
Brooks (Smith Observatory, Geneva). He described it as bright, without a
noticeable condensation. Perihelion had occurred on October 21 (r= 0.84 AU),
and when discovered the comet was heading towards Earth. Although fading should
have been very slow, the comet had already declined to magnitude 8.5 by the end
of November. At the same time, the comet's rapid motion had carried it into
Cancer. Perigee came on December 4 (0.28 AU), and although the predicted
magnitude should have been very near the discovery magnitude, the actual brightness was near 9. The coma was then estimated as near 5 arcmin in diameter.
By December 12, the comet was described as faint (magnitude near 10) and poorly outlined. By the 15th, estimates indicated a magnitude near 10.5 and a
coma diameter near 4 arcmin. The comet was last seen on December 20, near
Alpha Camelopardalis. It was then described as a faint, large object near
magnitude 11.5.
(Brooks)

1895 IV   Discovered: November 17   Discovery Magnitude: 6.5
     Discovered in the morning sky between Zeta and Nu Virginis by Charles
Dillon Perrine (Lick Observatory), this comet was described as bright with a
tail. As it approached both the sun and Earth, the comet brightened and by
November 20, it was near magnitude 6 and possessed a tail about 15 arcmin long.
By the end of November, a magnitude of 5.5 was being estimated and the tail was

found to be 1 degree in length. As December began, the comet rapidly became
a difficult object to observe as it headed southeastward into morning twilight
and was last detected on the 11th.

Perihelion came on December 18 (r= 0.19 AU), and the comet made a brief
appearance in the Southern Hemisphere between the 21st and 25th. Observers
in Australia and Africa described the comet as a bright object with a tail
one-half degree long. The comet was only visible for a short time after sunset and has been estimated as being brighter than 1st magnitude. The unexpected brightness led several observers to question whether this comet was
the same comet discovered by Perrine, but calculations confirmed the identity. After the 25th, the comet was invisible as it headed for conjunction
with the sun.

The comet reappeared in the morning sky during the first half of February 1896, and was described as near magnitude 9 with a coma 2 arcmin across
and a tail extending 1 degree on photographs. As February progressed, the
tail grew, and on the 21st, it was described as 2 degrees long. One month
later, the comet had faded to magnitude 11[*] and by May, it was described as
small and concentrated with a magnitude near 12.5[*]. The comet remained under
observation for most of the summer and was last observed on August 10, at a
magnitude near 14[*]. The orbit is hyperbolic with an eccentricity of 1.0001.
(Perrine)

1896 I   Discovered: February 15   Discovery Magnitude: 7.5

The discovery of this comet came about as a result of an erroneous telegram. On February 13, Dr. Lamp (Kiel, Germany) recovered comet 1895 IV and
sent a telegram to Boston, Massachusetts, to notify interested observers;
however, upon redistributing the telegram to American observatories, Lamp's
recovery position was accidentally altered. C. D. Perrine had also recovered
1895 IV a few days earlier and had then found it to be almost exactly in the
position expected; so, when he received the telegram from Boston, he assumed
Lamp had actually discovered a new comet since the position given differed
by a few degrees from the actual position of 1895 IV. Upon turning the Lick
Observatory telescope upon the position indicated, Perrine actually found a
comet nearly one magnitude brighter than 1895 IV and immediately notified
Boston that Lamp had actually found a new comet. Like Perrine, Lamp had
found 1895 IV in exactly the expected position and did not know what Perrine
was implying; so, like Perrine, he assumed a new comet had been found and
soon found Perrine's new comet. It was several weeks before Perrine learned
of the accidental discovery he had made.

The new comet had passed perihelion on February 1 (r= 0.59 AU), and when
discovered was moving rapidly to the north as it neared Earth. Perigee came
on February 23 (0.39 AU), and the comet was described as of magnitude 7 with
a coma 4 arcmin across and a tail which extended 3 degrees on photographs.
Afterwards, despite increasing distances from both the sun and Earth, the
comet brightened until March 3, when the magnitude reached 5.5 and then faded
thereafter. By March 12, the total magnitude had faded to 7.5[*] and, although
the tail was still nearly 2 degrees long on the 3rd, virtually no trace of
the tail could then be detected. By the 1st of April, the comet appeared as
an indefinite outline with a total magnitude near 9.5 and when last observed
on April 16, it had faded to nearly 11th magnitude[*]. Although a parabolic
orbit is generally given in catalogs, the orbit may be hyperbolic.
(Perrine-Lamp)

1896 III   Discovered: April 14   Discovery Magnitude: 6.5
     While searching for comets in the evening sky, Dr. Lewis Swift (Lowe
Observatory, California) discovered this comet just below the Pleiades. Then
describing it as bright, with a trace of tail, Swift was unable to detect a
motion since clouds soon interfered. Cloudy skies prevailed until April 16, and
shortly after sunset, Swift began searching the area again. After the skies
had darkened, he spotted the comet to the north of its position of two days
earlier and could definitely detect a short tail. Perihelion came on April
18 (r= 0.57 AU), and the comet was described as 2 arcmin in diameter, with
a nucleus near magnitude 10. On the 24th, despite moonlight, the comet was
easily seen and, in addition to the short tail extending southeast, there was
a very faint secondary tail, 4 arcmin long, which extended to the northeast.
The comet faded as it headed away from both the sun and Earth and also moved
rapidly to the north. By the end of April, the comet was located in Perseus
and possessed a magnitude near 7, with a nucleus dimly shining at magnitude
9. The coma was then nearly 3 arcmin across and the tail was estimated as
8 arcmin long. By mid-May, the total magnitude had declined to 8.5 and the
coma had shrunk to only 1.5 arcmin in diameter. Observations continued into
June, and when last seen on the 21st, the total magnitude was near 13.5[*]. The
comet's orbit is hyperbolic with an eccentricity of 1.0005.
(Swift)

1896 IV   Discovered: September 1   Discovery Magnitude: 11.0
     Having passed perihelion on July 11 (r= 1.14 AU), this faint comet was
discovered nearly two months later near Zeta Ursae Majoris by Sperra (Randolph,
Ohio). The lack of a nearby telegraph office caused Sperra to announce the
discovery by mail and before notice reached the proper authorities, Brooks
(Smith Observatory, New York) independently found the comet on September 5.
Brooks then described the comet as faint, with a coma diameter of 1.5 arcmin
and no condensation. On September 8, the comet was near magnitude 11.5 with
a coma 1.5 arcmin across and by mid-month, the magnitude had dropped to 12[*].
Moonlight interfered during the latter half of September, but observations
resumed on October 2, when the comet was described as exceptionally faint.
When last seen on the 6th, the total magnitude was near 13[*]. Although a parabolic orbit is generally given in comet catalogs, there is a possibility that
the orbit is hyperbolic.
(Sperra)

1897 I   Discovered: November 3, 1896   Discovery Magnitude: 11.0
     Perrine (Lick Observatory) discovered this comet in Vulpecula as a faint,
tailless object 2 arcmin across. The comet moved southward and, although nearing the sun, its increasing distance from Earth caused it to brighten slowly.
Several observations over the next few days confirmed Perrine's initial description, but, shortly thereafter, the comet was rendered invisible by moonlight. Observations resumed near the end of November, and the comet was described as near magnitude 10.5, with a slightly diffuse nucleus of 11th magnitude and a coma diameter of 1.5 arcmin. Shortly after December began, the
total magnitude attained a value of 10, while the coma increased to 2 arcmin.
As the month progressed, the comet neared evening twilight and by the 30th, it
could be observed only with great difficulty at a magnitude near 9.5. The
comet was not observed after the 31st.
     Perihelion came on February 8, 1897 (r= 1.06 AU), and the comet finally

exited the sun's glare on February 23. Then only visible from the Southern Hemisphere, it was described as near magnitude 9. Although the comet was heading away from the sun, the comet-Earth distance decreased rapidly enough to allow an increase in brightness during March. The brightness leveled off at magnitude 8.5 in mid-March, and remained at that brightness until mid-April. Thereafter, with the comet heading away from both the sun and Earth, the brightness rapidly faded and when last detected on May 6, the total magnitude had declined to near 10.5[*]. The orbit is hyperbolic with an eccentricity of 1.0010. (Perrine)

1897 III   Discovered: October 17   Discovery Magnitude: 8.0
     Perrine discovered this comet in Camelopardalis and described it as bright, with a coma 2 arcmin across and a star-like nucleus near magnitude 12. A tail extended 10 arcmin and from the nucleus a ray extended nearly 5 arcmin. Although the comet was heading towards perihelion, an increasing distance from Earth was calculated to cause only a slight variation in brightness throughout October and November; however, soon after the comet's discovery, a rapid fading began. By October 20, the total magnitude was estimated as near 8.5 and by October 24, the estimate was closer to 9. Curiously, beginning on October 18, the nucleus was reported as sometimes appearing double or oblong to Karl Mysz (Pola, Yugoslavia), who was using a 15-cm refractor. This appearance was again observed by Mysz on the 19th. On October 20, J. Moller (Kiel, Germany) was observing the comet with a 20-cm refractor when he noticed the nucleus to possess an oblong shape. Both men estimated the nuclear magnitude as near 10. Thereafter, the nucleus was reported to rapidly fade and was invisible to most observers by October 24; however, observers with larger telescopes estimated magnitudes near 13.5 on the 25th, and 14 on the 27th. By October 30, the total magnitude had declined to nearly 10th magnitude, while the nucleus was barely visible in large telescopes. The comet appeared as an exceptionally blurred object when observed in moonlight on November 2, and in the absence of moonlight on the 17th, it was described as exceptionally faint with a circular coma 1 arcmin across. No condensation was then visible. Observations on November 25 and 26 revealed the comet as nothing more than an indefinite outline with a magnitude near 14 and the comet was last detected on the 28th, with the 68-cm refractor at McCormick Observatory. On the latter date, the comet's magnitude should have been equal to the magnitude at discovery, however, the comet was at least 6 magnitudes fainter and was not detected thereafter. Perihelion came on December 9 (r= 1.36 AU), and no trace of the comet was found during subsequent searches.
(Perrine)

1898 I   Discovered: March 20   Discovery Magnitude: 7.0
     Having passed perihelion on March 17 (r= 1.10 AU), this comet was discovered by Perrine (Lick Observatory) three mornings later. It was then described as bright, with a tail 1 degree long which was brighter on the northern side than on the southern. Then located in Pegasus, the comet moved northeast.
     The comet brightened during the remainder of March, despite increasing distances from both the sun and Earth, and the magnitude reached 6 on the 28th. During the same time, the coma was estimated as 2 to 3 arcmin across and the tail was consistently noted as 1 degree long. Thereafter, the comet faded as predicted and by the end of April had attained a magnitude of 7, with a coma 2 arcmin across and a tail 20 arcmin long. By the end of May, the comet was

estimated as between magnitude 9 and 9.5 and, one month later, when located in Camelopardalis, it had faded to magnitude 10.5[*]. The comet was followed throughout the summer and into autumn and was seen for the last time on November 16, when Perrine estimated a total magnitude near 16.5. The orbit is elliptical with a period of 419 years.
(Perrine)

1898 V   Discovered: June 19   Discovery Magnitude: 10.5
    While searching through Capricornus, Giacobini (Nice Observatory, France) discovered this comet and described it as faint with a condensation 15 arcsec across and a coma between 30 arcsec and 1 arcmin in diameter. Although he estimated the total magnitude as between 11 and 12, this may actually have been an estimation of the condensation, since observers during the next few days estimated the total magnitude as near 10, with a coma diameter of 2 arcmin and a nucleus of magnitude 12. The comet-Earth distance changed little through the end of June, but a decreasing distance from the sun brightened the comet to near magnitude 9.5; however, after the first week of July, the comet began to fade rapidly as the distance from Earth began to increase. By July 16, the coma was estimated as near 12th magnitude and possessed a nucleus near magnitude 15. Perihelion came on July 26 (r= 1.50 AU), and the comet was followed until August 16, when the magnitude had dropped to nearly 13[*].
(Giacobini)

1898 VI   Discovered: June 15   Discovery Magnitude: 10.0
    Discovered by Perrine (Lick Observatory) near the Camelopardalis-Perseus border, this comet was described as a faint, circular object 2 arcmin across without a nucleus. As it neared both the sun and Earth, the comet brightened and was commonly estimated as near 9th magnitude shortly before the end of June. At the same time, although no nucleus was visible, a condensation was noticed as becoming more condensed and by the 23rd, it was near magnitude 11.5. The comet brightened to magnitude 8.5[*] by mid-July, and at the end of the month had attained a magnitude near 7.5[*]. As August began, the comet moved deeper into evening twilight, and soon became a difficult object to observe. It was last seen on August 11. Perihelion came on August 16 (r= 0.63 AU), and the comet was not recovered in the Southern Hemisphere after perihelion.
(Perrine)

1898 VII   Discovered: June 10   Discovery Magnitude: 7.0
    On June 11, Mr. Coddington (Lick Observatory) was examining the negative plate of a photograph taken of the Antares region on June 10 when he noticed a faint streak on the plate. He at once thought it could have been caused by an undiscovered comet and during a telescopic examination of that region that evening, his suspicions were confirmed. On June 14, Pauly (Bucharest, Rumania) independently discovered the comet with a 7.5-cm telescope while observing the globular cluster M4. Both men reported the comet as bright and observations by other observers during the next few days consistently described the total magnitude as near 7, with a nucleus near magnitude 8. The coma was estimated as 4 arcmin in diameter and a short, broad tail was occasionally seen.
    The comet moved southwest after discovery, but changed little in brightness as its decreasing distance from the sun was countered by an increasing distance from Earth. During the first days of July, observations ceased for Northern Hemisphere observers as the comet passed below the southern horizon; however, southern observers continued to follow the comet as it climbed high-

er into their skies. By the end of July, the magnitude was near 6.5[*]. Unfortunately, southern observers rarely described the comet's appearance, but
orbital conditions should have caused a slow fading during August, and during
September the magnitude should have been near 7. Perihelion came on September
14 (r= 1.70 AU), and the comet was simply referred to as small, but centrally
condensed through the end of the year. In February 1899, observations indicated a magnitude near 10 and when observed on March 3, it had faded to about
10.5[*]. The comet was lost in evening twilight soon afterward, but late in the
year, after conjunction with the sun, it was again observed as a very faint
object between magnitude 13 and 14[*]. It was last seen on December 7, 1899.
The orbit is hyperbolic with an eccentricity of 1.0009.
(Coddington-Pauly)

1898 VIII   Discovered: November 15   Discovery Magnitude: 12.0
     Between November 13 and 17, Dr. Frederick L. Chase (Yale College Observa-
tory, Connecticut) made a detailed photographic record of the Leonid meteor
shower. In all, 60 plates were taken and on November 21, the tedious study
of these plates began. After only a short time, Chase's attention was captured
by a hazy trail which appeared on a plate exposed on the 15th. Further examination revealed 3 additional plates which indicated movement and the possibility of a new comet. During the morning hours of the 22nd, Chase and his
colleagues decided to confirm the object's existence. Chase failed to locate
it visually with a 20-cm refractor, but photographs again revealed its presence.
     The comet was found in Leo and moved slowly northeast after discovery.
Perihelion had occurred on September 20 (r= 2.28 AU), but the distance from
Earth was decreasing after discovery as the comet neared opposition and the
comet brightened slightly. Observations in December indicated a magnitude
between 11 and 11.5. During January 1899, the brightness seemed firmly set
at 11 and one estimate of the coma diameter early in the month gave a value
of 1 arcmin. The comet was at opposition in late January, and, thereafter,
it slowly faded. By early March, the total magnitude was estimated as near
12 and by early April, it was near 12.5[*]. On the latter date, the coma was 1
arcmin across and a nucleus of 14th magnitude was visible. The comet was followed until June 27, 1899, when the magnitude must have been near 13.5[*]. The
orbit is elliptical with a period near 315,000 years.
(Chase)

1898 IX   Discovered: September 13   Discovery Magnitude: 8.0
     This comet was discovered by Perrine (Lick Observatory) as it moved south-
east through Leo. It was then described as bright, with a coma about 4 arcmin
in diameter and a prominent nucleus. Two days later, an independent discovery
was made by P. Chofardet (Besancon, France). During the next few days, observers
described the comet as near magnitude 7.5, with a nucleus near 9th magnitude
and a coma between 2 and 3 arcmin across. Depending on the magnification used
for studying the comet, observers variously estimated the tail length as between
5 arcmin and 30 arcmin. As September progressed, the comet drew closer to both
the sun and Earth, and by month's end it had brightened to nearly 6th magnitude.
The comet moved into morning twilight as October began and became more difficult to observe as each day went by. It was considered brighter than magnitude 6 by the 2nd, and when last detected on the 10th, it had increased to
about magnitude 5. Perihelion came on October 21 (r= 0.42 AU), and the comet

was not detected by observers in the Southern Hemisphere after its perihelion passage.
(Perrine-Chofardet)

1898 X   Discovered: October 21   Discovery Magnitude: 7.0
     Brooks (Smith Observatory, New York) discovered this comet near Alpha Draconis while using a 25-cm refractor. He described it as bright, with a coma 4 arcmin across and a central condensation. Although Brooks estimated the total magnitude as near 7, estimates by other observers during the next few days indicated a magnitude closer to 8. The comet moved rapidly southeast after discovery and brightened to magnitude 7.5*by the end of October as it approached perihelion. After October, the distance from Earth began to increase more rapidly and the comet began to fade. By November 19, estimates of the total magnitude were near 8.5 and the condensation was about magnitude 10. The coma diameter was then estimated as between 30 arcsec and 1 arcmin. Perihelion came on November 23 (r= 0.76 AU), and the comet was last observed in Scutum on the 26th, at low altitude and in twilight. The comet was not observed in the Southern Hemisphere. During the first half of November, Lick Observatory conducted a photographic study of the comet to determine accurate positions. On a November 10th photo, J. E. Keeler noted the nucleus as split into two components--a feature not noted visually on that morning. No additional observations of this double nucleus were obtained and it now seems likely that the supposed companion was merely a plate flaw. Although comet orbit catalogs generally give a parabolic orbit for this comet, there is a possibility that the comet may travel in an elliptical orbit of very long period.
(Brooks)

1899 I   Discovered: March 4   Discovery Magnitude: 6.0
     L. Swift (Lowe Observatory, California) discovered this comet in Eridanus as a bright object with a star-like nucleus and a short tail. The comet was then nearing morning twilight and was approaching perihelion. Observations during the next few days indicated a total magnitude near 6 and a coma diameter near 3 arcmin. The comet steadily brightened as March progressed, and by the 17th, it was being estimated as near 5th magnitude. Five days later, the magnitude reached 4.5 and in addition to the long, wide tail, observers reported a faint secondary tail. The comet was lost in twilight after March 26.
     Perihelion came on April 13 (r= 0.33 AU), and the comet remained lost in the solar glare until the first days of May. Thereafter, a decreasing distance from Earth should have caused the comet to brighten as May progressed; however, the comet faded from a magnitude of 3 to 4 between the 7th and 13th of May. On the latter date, the tail extended 1 degree and the magnitude of the central condensation was near 5. By the end of the month, observers were estimating a total magnitude near 6 and a coma diameter of 10 arcmin. The comet was then near perigee (0.54 AU) and photographs revealed a tail 10 degrees long. Also, during May, a secondary nucleus was visible beginning on the 12th. On that date, Perrine (Lick Observatory) detected a 9.5-magnitude nucleus 12.5 arcsec from the 8.0-magnitude main nucleus. Barnard (Yerkes Observatory, Wisconsin) also detected the nuclei and continued observations until May 21. Perrine last detected the companion on May 24, when the separation was 38.5 arcsec. Later studies of the separation have led researchers to calculate the date of splitting as April 25, 1899.

During the first days of June, observers reported a total magnitude near 5.5, a nuclear magnitude near 9.5 and a coma 6 arcmin in diameter; however, on June 4, the comet suddenly brightened and by the 5th, it had reached a magnitude of 4.5, with a nucleus near magnitude 7 and a coma 7 arcmin across. Thereafter, as the comet moved away from both the sun and Earth, the magnitude decreased rapidly. On the 8th, it had recovered from the outburst and was then shining at a magnitude near 6, with a nucleus near magnitude 9.5; however, by the 17th, it had dropped to 7th magnitude and on the 21st, it was near 8. By the end of the month, the total magnitude had declined to 9th magnitude, the nuclear magnitude had dropped to near 11.5 and the coma was estimated as 2.5 arcmin across. Curiously, another secondary nucleus became visible on June 7. Perrine was the only observer this time and managed to keep it under observation until the 11th. Researchers believe the division occurred on May 28, 1899.

Fading continued to be rapid throughout July and into August, as the comet's magnitude declined from a value of 9 on July 1 to 12.5 on August 1. During the same interval, the coma shrank from a diameter of 2 arcmin to 1 arcmin. The comet was last observed on August 12, 1899. The orbit is hyperbolic with an eccentricity of 1.0004.
(Swift)

1899 V   Discovered: September 29   Discovery Magnitude: 11.0
Having passed perihelion on September 15 (r= 1.79 AU), this comet was discovered in Ophiuchus by Giacobini (Nice Observatory) and was described as circular, with a diameter of 1 arcmin and a condensation near magnitude 13. Thereafter, it moved slowly away from both the sun and Earth and by the end of October, it had faded to magnitude 11.5. One month later, the comet was near magnitude 12 and when last seen on December 24, observers at Lick Observatory indicated a magnitude near 12.5. Throughout its duration of visibility, the comet remained a tailless, circular coma with a diameter of 1 arcmin--only the brightness changed. No nucleus was ever visible, though a weak condensation was frequently noted.
(Giacobini)

1900 I   Discovered: January 31   Discovery Magnitude: 11.0
Discovered by Giacobini (Nice Observatory) in Eridanus, this comet was described as a faint telescopic object with a 13th-magnitude nucleus and a stubby tail. Although approaching the sun, an increasing distance from Earth kept the comet from brightening noticeably during February and March. The total magnitude remained near 11 and the nuclear magnitude was frequently noted as between 11.5 and 12. A possible outburst may have occurred near the end of February, when Marchetti (Pola, Austria) detected the comet while using a 15-cm refractor and described it as possessing a nucleus near magnitude 10.5. Similarly, G. Abetti (Arcetri) reported fluctuations of the nuclear condensation during the same period of time. The comet was followed until March 24, when it became lost in the solar glare.

Perihelion came on April 29 (r= 1.33 AU), and the comet was recovered as it exited the sun's glare on May 25. The total magnitude was then estimated as near 10.5--almost exactly as expected--and the brightness did not change during the next 2 months as the increasing distance from the sun countered the effects of a decreasing distance from Earth. The coma remained near 1 arcmin across and the condensation was estimated as near 13th magnitude. At the end

of July, the comet passed perigee (1.15 AU) and, thereafter, faded rapidly. It
was last detected on August 18, by Perrine at a magnitude near 14.5. The orbit
is hyperbolic with an eccentricity of 1.0011.
(Giacobini)

1900 II   Discovered: July 24   Discovery Magnitude: 6.5
      This comet was independently discovered near the Aries-Cetus border by
Borrelly (Marseilles, France) and Brooks (Smith Observatory, New York) only 5
hours apart, with the former observer describing it as 2 arcmin in diameter
with a nucleus of magnitude 9.5. During the next few days, observers reported
a visible tail 1 degree long, while photographs revealed it closer to 6 degrees
in length. The comet was at perigee around July 30 (0.44 AU), with estimates
of the coma diameter then ranging from 10 to 30 arcmin. Perihelion came on
August 3 (r= 1.01 AU), and the comet faded fairly rapidly thereafter. At
perihelion the total magnitude had been near 7, with a nucleus near magnitude
9.5, and by the 31st, the total magnitude had dropped to 8, while the nucleus
had declined to slightly fainter than 10. The comet had passed 4 degrees from
the north celestial pole on August 25.
      By mid-September, observers estimated the total magnitude as near 9 and
measures of the coma diameter gave values near 1.5 arcmin. During October,
the last traces of tail were seen as the comet faded to near magnitude 11.5*
by mid-month. As October ended, the comet was again heading for the north
pole, though at a very slow pace. Observations continued through November,
and the comet was last seen on December 23, when Robert Grant Aitken (Lick
Observatory) estimated the total magnitude as near 15. The orbit is hyperbolic with an eccentricity of 1.0004.
(Borrelly-Brooks)

1901 I   Discovered: April 12   Discovery Magnitude: 1.5
      First discovered by Viscara (Paysandu, Uruguay) as a bright object with a
noticeable tail in the morning sky, this comet brightened rapidly and, since
its elongation from the sun decreased slowly, it was seen with greater ease
when independently discovered on April 24, by Halls (Queenstown, Tasmania)
and Tattersall (Cape Leeuwin, Australia). On this date, the comet was at perihelion (r= 0.24 AU), and was described as bright with a nucleus near magnitude
0 and a tail which extended 10 degrees. The comet seemed yellow in appearance
and remained visible for some time after sunrise.
      After the 24th of April, the comet moved into the evening sky and appar-
ently remained lost in the sun's glare until May 2, when the nucleus was described as rivaling Sirius in brightness (hence, a nuclear magnitude of -1.5).
A reported observation made on April 27, by observers at Yerkes Observatory in
Wisconsin, seems to definitely be a false report apparently generated by local
newspapers since the direction from the sun nowhere matches the comet's true
position on that date. As the comet steadily moved away from both the sun and
Earth it rapidly faded. On May 5, the total magnitude was estimated as near
1.0 and by the 24th, estimates were near 6. The comet's tail reached a maximum length of 10 degrees on May 6; however, a secondary tail which had first
appeared on the 3rd was estimated as extending 30 degrees on the 6th. By the
middle of May, the tail lengths were virtually unchanged, but as the comet
faded, the secondary tail rapidly vanished and by month's end only the main tail
was visible, with an estimated length of barely 1 degree. Curiously, R. T. A.
Innes (Transvaal Observatory, South Africa) observed two nuclei on May 15,

which were separated by 1 arcsec and differed in brightness by 1 magnitude.
Although a date of splitting has been calculated by one researcher (May 1,
1901), the existence of the secondary nucleus is somewhat in doubt since Innes,
as well as other observers, failed to detect it after that date. The comet was
followed until June 14, when Innes indicated a magnitude near 10.5. With re-
gard to the comet's discovery magnitude and orbit, E. M. Pittich concluded in
1969 that this comet was a good candidate for a prediscovery outburst in
brightness. He found the comet would have been bright enough for a discovery
95 days earlier than April 12. Even after eliminating unfavorable conditions
such as moonlight and bad weather, he still came up with 47 days which would
have been good for discovery. Other researchers have found that the comet's
absolute brightness was nearly 3 magnitudes brighter before perihelion than
afterwards.
(Great Comet)

1902 I   Discovered: April 15   Discovery Magnitude: 7.5
     While searching for comets in the eastern morning sky, Brooks (Smith
Observatory, New York) discovered this comet near Beta Pegasi and described it
as bright (magnitude near 7.5), with a stellar nucleus and a short tail. The
comet was then moving at a daily rate of 2 degrees to the south-southeast
toward morning twilight. On April 17, Hartwig (Bamberg, Germany) and Aitken
(Lick Observatory, California) independently estimated the comet's total magnitude as 8.5 and measured the straight tail as extending between 25 and 30
arcmin. Neither observer remarked on a stellar nucleus, although Hartwig did
estimate the magnitude of the condensation as near 9. Observations on April
18 and 19 indicated the comet was faint, with a diffuse nucleus between magnitude 10 and 10.5. On the same morning, Max Wolf (Heidelberg, Germany) took
several photographs of 7-minute exposure which revealed a wide, faint tail, of
which the southern, concave edge was noticeably brighter than the northern
edge. The comet was last detected on April 20. The few observations available indicate the comet faded as it neared the sun, which is quite contrary
to the expected brightening of nearly one magnitude based on the comet's
decreasing distance from the sun. Perihelion came on May 7 (r= 0.44 AU), and
attempts to locate the comet after perihelion at Lick Observatory ended in
failure despite the use of both 30-cm and 91-cm refractors.
(Brooks)

1902 III   Discovered: September 1   Discovery Magnitude: 9.0
     This comet was discovered in southern Perseus by Perrine (Lick Observatory),
who described it as of the 9th magnitude with a well-defined, but non-stellar
nucleus between magnitude 10.5 and 11. He added that the coma was 4 to 5
arcmin across and possessed a short, brushy tail which extended 8 to 10 arcmin.
The next evening, Borrelly (Marseilles Observatory) made an independent discovery. The comet was then approaching both the sun and Earth and rapidly brightened as the month progressed. On September 20, the total magnitude was near
7.5 and by the 30th, it was near magnitude 5. Also, on the latter date, the
comet's tail was visually estimated as 15 arcmin long, while photography revealed two tails--one straight and the other curved--with the longest being
1 degree in length. Perigee came on October 4 (0.36 AU), and the increasing
distance from Earth finally prevailed over the decreasing solar distance in
mid-October, when the magnitude peaked at 4. Thereafter, the comet slowly
faded, although the tail continued to grow until later in the month, when the

length peaked at 11 degrees. As November began, the comet was moving into
evening twilight and when last seen on the 17th, it was near magnitude 5.5,
with a coma 1.5 arcmin across and a tail 20 arcmin long.

Perihelion came on November 24 (r= 0.40 AU), and the comet remained lost
in the sun's glare until December 11, when recovered in the Southern Hemisphere at a magnitude near 4.5. Then moving southward, the comet picked up a
westward movement in early January 1903, and finally moved northward near the
end of January. It was relocated by Northern Hemisphere observatories on the
29th, at a magnitude near 9.5. By late February, it was described as very
faint (magnitude near 11) with an irregularly shaped coma and no nucleus. The
comet was last seen on March 31, 1903. The orbit is elliptical with a period
near 1.4 million years.
(Perrine)

1903 I   Discovered: January 15   Discovery Magnitude: 10.0
Giacobini (Nice Observatory) discovered this comet in the evening sky
near Beta Piscium. He then described it as small and circular, without any
nucleus. The comet was then nearing both the sun and Earth and developed
fairly rapidly with the first traces of tail becoming visible on January 27.
In mid-February, the comet brightened to magnitude 9 and observers estimated
the coma as 1.5 arcmin across. A nucleus of magnitude 10 was occasionally
seen and the tail was still very short. Although the tail continued to ex-
hibit a short size, photographs during the last days of February revealed
it to be a respectable size with measures of 1 degree on the 25th, and 4 degrees on the 28th. As March began, the comet was near magnitude 7.5 and tail
lengths were restricted to visual estimates due to a decreasing altitude and
twilight. On the 4th, the tail extended 3 arcmin and by the 16th, it was estimated as near 7 arcmin. On March 10, the total magnitude was near 5.5, and
by the 16th, it had brightened to 4. Perihelion occurred on the 16th (r= 0.41
AU), and the comet was lost in the sun's glare until recovered in the Southern
Hemisphere on April 6. It was then described as bright and visible to the
naked eye (magnitude between 4 and 5). Fading was rapid thereafter, and the
comet was last detected on May 5. The orbit is elliptical with a period near
43,900 years.
(Giacobini)

1903 II   Discovered: December 2, 1902   Discovery Magnitude: 11.0
This comet was discovered by Giacobini near Delta Monocerotis and was described as a faint, circular object with a sharp nucleus, but no tail. The comet was then located 3.00 AU and 2.25 AU from the sun and Earth, respectively,
with both distances decreasing. Subsequently, the comet brightened very slowly
with observers indicating an increase of only 0.5 magnitude by mid-January. At
that time, the coma was estimated as about 1 arcmin across. Perigee came on
January 24 (1.90 AU), and the comet began to slowly fade thereafter. In mid-
February, the brightness had declined to 11.5, with a nuclear magnitude near
12.5, and by mid-March, the total magnitude was 12, while the nuclear magnitude
was near 13. Perihelion came on March 24 (r= 2.77 AU), and by April 21, the
total magnitude had dropped to 12.5[*]. The comet was followed until June 27,
when Aitken (Lick Observatory) detected it with the 91-cm refractor at a total
magnitude of 13.5. The nucleus was then near magnitude 16 and the coma was 2
arcmin across. The orbit is hyperbolic with an eccentricity of 1.0006.
(Giacobini)

1903 III   Discovered: April 17   Discovery Magnitude: 9.0
     Having passed perihelion on March 26 (r= 0.50 AU), this comet was discov-
ered in Eridanus by John Grigg (Thames, New Zealand), who described it as faint
in his 7.6-cm refractor, with no visible nucleus. The comet moved away from
the sun and faded very rapidly. Few descriptive observations were obtained
by Southern Hemisphere observers and the comet was last seen on May 28, when
observers in Australia and South Africa indicated a total magnitude near 12.5.
(Grigg)

1903 IV   Discovered: May 29   Discovery Magnitude: 8.0
     Borrelly (Marseilles Observatory) discovered this comet on June 21
about 5 degrees west of Theta Aquarii and described it as fairly bright, with
a 10th-magnitude nucleus and a fan-shaped tail extending 10 arcmin. Subsequently, after the first orbital calculations, images of the comet were located on photographic plates exposed at Harvard Observatory on May 29 and 31.

     The comet rapidly approached both the sun and Earth, and, thus, rapidly
brightened. On June 24, the total magnitude and nuclear magnitude were estimated as near 7 and 9.5, respectively, while observations on the 28th gave
values of 5 and 9.7, respectively. By July 14, the total magnitude was near
4 and the comet reached its maximum brightness of 3 on July 18. Between these
two dates, the comet had arrived at perigee (0.25 AU). Despite a decreasing
distance from the sun, the increasing distance from Earth caused the comet to
fade slowly after perigee with reported total magnitude values of 3.5 at the
end of July and 4 in mid-August. The tail was of most interest during this
time and photographs revealed a maximum length of about 17 degrees at the
beginning of August. Curiously, the comet ejected its tail on July 24--an
event which was well photographed at one French and two American observatories.
The comet was followed until August 19, with observations thereafter being
affected by the sun's glare. Perihelion came on August 28 (r= 0.33 AU), and
the comet remained hidden until September 30, when observers at the Cape of
Good Hope recovered it as a relatively faint object. Observations in the
Southern Hemisphere continued until October 23, when the magnitude had faded
to near 11.5[*]. The orbit is hyperbolic with an eccentricity of 1.0004.
(Borrelly)

1904 I   Discovered: May 15, 1903   Discovery Magnitude: 11.5
     Brooks (Smith Observatory, New York) discovered this comet a few degrees
northwest of M92 in Hercules on April 17, 1904. He described it as a 9th-
magnitude object, with a 10th-magnitude nucleus, a 1.5 arcmin diameter coma
and a 4 arcmin long, fan-shaped tail. The comet was independently discovered
by Lucien Rudeaux (Manche, France) the next day while examining a photo he had
exposed on the M92 region at his private observatory 4 hours prior to Brooks'
discovery; however, due to cloudy weather, he was unable to confirm the unknown nebulous patch as a comet. Shortly after the first detailed calculations of the comet's orbit, astronomers at Harvard Observatory found prediscovery images of the comet on 5 photos exposed at the Arequipa, Peru, station
between May 15 and June 25, 1903.

     The comet had passed perihelion on March 4, 1904 (r= 2.71 AU), and, when
discovered, was located 2.74 AU and 2.25 AU from the sun and Earth, respectively, with both distances increasing. The total magnitude subsequently faded, though very slowly, and the comet was well observed due to its northern
location in the sky. The comet remained brighter than magnitude 9.5 until

early June, and was frequently measured as 1.5 arcmin in diameter with a tail about 10 arcmin long. By mid-July, it had faded to magnitude 10 and in mid-November, it had declined to magnitude 10.5[*]. As 1905 began, the comet was described as near magnitude 11 and by the end of February, it had faded to near 12. During this time the coma was estimated as 1 arcmin in diameter. In April, the total magnitude was estimated as about 12.5, while the coma diameter had shrunk to 30 arcsec. The comet was last detected on June 6, 1905, when the total magnitude was indicated as near 15.5. The orbit is hyperbolic with an eccentricity of 1.0013.
(Brooks)

1904 II   Discovered: December 18   Discovery Magnitude: 11.0

Giacobini (Nice Observatory) discovered this comet in the morning sky in southeastern Corona Borealis. He then described it as a small, faint object with a seemingly granular structure. The comet had passed perihelion on November 4 (r= 1.88 AU), and, when discovered, it was slowly nearing a late-January perigee (2.19 AU).

During the first days following discovery, observers agreed with Giacobini's initial description--including the apparent granular structure of the coma--but by mid-January, a nuclear condensation had appeared and the magnitude had declined to 11.5. The coma was then estimated as 30 arcsec in diameter. By mid-February, the total magnitude was near 12 and the condensation had become a well-developed feature. In mid-April, the comet was described as very faint, and, while using the 67-cm refractor at the Vienna Observatory, Palisa reported a star-like nucleus of the 14th magnitude. The comet was last seen on June 1, 1905, when it was described as almost invisible. The orbit is hyperbolic with an eccentricity of 1.0007.
(Giacobini)

1905 III   Discovered: March 26   Discovery Magnitude: 11.5

This comet was discovered nearly 4 degrees north of Alpha Orionis in the evening sky by Giacobini (Nice Observatory). It was then described as 3 arcmin in diameter with a condensation 5 to 6 arcsec across. Other observers during the next few days confirmed Giacobini's initial description, although a possible tail was reported on March 30. Perigee came on April 1 (0.70 AU), and observations on that date indicated a total magnitude between 11 and 11.5. The coma was still estimated as 3 arcmin across, with a noticeable condensation, and a fan-shaped tail extended to the northwest. The comet passed perihelion on April 4 (r= 1.11 AU), and, thereafter, faded slowly. On April 6, it was near magnitude 11.7 and by the 10th of May, it had faded to 12.5. On the latter date, the coma was 1 arcmin across. The comet was followed until June 22, when it had faded to nearly 15th magnitude[*]. The orbit is elliptical with a period of 226 years.
(Giacobini)

1905 IV   Discovered: January 10, 1904   Discovery Magnitude: 15.5

A. Kopff (Heidelberg, Germany) discovered this comet on March 3, 1906, on a photograph he had taken of the Leo-Virgo region. He described it as well-condensed, with a coma 40 arcsec across and a nucleus of magnitude 11.5. The tail extended one-half degree; however, a visual examination by Kopff shortly thereafter revealed no tail. Several months later, a more accurate orbital calculation allowed Wolf (Heidelberg) to find the comet as a 12th-magnitude object on a photo he had exposed on January 14, 1905--413 days prior to Kopff's discovery. A new orbit and ephemeris then helped to locate the comet as an object near magnitude 15.5 on a photo exposed on January 10, 1904--782 days prior to Kopff's discovery.

Perihelion had occurred on October 18, 1905 (r= 3.34 AU), with the solar distance then surpassing every other observed comet except that of 1729. When discovered, the comet was located 3.57 AU and 2.60 AU from the sun and Earth, respectively, with both distances increasing. The comet therefore faded very slowly, with the total magnitude dropping to 11 at the end of March, and 12 by the end of April. Interestingly, while using the largest refractor in the world (101-cm) on March 18, Barnard (Yerkes Observatory, Wisconsin) discovered a faint, star-like condensation preceding the comet's nucleus. He was then unsure whether the object was a star or a secondary nucleus, but when he next observed the comet on March 25, he found the latter title to be correct. He then described the secondary nucleus as near magnitude 14 or 15 and measured the separation as 4.68 arcsec. He next observed it on April 1, and measured the separation as 5.67 arcsec. This nucleus was not observed elsewhere and was not seen by Barnard after April 1. Although the lack of observations does not allow a detailed look at how the secondary nucleus evolved, researchers believe the split occurred in December 1905.

The comet was followed throughout May, and by mid-June, the total magnitude had dropped to 13.5.[*] The comet was not observed after June 23. At opposition in 1907, the comet was recovered in mid-March, and was described as near magnitude 14, with a coma 1 arcmin across. By mid-April, the total magnitude was unchanged, although a nucleus near magnitude 15.5 was detected. The comet was followed by Aitken (Lick Observatory) throughout May and June, and was last seen on July 4, 1907, when the total magnitude was between 15 and 16.[*] The orbit is hyperbolic with an eccentricity of 1.0015.
(Kopff)

1905 V   Discovered: November 17   Discovery Magnitude: 7.0

Having passed perihelion on October 26 (r= 1.05 AU), this comet was discovered a few days before perigee by E. Schaer (Geneva, Switzerland) when only 4 degrees from the north celestial pole. The comet was then moving at a rate of 9 degrees per day to the south and was described the next evening as bright, with a coma diameter of 8 arcmin, but no tail. Perigee came on November 20 (0.25 AU), and, thereafter, the comet faded rapidly. On November 21, the total magnitude was estimated as 7, while the nucleus was near magnitude 11 and the coma diameter was found to be 7 arcmin. A photograph by Wolf (Heidelberg) showed several rays extending from the comet's head and a narrow, faint tail 1.5 degrees long. By November 26, the total magnitude had declined to 8, the nucleus was estimated as near 10 and the coma was still 7 arcmin across. During the first week of December, the comet reached 10th magnitude and by mid-month, it was near 11, with a coma diameter of 1.5 arcmin. The comet was followed until December 30, when observers at Lick Observatory and Nice Observatory indicated a total magnitude near 12.5. After the moon had left the region of the comet, searches were conducted in mid-January 1906, but no trace was found. Although the orbit is generally listed as parabolic, there is a possibility of it being hyperbolic.
(Schaer)

1905 VI   Discovered: January 27, 1906   Discovery Magnitude: 9.0
Brooks (Smith Observatory, New York) discovered this comet in Hercules

and described it as circular, with a diameter of 8 arcmin and a faint nucleus.
Having passed perihelion on December 25, 1905 (r= 1.30 AU), the comet was
nearing Earth when discovered and faded very little during the next two weeks.
On January 31, the total magnitude was 9.4, while the nuclear magnitude was
estimated as near 12.5. By mid-February, the total magnitude was near 9.5,
while the nuclear magnitude was near 12. During this interval, the coma was
consistently estimated as between 8 and 10 arcmin. Perigee came on February
10 (0.90 AU), and the comet faded more rapidly thereafter. By the end of
February, the total magnitude was near 10 and by mid-March, it had dropped
to 11. On March 28, when the brightness had dropped to near 12.5[*], C. W.
Wirtz (Strasbourg, Germany) detected a double nucleus with a separation of
12 arcsec. This secondary nucleus was not detected again and although one
researcher has calculated the date of splitting as March 14, the observation
is considered to be of doubt. The comet was followed until April 25, 1906,
when observers at Lowell Observatory (Arizona) indicated a total magnitude
near 15. The orbit is hyperbolic with an eccentricity of 1.0002.
(Brooks)

1906 I   Discovered: December 7, 1905   Discovery Magnitude: 8.0
       Giacobini (Nice Observatory) discovered this comet about 2 degrees from
Alpha Boötes in the morning sky. He then described it as circular, with a
diameter of 1 to 2 arcmin and a total magnitude of 8. Observers during the
next few days confirmed Giacobini's coma diameter estimate and added that
the nucleus was between 11th and 12th magnitude; however, estimates of the total
magnitude were consistently near 9, thus making Giacobini's initial estimate
of 8 appear slightly exaggerated. At discovery, the comet was nearing both
the sun and Earth and became brighter. In mid-December, the total magnitude
and the nuclear magnitude were estimated as 8.5 and 9.5, respectively, while
the coma diameter was measured as 2 arcmin across. By December 25, the total
and nuclear magnitudes had brightened to 8 and 9.0, respectively, and a tail
5 arcmin long was also observed.
       On December 30, observers found the comet much brighter than expected
with a total magnitude near 5 and a tail 30 to 45 arcmin long. During the
first week of January, the total magnitude brightened to 4, with a tail 40
arcmin long and a coma 4 arcmin across. The comet arrived at perigee on
January 10 (1.10 AU), and, as it neared perihelion and morning twilight, it
became more and more difficult to observe. It was last seen on January 14,
when Wirtz (Strasbourg, Germany) estimated the total magnitude to be 0.7. He
added that the nucleus appeared as a bright disk 9 arcsec in diameter and the
tail was split into two tails with lengths of 5 arcmin and 4 arcmin.
       Perihelion came on January 23 (r= 0.22 AU), and the comet remained lost
in the sun's glare until recovered in the Southern Hemisphere on February 5.
On February 15, observations resumed in the Northern Hemisphere, and observers
on the 22nd estimated the total magnitude as about 9 and reported only a faint
trace of tail. Fading was rapid as the comet moved away from both the sun and
Earth and by March 3, the total magnitude had dropped to 10. The comet was
followed until March 23, when observers estimated a total magnitude of about
13.
(Giacobini)

1906 II   Discovered: March 18   Discovery Magnitude: 8.0
       Ross (Melbourne, Australia) discovered this comet in Cetus while sweeping

through the evening sky. He described it as circular, with a diameter of 3
arcmin and a central condensation. Perihelion had occurred on February 21 (r=
0.72 AU), and with the distances from both the sun and Earth increasing, the
comet faded rapidly. By March 21, the total magnitude had dropped to 10 and
the coma diameter had shrank to 2.5 arcmin. Thereafter, the comet rapidly
became a difficult object as it moved deeper in evening twilight and, by the
end of March, observations could not be made in a dark sky until the comet had
dropped to an altitude of only 5 to 8 degrees. Observations were no longer
possible after April 4.
(Ross)

1906 VII   Discovered: November 11   Discovery Magnitude: 8.5
     Holger Thiele (Copenhagen, Denmark) discovered this comet shortly after
midnight in the eastern part of Cancer. He described it as round, with a coma
diameter of 4 arcmin and a central condensation near 10th magnitude. With the
distances from the sun and Earth decreasing, the comet brightened, and between
the 17th and 20th, it reached a maximum magnitude of 7.7. At the same time,
the nucleus was about magnitude 9.5 and the coma had a diameter of at least
5 arcmin. Perihelion came on November 22 (r= 1.21 AU), and the comet arrived
at perigee on the 28th (0.63 AU). On the latter date, the magnitude was near
8 and the coma was 5 arcmin across. Thereafter, the comet faded and by mid-
December, the total magnitude was about 9. During the first week of January
1907, observers estimated a magnitude near 12 and when last seen on the 19th,
it had faded to about 13.5[*] The orbit is elliptical with a period of 583
years.
(Thiele)

1907 I   Discovered: March 9   Discovery Magnitude: 11.0
     Giacobini (Nice Observatory) discovered this comet 6 degrees east and 1
degree south of Alpha Canis Majoris. He then described it as faint, with a
coma diameter of 30 arcsec. Observations by other observers during the next
few days confirmed Giacobini's initial description. Perihelion came on March
19 (r= 2.05 AU), and perigee came at about the same time (1.44 AU). Thereafter, the comet slowly faded, and was observed throughout April, as the total
magnitude declined from 11.5 to 12.5. Because of the comet's proximity to the
sun and its ever-fading light, observations ceased after May 11.
     The comet was recovered by Wolf (Heidelberg) on December 4, when near
opposition. It was then described as near magnitude 12.5. An observation on
December 7 gave a nuclear magnitude of 13.5 and a coma diameter of 12 to 18
arcsec. The comet was followed into the new year, and was last observed at
magnitude 15[*]on February 26, 1908. The orbit is hyperbolic with an eccentricity of 1.0010.
(Giacobini)

1907 II   Discovered: April 8   Discovery Magnitude: 6.5
     This comet was discovered in Caelum by Grigg (Thames, New Zealand) and
was described as 15 arcmin in diameter with no clearly visible nucleus or tail.
Perihelion had occurred on March 28 (r= 0.92 AU), and, at discovery, the comet
was at perigee (0.24 AU). On April 14, John E. Mellish (Madison, Wisconsin)
independently discovered the comet and described it as faint and diffuse, with
a rapid movement to the northeast. Interestingly, Mellish sent his telegram
of announcement to Yerkes Observatory, where, upon receiving it, his friend
Barnard proceeded to examine a photo of the region he had exposed for comet

1907 I only a few hours earlier.   The new comet was located after only a short
search near the corner of the photo.

After discovery, the comet faded rapidly.   In mid-April, the total magnitude was near 8.5 and the coma was estimated as 2 arcmin across in moonlight.
By the end of the month, in dark skies, the total magnitude was near 10 and
the coma was 3 to 4 arcmin in diameter.   Most physical observations of the
comet in May were made by Wirtz (Strasbourg).   On May 7, he described the
comet as a diffuse nebulous spot of magnitude 10.9.   It was measured as 3.3
arcmin across and possessed no nucleus.   On May 14, he estimated a total magnitude of 12.0 and a coma diameter of 2.5 arcmin.   Observations ceased thereafter, due to moonlight, and on the next favorable, dark night, no trace of
the comet could be found.

Up until recent times, this comet's orbit was generally believed to be
elliptical with a period of 165 years--based on the assumption that it was
identical with the comet of 1742.   However, in 1974, Marsden recalculated the
orbit and found the link between these two comets to be impossible.   Only a
parabolic orbit is given today.
(Grigg-Mellish)

1907 IV   Discovered: June 10   Discovery Magnitude: 9.5
Zaccheus Daniel (Princeton University Observatory, New Jersey) discovered
this comet about 5 degrees south of Beta Piscium in the morning sky.   He was
then using a 14.6-cm comet-seeker and described the comet as 2 arcmin across
with a 10th-magnitude condensation.   Being unable to detect motion before sunrise, Daniel telegraphed word of his suspected discovery to Harvard College
Observatory.   Cloudy weather prevented confirmation at Princeton for the next
several days, but on June 12, Brooks (Smith Observatory, New York) succeeded
in locating the new object only a short distance from its discovery position.

At discovery, the comet was heading toward both the sun and Earth and
brightened rapidly.   On June 20, observers gave a total magnitude of 8 and a
coma diameter of nearly 7 arcmin.   A trace of tail also extended to the west.
By July 10, the total magnitude had become slightly brighter than 5 and visual
estimates of the tail length gave values of 1 to 1.5 degrees.   At the end of
July, the total magnitude was near 3, the coma diameter was between 7 and 9
arcmin and the tail extended nearly 3 degrees.   Perigee came on August 1 (0.75
AU), and, thereafter, the comet brightened slowly as it continued to approach
the sun.   At mid-month, the total magnitude was 2 and the tail extended about
8 degrees.   By August 31, the brightness was unchanged, but visual estimates
of the coma diameter and tail length gave values of 2 arcmin and 3 degrees,
respectively.

This comet was photographed more than any other before it and proved to
be a much more unusual object than visual observations had led astronomers to
believe.   During August, photographs of the comet's head showed that instead
of the tail flowing uniformly out of the coma, it first appeared as a narrow
cluster of 15 to 20 rays, which, a short distance from the coma, diffused to
form the tail.   The tail was longest shortly after mid-August, when photographic estimates gave values as great as 17 degrees.   Photographs also indicated a possible separation of the tail from the coma on two occasions: July
11 and August 18.

Observations continued into September, as the comet moved into Leo and
became involved in morning twilight.   Perihelion came on the 4th (r= 0.51 AU),

and the comet was then near magnitude 1.5 with a nucleus near magnitude 4.5.
Therafter, the comet faded and when last seen in strong twilight on September
24, it was near magnitude 3. The comet remained lost in the sun's glare until
November 14, when it was recovered as an object of magnitude 8.5, with a coma
diameter of 1 arcmin. In mid-December, the comet was described as an out-of-
focus 9th-magnitude star and by the beginning of February 1908, the total magnitude had faded to 10. Observations continued until June 30, when the comet
was described as an ill-shaped object about 24 arcsec in diameter, with a
total magnitude near 14. Interestingly, Wolf (Heidelberg) spotted a nebulous
object near the expected position on two plates exposed on April 19, 1909.
The total magnitude was then between 16 and 17; however, the identification
with 1907 IV is still questionable since the object is at the extreme limit of
the photographic plate. The orbit is elliptical with a period near 8,700 years.
(Daniel)

1907 V   Discovered: October 14   Discovery Magnitude: 9.0
    Mellish (Madison, Wisconsin) discovered this comet in the morning sky near
the Hydra-Puppis border. He then described it as visible with an opera-glass
and moving slowly northwest. Perihelion had occurred on September 15 (r= 0.98
AU), but, after discovery, the comet brightened slowly as its distance from
Earth decreased.
    A few days after discovery, observers were estimating a total magnitude
between 9 and 9.3 and found a coma diameter of 2 to 3 arcmin. During the first
week of November, the brightness had increased to about 8.7, while the coma
had grown to 4 arcmin across. Perigee came on November 10 (0.4 AU), and, thereafter, the comet faded rapidly. By the end of November, the total magnitude
was slightly brighter than 12, with a coma diameter near 3 arcmin, and by mid-
December, it had faded to about 12.5 and possessed a coma diameter of 2.5 arcmin. Observations continued into the new year and the comet was last seen on
January 22, 1908, when the photographic magnitude was near 16.5.
(Mellish)

1908 III   Discovered: September 2   Discovery Magnitude: 9.0
    D. W. Morehouse (Drake University, Iowa) discovered this comet photograph-
ically near the Camelopardalis-Cassiopeia border and described it as conspicuous with a long tail. On September 3, Borrelly (Marseilles Observatory) made
an independent.discovery. Thereafter, the comet's distance from both the sun
and Earth decreased and by the end of the month, it had brightened to magnitude
6. Also during the month, the comet showed signs of unusual activity, with
photographs on the 14th and 20th revealing prominent tail condensations which
moved away from the head in the following days. By month's end, the comet displayed no further evidence of unusual activity and during the final hours of
the 30th, photographs revealed a tail 8 degrees long emanating from the coma.
During the early hours of October 1, photographs began to reveal something
different as the dense matter of the tail began to separate from the coma. The
next evening, Barnard commented that the evidence of the separation was very
strong as "the wreck of the tail was shown as a very large, long mass some two
degrees from the comet, and apparently attached to it by one or two slender
threads or streams of matter." This matter remained visible on photographs for
several days as it slowly dissipated into space.
    The comet arrived at perigee after the first week of October (1.06 AU),
with no additional activity being noted. The magnitude had nearly brightened

to 5.5 and the tail had recovered half of its pre-October length of 8 degrees. By October 14, the tail had grown to a length of 7 degrees; however, the next night, Barnard's photographs showed "that the tail for one-half degree out was made up of a slender stream which joined on to great cloudlike masses which were nearly at right angles to the tail. From these (masses), a broadening tail, curved and irregular on its north side, ran eastward for six or eight degrees." Barnard added that the disconnected tail was observed on photographs for the next several days. During the last half of October, the comet underwent a different type of change as it began to fluctuate in brightness. Between October 14 and 21 the comet suddenly faded and then brightened to its expected brightness and between the 28th and 30th, the opposite happened, with the comet being suddenly brighter on the 29th, but back to normal by the 30th. These fluctuations amounted to only 1 magnitude, according to several observers.

At the beginning of November, the comet was near magnitude 5 and, due to the increasing distance from Earth, the brightness remained near 5 throughout the month as the comet neared the sun. The tail took on a slightly different form as the tail structures became very complex and long slender rays emanated from the coma. The brightness continued to fluctuate, with the greatest being an increase between the 17th and 20th, but the tail separations noted in October were absent. The coma was consistently estimated as between 3 and 4 arcmin across, while the tail seemed to reach a maximum length of nearly 15 degrees on photographs taken shortly after mid-month. Observations continued until December 14, when the total magnitude was near 5.5, and twilight hid the comet thereafter.

Perihelion came on December 26 (r= 0.95 AU), and the comet remained lost in the sun's glare until January 22, when it was recovered in the Southern Hemisphere at a magnitude near 6.5. Photographs during February and March 1909 again revealed the complicated tail structure observed before perihelion, though nothing as dramatic as seen on photos in October and November. Although the comet-Earth distance was decreasing during this time, the increasing distance from the sun caused it to fade rapidly. As May began, the comet was again visible in the Northern Hemisphere as a 9th-magnitude object and it was last seen on the 11th. The orbit is hyperbolic with an eccentricity of 1.0007. (Morehouse)

1909 I   Discovered: June 15   Discovery Magnitude: 9.0

This comet was discovered near Alpha Trianguli by Borrelly (Marseilles Observatory) and was described as visible in a small telescope with a rapid movement northeast. The next evening, an independent discovery was made by Daniel (Princeton University Observatory), who described the comet as round, with a slight elongation away from the sun and a faint nucleus. Observers during the next few days confirmed these initial remarks, but added that the coma had a diameter of 1.5 to 2 arcmin and possessed a nuclear condensation near magnitude 11.

With perihelion having occurred on June 6 (r= 0.84 AU), the comet faded as it headed away from both the sun and Earth. By the beginning of July, the total magnitude had declined to 10 and the coma was near 1.5 arcmin. By the end of the month, the total magnitude was near 11.5. The comet was last detected on August 19, with the use of photographs, when the total magnitude was fainter than 13[*](Wolf's indication was that it was near 16--indicating a very rapid decline). The orbit is elliptical with a period near 2,000 years. (Borrelly-Daniel)

1910 I   Discovered: January 13   Discovery Magnitude: 1.0

The "Great January Comet" was first seen in the early morning sky by work-
men at the Transvaal Premier Diamond Mine in South Africa. It was described as
looking like an ordinary star, but with a tail attached, and was situated slightly to the right of where the sun rose. Word of the comet actually did not
spread until January 15, when R. T. A. Innes (Director of the Transvaal Observatory) received a telephone call from the "Leader" newspaper of Johannesburg
telling of a telegram received from the railway station master at Kopjes. The
telegram said three men at the station had seen Halley's Comet that morning for
about 20 minutes. Since Halley's Comet was not in that portion of the sky, nor
bright enough to be so conspicuous, Innes believed an unexpected comet had been
found. He and W. M. Worssell searched the sky the next morning, only to be
clouded out. Similar circumstances appeared to be prevailing on the morning
of the 17th, until a break in the clouds above the place of sunrise revealed
the comet to the two astronomers. Telegrams were immediately sent to the appro-
priate authorities around the world.

By midday on the 17th, Innes had a fine view of the comet when it was only
4.5 degrees from the sun's limb. Its appearance to the naked eye was that of
a snowy-white object with a tail 1 degree long. At the Santiago Observatory
the comet was picked up at 11:15 a.m. (local time) on January 19, when it was
7 degrees east and 3.5 degrees north of the sun, and was kept under observation
until 6:00 p.m. Further daylight observations were made at the Milan Observatory and at the Lick Observatory.

As January neared its end, the comet moved into the evening sky and became
a brilliant object visible in the Northern Hemisphere. Even with the interference of moonlight, the tail was estimated as 18 degrees long on January 26, and,
afterwards, with the absence of moonlight, it reached a maximum of 30 degrees
on January 30--although some estimates were as high as 50 degrees.

The comet gradually faded after its perihelion date of January 17.6 (r=
0.13 AU), when about magnitude -4[*], until January 27, when it appeared to be near
magnitude 2. After the 27th, a rapid fading occurred which dropped the total
magnitude to 7.6 by February 4. A gradual fading thereafter persisted until
July 15, when the comet was seen for the final time. Wolf (Heidelberg) then
reported a magnitude of 16.5. The comet was then located 3.4 AU from the sun.
The orbit is elliptical with a period near 4 million years.
(Great January Comet)

1910 IV   Discovered: August 9   Discovery Magnitude: 8.5

Using a broken-backed comet-seeker, Joel H. Metcalf (South Hero, Vermont)
discovered this comet 1.5 degrees northwest of Omega Herculi. Although the
sky clouded 15 minutes after this sighting, Metcalf's familiarity with this
region of the sky ruled out every alternative identification except for a
comet. Metcalf then described the comet as 2 arcmin across and a few days later,
observers were estimating a nuclear magnitude of 9.5 to 10. Physical descriptions varied throughout August, depending on the size of optical instrument
used, with the total magnitude being estimated as 8 to 11 and the nucleus
being described as stellar and extremely diffuse. The only consistent aspect
was the tail, which extended 1.5 arcmin on the 11th. Perihelion came on September 16 (r= 1.95 AU), and the comet was then described as near magnitude 10,
with a coma diameter of 2 arcmin. In early October, observers reported the
tail as 18 arcmin long and the total magnitude as still near 10. This general

appearance continued throughout November and December, except for a very slow
fading, which placed the total magnitude near 10.5 at the end of December.
During the last days of January 1911, an outburst in brightness occurred, with
the total magnitude rising to 9.5 on the 29th, and 9.3 by February 2. The
coma was then near 1.5 arcmin across and the tail extended 5 arcmin. In late
March, the total magnitude was near 11, while the nucleus was slightly brighter than magnitude 13. In mid-April, the magnitude was estimated as 11.3 and
by late May, it was near 13. The comet was last observed on June 23, near a
magnitude of 14.5. The orbit is elliptical with a period near 940,000 years.
(Metcalf)

1911 II   Discovered: July 7   Discovery Magnitude: 6.0
     Discovered by C. C. Kiess (Lick Observatory) on the edge of a photo-
graphic plate he had taken with the Crocker Photographic Telescope this comet
was described as a distorted, nebulous object with a faint tail-like streamer
of light attached to it. Visual observations the next morning with a 30.5-cm
telescope confirmed the object's existence.
     Photographs during the next two days revealed a coma diameter of 2 arcmin
and a tail length of 30 arcmin. It soon became obvious from orbital calculations that the comet had passed perihelion on June 30 (r= 0.68 AU), and would
be closest to Earth on August 19 (0.25 AU). During the last days of July, the
coma diameter had increased to about 5 arcmin, while the total magnitude was
estimated as near 5.8. On August 5, Raimond Moravansky (Moravia, Czechoslovakia)
independently discovered the comet at a magnitude near 5.5 and with a coma diameter of nearly 6 arcmin.
     The comet was at its best on August 22, when it was described as near mag-
nitude 4 with a coma diameter of 15 arcmin. Two days later, the magnitude had
faded to 4.5 and by the 25th, it was near 6. The rapid fading continued as
the comet's retrograde orbit took it quickly away from Earth, and by August 27,
it was described as near magnitude 9--though moonlight and haze were interfering. A combination of the comet's low altitude, increasing southern declination and decreasing brightness ended observations on September 18, 1911. The
orbit is elliptical with a period near 2,500 years.
(Kiess)

1911 IV   Discovered: September 29   Discovery Magnitude: 3.0
     This comet was discovered before sunrise between Chi and Rho Leonis by
S. I. Beljawsky (Simeis Observatory) as a bright object with a very bright tail
at least 2 degrees long. Two days later, the total magnitude was estimated as
2 to 2.5 and the tail was described as near 15 degrees in length. In addition,
Professor A. A. Nijland (Utrecht, Netherlands) described two dense streamers
flowing from the head. Numerous independent discoveries were made around the
world by October 4--with at least 7 in the United States--and the comet continued to brighten as it neared its October 11 perihelion (r= 0.30 AU). Maximum brightness was achieved on the 15th, with observers estimating the magnitude as between 1 and 1.5. The comet faded thereafter, as it headed away from
both the sun and Earth, with the total magnitude dropping from 2 to 2.5 between
the 18th and 19th. During this same time, the tail shrank from a length of
10 degrees to 6 degrees. By October 28, the comet was located at such a low
altitude that observations were no longer possible in the Northern Hemisphere
thereafter. The magnitude was then near 5. The comet was recovered in the
Southern Hemisphere in November by Perrine (Cordoba Astronomical Observatory).

On January 28, 1912, the total magnitude was estimated as 11, with a coma
diameter of 3 arcmin and a trace of tail, and when seen for the final time
on February 17, the comet had faded to near 12[*]and possessed a barely noticeable condensation. The orbit is hyperbolic with an eccentricity of 1.0001.
(Beljawsky)

1911 V   Discovered: July 21   Discovery Magnitude: 10.0

   While sweeping the eastern sky on the morning of July 21, Brooks (Smith
Observatory, New York) discovered a faint comet in Pegasus, which moved slowly
northwest. The next evening, observers described the comet as near magnitude
10, with a nucleus near magnitude 11 and a coma 5 arcmin across. The comet
brightened rapidly as it approached both the sun and Earth--reaching a total
magnitude of 8 on August 1, and 5.5 by August 20. On the latter date, the
coma was measured as 10 arcmin in diameter. The comet's increasing northern
declination halted on September 12, at 57 degrees, and, thereafter, it moved
rapidly southward. Perigee came on the 19th (0.51 AU), and observers then
described the comet as near magnitude 4. Although the comet moved steadily
away from Earth in the following days, it also approached perihelion and
by the end of September, had brightened to 3rd magnitude, with a tail estimated
as 20 degrees long. On October 1, W. Doberck, an experienced double-star
observer, detected a double nucleus with a separation of 2 arcsec. No other
observations were made and the secondary nucleus is considered doubtful.

   Perihelion came on October 28 (r= 0.49 AU), and the comet was then described as near 2nd magnitude, with a tail extending nearly 30 degrees. Fading
was rapid thereafter, and by November 21, the total magnitude had dropped to
5. Also, on the 21st, the coma was estimated as 1 arcmin across and the tail
had a length of 2 degrees. At the end of December, the total magnitude had
declined to 8 and, when last detected on February 28, 1912, it was near 11.
Interestingly, the comet seems to have been subject to minor variations in
brightness between August 20 and October 4. The orbit is elliptical with a
period near 2,100 years.
(Brooks)

1911 VI   Discovered: September 23   Discovery Magnitude: 7.5

   While making binocular observations of the region around Beta Ursae Minoris, F. Quenisset (Juvisy Observatory, France) discovered this comet as a slightly elongated object, with a nucleus and a trace of tail. The coma was then 4
arcmin in diameter. That same evening, Quenisset and F. Baldet photographed
a tail 1 degree long. F. Brown (England) independently discovered the comet
a short time later.

   The comet travelled rapidly southward as it approached its November 12th
perihelion (r= 0.79 AU), and by mid-October, it became visible to the naked
eye at a magnitude near 5.7. On this same date, observers reported a tail 30
arcmin long, a nucleus near magnitude 7 and a coma 4 arcmin in diameter. Thereafter, the comet-Earth distance increased more rapidly than before and the comet
faded slowly as it neared the sun. By the end of October, it was situated near
the Corona Borealis-Serpens border and was estimated as near 6th magnitude.
On November 8, the total magnitude was near 6.7 and the comet was lost in the
solar glare after the 16th. After conjunction with the sun, the comet was
recovered on December 18 by F. Gonnessiat (Algiers, Algeria) at a declination
of -20 degrees and at a magnitude of 7 or 8. Further southern movement ended
all chances of observation from the Northern Hemisphere at the beginning of the

new year. Perrine (Cordoba Astronomical Observatory) continued observing the comet from the end of December until February 17, 1912, when the comet was described as very diffuse and uncondensed, with a magnitude near 11. The comet was not detected thereafter. The orbit is elliptical with a period near 9,000 years.
(Quenisset)

1912 II   Discovered: September 9   Discovery Magnitude: 5.0

Gale (Sydney, Australia) discovered this comet while examining the northern part of Centaurus with a 5-cm field glass. He then described it as obviously cometary with two tails and a coma 3 to 4 arcmin across. On the 12th, J. F. Skjellerup (Rosebank, South Africa) independently found the comet near g Centaurus while using binoculars. The next evening, observers at Johannesburg described the comet as 4 arcmin across with a nearly stellar nucleus. Their photographs revealed two tails with lengths of 4 degrees and 40 arcmin. Although moving away from the Earth, the comet brightened slightly as it neared the sun, and, by the end of September, it was near magnitude 4.5.

Perihelion came on October 5 (r= 0.72 AU), and, thereafter, the comet faded slowly. On October 13, the total magnitude was near 5.5 and, by month's end, it had faded to 6. On October 7 and 12, Barnard (Yerkes Observatory) photographed a slender tail 7 degrees long, but on the 14th, he found it bifurcated at a distance of one-half degree from the head. Photos by Quenisset on the same night revealed a similar feature and his photos on the 16th showed the main tail to be straight and over 6 degrees long, while the secondary tail was curved to the south and measured only 1 degree. In early November, observers still reported the comet to be barely visible to the naked eye with a tail 6 degrees long. On the 8th, the total magnitude was 6.3, with a coma diameter of 4 arcmin and a tail 2 degrees long. By late December, the comet had faded to near 8.5 and possessed a coma 2 arcmin across. Observations continued into 1913, with magnitude estimates of 9 in early January, 11.5 in mid-March, and 12.5 at the end of April. The comet was last observed on May 26, at a total magnitude near 14 and with a coma 30 arcsec across. The orbit is hyperbolic with an eccentricity of 1.0005.
(Gale)

1912 III   Discovered: November 2   Discovery Magnitude: 7.5

Borrelly (Marseilles) discovered this comet 2 degrees northwest of Theta Herculis and described it as of magnitude 10. This estimate seems to indicate the magnitude of the nuclear condensation, since magnitude estimates made the next evening by other observers indicated a total magnitude near 7.5. The comet's retrograde orbit quickly took it away from the Earth and the brightness, as well as coma and tail sizes, subsided rapidly. On November 4, Barnard photographed a thin, slightly widening tail which extended over 1 degree and, on the 8th, Frederick C. Leonard (Chicago, Illinois) described the coma as 3 arcmin across. On November 14, Barnard estimated the magnitude as near 9 and by the 3rd of December, observers estimated the brightness as near 11 or 12. The comet was last detected on December 13, as a very faint object near magnitude 12 or 13. Perihelion had occurred on October 21 (r= 1.11 AU).
(Borrelly)

1913 II   Discovered: May 7   Discovery Magnitude: 9.5

Schaumasse (Nice Observatory) discovered this comet near the Equuleus-Delphinus border as a faint object moving in a northeasterly direction in the

morning sky. The next morning, Gonnessiat (Algiers Observatory) estimated the
total magnitude as 9.5 and measured the coma diameter as 3.5 arcmin. The comet
passed perihelion on May 15 (r= 1.46 AU), and, although a sharp fading was noted on May 17 (magnitude 10.5) by two observers, it slowly brightened as the
comet-Earth distance decreased. By early June, the comet was at perigee (0.65
AU) and then reached a maximum magnitude of 8.5. Fading was rapid thereafter,
with the comet dropping to magnitude 9 by June 15, and 10 by July 2. Observing
on August 6, Gonnessiat estimated the total magnitude and nuclear magnitude as
12.5 and 14, respectively. He added that the coma was 4 arcmin across. The
comet was last seen on August 23, when it was described as very faint, with a
diameter of 1 arcmin. Observers never detected a stellar nucleus during the
comet's time of visibility and the tail was not observed for a very long period
of time. The orbit is elliptical with a period of 5,440 years.
(Schaumasse)

1913 IV   Discovered: September 2   Discovery Magnitude: 9.5
     Metcalf (South Hero, Vermont) discovered this comet moving slowly north-
ward in Lynx and during the next few days, observers estimated the total magnitude as near 9.5. The suggestion was immediately made that the comet might be
the expected periodic comet Westphal, for which numerous searches were then
being made, but further observations quickly disproved this. On September 9,
the comet had brightened to near 8.5, while the coma was measured as near 5
or 6 arcmin in diameter.
     The comet passed perihelion on September 14 (r= 1.36 AU), and, thereafter,
it approached Earth. On September 23, it was described as 10 arcmin in diameter
with a total magnitude near 7.5 and a few days later, the comet passed 13 degrees
from the north celestial pole. Perigee came on October 8 (0.62 AU), and the
comet faded rapidly thereafter as the retrograde orbit took it quickly away
from Earth. On October 10, the total magnitude was estimated as between 7.5
and 8 and, although it experienced a sharp, temporary drop in brightness on
October 17 (magnitude 9.5), the comet faded fairly quickly to near 8.5 on the
19th. By early November, the comet was near 10th magnitude and, by the 21st, it
was near 11.5. The final observation was made on November 30. The orbit is
elliptical with a period near 13,100 years.
(Metcalf)

1914 I   Discovered: May 15   Discovery Magnitude: 4.0
     Having passed perihelion on May 8 (r= 0.54 AU), this comet was discovered
by Zlatinsky (Latvia, Russia) near Eta Persei. The comet was moving toward
Earth when discovered, and during the next few days, observers described it as
near magnitude 4, with a rapid southeasterly motion. The total magnitude
varied little in May, except for minor variations in brightness noted by several
observers, and the comet passed perigee (0.54 AU) on the 23rd. After discovery,
the tail grew significantly, with estimates being near 1 degree on the 17th,
and up to 12 degrees by May 23rd. On the 24th, it had declined to 9 degrees
and, by month's end, it was considerably shortened due to twilight. The total
magnitude on the latter date was near 5.5. Observations during June, were described as difficult due to evening twilight and low altitude and the comet was
not observed after the 12th.
(Zlatinsky)

1914 II   Discovered: March 24   Discovery Magnitude: 9.5
     This comet was discovered by Dr. Kritzinger (Bothkamp, Germany) as a faint

object with a coma diameter of 6 arcmin and a tail nearly 40 arcmin long. Kritzinger had actually seen the comet on March 24, but had taken it for a new nebula. It was then located near Psi Scorpii and was moving northeast. As the comet neared the sun and Earth, it slowly brightened, and by April 15, it was estimated as near magnitude 8.5, with a coma 10 arcmin across. Interestingly, the comet varied little in brightness from that date until mid-June, despite passing perigee on May 17 (0.5 AU). Perihelion came on June 4 (r= 1.20 AU), and the comet faded slowly thereafter. During the first days of July, it was estimated as near 10 and by mid-August, it had dropped to 11.5. The comet was followed until December 14, when the total magnitude was found to be near 15. The orbit is hyperbolic with an eccentricity of 1.0001.
(Kritzinger)

1914 III   Discovered: June 24   Discovery Magnitude: 14.0
During the course of a regular asteroid survey, G. N. Neujmin (Simeis, Russia) discovered this comet on a photograph of the Gamma Scuti region. The initial magnitude estimate was photographic and observers during the next few days found a visual magnitude closer to 12. Descriptions of the comet during these first few days indicated it was small and round, with a trace of tail. The comet was then located 3.79 AU and 2.82 AU from the sun and Earth, respectively, and these large distances caused the comet to change little in brightness by the time it reached perihelion on July 30 (r= 3.75 AU). Thereafter, the comet faded very slowly with magnitude estimates of 13.5 and 15 being reached by mid-September and late December, respectively. The comet was again detected near opposition on August 10, 1915, when photographic magnitude estimates gave values near 16.5. The orbit is hyperbolic with an eccentricity of 1.0033.
(Neujmin)

1914 IV   Discovered: September 18   Discovery Magnitude: 3.5
This comet was discovered by Leon Campbell (Arequipa Station, Peru) in Doradus and was described as very bright, with a movement to the north. Only hours later, independent discoveries were reported by J. Lunt (Royal Observatory, South Africa) and Charles Westland (New Zealand). A fourth independent discovery was made by Bernard Thomas (Tasmania) on the 20th. The comet had passed perihelion on August 5 (r= 0.71 AU), and, when discovered, was nearing its closest distance from Earth. Since the comet was discovered in rapid succession by 4 observers when located 90 degrees from the sun, it has been considered as a very likely candidate for a prediscovery outburst in brightness.

On September 20, Lunt reported a widely separated double nucleus, which was confirmed by Innes and Wood the next evening. Observations continued until October 10. Several years later, photos by W. H. Pickering that had been exposed shortly after Campbell's discovery showed both nuclei clearly and are considered the first successful photos of a double comet. From all of these observations, Sekanina has calculated a probable separation date of August 25.

Perigee occurred on September 19 (0.25 AU), and, although the comet remained brighter than magnitude 4 through the end of the month, it rapidly faded after October began. On the 17th, Fabre (Barcelona, Spain) independently discovered the comet and reported a total magnitude near 8. Other observers on that date indicated a coma diameter of 1 arcmin and a tail length of 14

arcmin. By November 17, observers described the comet as of magnitude 12.5, with a circular coma 4 arcmin across. Observations continued until February 16, 1915, when the magnitude was estimated as near 15.5. The orbit is ellip- tical with a period near 12,300 years.
(Campbell)

1914 V   Discovered: December 18, 1913   Discovery Magnitude: 11.0
    This comet was discovered slightly north of Zeta Eridani by Delavan (La Plata, Argentina) nearly 10 months before perihelion. Then located 4.24 AU and 3.52 AU from the sun and Earth, respectively, the comet was des- cribed as tailless with a coma diameter of 1 arcmin. The comet changed little during the remainder of the month and into January 1914, but at the beginning of February, it began to brighten more rapidly. At that time, it was near magnitude 10 and, by mid-month, it had brightened to 9.5 and possessed a coma 2 arcmin across which was slightly elongated as the tail began to develop. By the end of March, the total magnitude had increased to 8.5 and the comet vanished in morning twilight shortly thereafter as it neared conjunction with the sun.
    The comet was recovered in late June, and by July 15, the total magnitude was estimated as near 6.5 and the coma was 2 arcmin across. Development was rapid thereafter, and by the end of July, the magnitude had increased another magnitude, while the tail length and coma diameter attained dimensions of 1 degree and 2.5 arcmin, respectively. After the first week in August, the total magnitude was 5.2 and the tail length was estimated as 2 degrees. By late August, the comet was circumpolar and observers reported a total magni- tude of 3.5, a coma diameter of 4 arcmin and a tail length of 2 degrees.
    The comet was brightest during late September and early October at a magnitude near 2.8. It was then closest to Earth (1.5 AU) and possessed a double tail: one being narrow and straight with a length of over 10 degrees and the other being curved and 5 degrees long. Fading was slow thereafter as the comet continued to approach the sun, and when it arrived at perihelion on October 26 (r= 1.10 AU), the total magnitude was only slightly fainter than 3. By November 5, the brightness had dropped to 3.5 and, one month later, it had only declined to 3.9 and possessed a tail 2 degrees long.
    Observations were still numerous as the new year began and in mid-January, the total magnitude was still near 6. One month later, the comet had faded to magnitude 6.5 and as March began, the comet became a difficult object to ob- serve as it neared conjunction with the sun. It was recovered in Scorpius in late April, as a bright nebula with a nucleus near magnitude 10. By August, the total magnitude had decreased to 11 and the comet was last seen on Septem- ber 8, 1915. The orbit is hyperbolic with an eccentricity of 1.0002.
(Delavan)

1915 II   Discovered: February 10   Discovery Magnitude: 9.0
    While searching for comets in the morning sky, Mellish (Madison, Wiscon- sin) found an object near Sigma Ophiuchi. Though suspected to be a new comet, twilight interfered before any motion was detected. The next morning the comet was found a short distance southeast and was described as 3 arcmin in diameter, with a very short, fan-shaped tail. During the next few days, observers were reporting a total magnitude near 9, a nuclear magnitude near 10 and no tail. The comet brightened slowly and was well observed during the next few months. Perigee came in early June (0.35 AU) and the comet was then at its brightest,

with a total magnitude near 4 and a tail 6 degrees long. The comet changed little during the remainder of June and in the first half of July, as the comet's distance from the sun decreased and perihelion came on July 17 (r= 1.01 AU). Thereafter, fading was fairly slow, with the total magnitude being estimated as 6 in mid-August, and 8.5 in mid-November. On the latter date the coma was estimated as 4 arcmin across. By the end of the year, the comet was detected as a 10th-magnitude object. Observers in late January 1916 estimated the total magnitude as near 12.5 and the comet was soon afterward lost in the sun's glare as it neared conjunction. It was recovered in September and was followed until October 21, when it was near magnitude 16.

The most interesting feature about this comet was the appearance of subsidiary nuclei at the beginning of May. Several observers had remarked on the elongation of the nucleus during April, but on May 6, T. Banachiewicz (Englehardt Observatory, Russia) and G. Abetti (Arcetri, Italy) independently reported a double nucleus. On May 12, Barnard discovered a third nucleus, which lasted for over two weeks, and between June 5 and 8, photos at Sydney Observatory (Australia) revealed a fourth nucleus. The May 6th nucleus was the longest-lived of the three nuclear companions and was last seen on June 12. Its fading was rapid in relation to the primary, being 1.5 magnitude fainter on May 14 and 3 magnitudes fainter on May 31. The comet's orbit is hyperbolic with an eccentricity of 1.0002.
(Mellish)

1915 IV   Discovered: September 14   Discovery Magnitude: 9.5
Mellish (Madison, Wisconsin) discovered this comet in Leo Minor while sweeping for comets in the morning sky with his 15-cm comet-seeker. Unfavorable weather hampered his observations over the next few days, but he finally decided the comet was moving southeast. Shortly after an announcement telegram was received at Yerkes Observatory from Mellish, George van Biesbroeck obtained an accurate position and reported a magnitude near 9 and a coma diameter of 2 arcmin. Due to the comet's rapidly approaching morning twilight, the comet was becoming a difficult object. Van Biesbroeck could no longer detect it after the 20th; however, the comet's only other observer, R. G. Aitken (Lick Observatory), made observations between the 19th and 23rd, while using a 30-cm refractor. No other observations were obtained thereafter. The comet arrived at perihelion on October 13 (r= 0.44 AU), but remained on the far side of the sun until November. After November, it was located too far south for observations from Northern Hemisphere observatories and was too faint to be observed by the largest Southern Hemisphere telescopes.
(Mellish)

1917 II   Discovered: April 26   Discovery Magnitude: 9.5
M. A. Schaumasse (Nice Observatory) found this comet below Alpha Pegasi as a relatively faint object moving rapidly northeast. Shortly thereafter, observers were describing it as circular, with a yellowish tint and a central condensation. The comet became circumpolar in May and brightened from a magnitude of 9 on the 3rd to 7.3 on the 18th. On the latter date, the comet was at perigee (0.4 AU) and perihelion (r= 0.76 AU), and possessed a stubby tail and a coma 7 arcmin across. The retrograde orbit, as well as the movement away from Earth, caused the comet to fade rapidly during June. On the 9th, it was near magnitude 9, and by the 15th, it was 9.5. The final observation came on the 22nd, when the southeast movement was taking the comet into the

sun's glare.
(Schaumasse)

1917 III   Discovered: April 4, 1916   Discovery Magnitude: 13.0

Dr. Wolf (Königstuhl Observatory, Germany) discovered this comet on a
photograph exposed during a search for asteroids. In fact, the comet was
given the asteroid designation 1916 ZK, as it was then distinctly stellar
in appearance. This stellar appearance continued throughout April, although
observers with large telescopes occasionally reported a possible weak coma
around the object and, this in turn led astronomers to believe they had a
very unusual asteroid on their hands. On April 27, Dr. Johann Palisa (Vienna)
observed a very distinct halo-like feature surrounding the stellar body and
added that his latest positional measurements indicated a change in daily
motion that "was not asteroidal in character." Photographs by other observers
thereafter confirmed Palisa's observation and orbital calculations gave the
comet a nearly parabolic orbit.

This comet was discovered 14 months before perihelion and was then located
5.20 AU and 4.20 AU from the sun and Earth, respectively. Thereafter, the
comet slowly faded as the Earth headed for the opposite side of the sun and
the comet was lost in the solar glare in July. The comet was recovered in
late December, and slowly grew brighter as the new year progressed. By April
1917, the total magnitude was estimated as near 11 and a tail 2 arcmin long
was visible on photographs. Perihelion came on June 17 (r= 1.69 AU), and
observers then estimated a total magnitude of 9, a nuclear magnitude of 9.8
and a coma 2.5 arcmin across. The comet continued to brighten as it neared
the Earth and reached its maximum magnitude (8.5) in late July. Perigee came
on August 21 (0.98 AU). By mid-September, the total magnitude was near 9.5,
the nuclear magnitude was near 10 and the coma was 1.5 arcmin across. By
November 5, these figures had declined to 10.5, 11 and 1 arcmin, respectively.
On November 4, the comet passed 0.14 AU from the asteroid Vesta. By late
December, the total magnitude had declined to 15 and the comet was last detected on January 29, 1918. The orbit is elliptical with a period near
151,000 years.
(Wolf)

1918 II   Discovered: June 12   Discovery Magnitude: 10.5

Discovered by William Reid (Cape of Good Hope, South Africa) using a
15-cm refractor, this comet was then located near Alpha Hydrae and was described as resembling a small, hazy patch, with a slight condensation. The
daily motion was established as 48 arcmin due south--thus restricting observation to all but southern observers. Perihelion had occurred on June 6 (r=
1.10 AU), and the comet was moving very slowly away from both the sun and
Earth. Nearly a dozen photos were taken at the Royal Observatory and the
comet was last detected on July 17. Observers on that date indicated a total
magnitude of 14, although the predicted magnitude was only 0.1 fainter than
at discovery.
(Reid)

1919 V   Discovered: August 23   Discovery Magnitude: 8.0

This comet was discovered by Metcalf (South Hero, Vermont) as a fairly
bright telescopic object in Boötes. The next day, an independent discovery
was made by Borrelly (Marseilles). Observations by other observers during
the next few days indicated a total magnitude closer to 9 and a coma diameter

of 2 arcmin. The comet brightened thereafter as it headed towards the sun
and Earth and, during September, the total magnitude was commonly estimated
as 8 to 8.5 and the nuclear magnitude was near 9.5. The coma was nearly 4
arcmin in diameter. By October, the total magnitude had brightened to 7.5
or 8 and, in November, it was between 7 and 7.5.

The comet's southerly motion made observations difficult in the Northern
Hemisphere during the latter half of November, and the comet was lost in the
sun's glare after the 23rd. Perihelion came on December 7 (r= 1.12 AU), and
the comet was recovered on the 19th, by observers in the Southern Hemisphere.
The comet remained visible to these observers until February 3, 1920. The
orbit is hyperbolic with an eccentricity of 1.0002.
(Metcalf)

1920 I   Discovered: December 19, 1919   Discovery Magnitude: 8.5
     J. F. Skjellerup (Cape of Good Hope, South Africa) discovered this comet
as a relatively bright telescopic object located 33 degrees from the sun. It
was then moving towards the sun and was lost in twilight after December 26.
The only other observations appear to have been made by R. Woodgate (Cape
Observatory), who managed to secure two photographs of the comet before it was
lost. The comet was discovered when near perigee (0.6 AU) and passed perihelion on January 3 (r= 0.30 AU). It remained behind the sun during January,
and upon emerging from twilight in February, it was too faint for further
observations.
(Skjellerup)

1920 III   Discovered: December 8   Discovery Magnitude: 10.0
     Discovered by C. J. Taylor (Cape Town, South Africa) on December 8, this
comet was near perigee (0.25 AU) and was described as diffuse and rather
faint. Perihelion came on December 11 (r= 1.15 AU), and the comet was independently discovered on the 13th, by Skjellerup (Cape of Good Hope). The
comet faded thereafter as it moved away from both the sun and Earth, and, during the remainder of December, observers described it as a circular nebulosity
8 arcmin across with a nuclear magnitude of 11. On January 5, 1921, the total
magnitude had decreased to 10.5 and a tail extended westward for 24 arcsec.
By February 17, the magnitude had dropped to 14, while the coma was estimated
as near 0.8 arcmin. The comet was last detected on March 11, at a total magnitude of about 15.5. The orbit is elliptical with a period near 2,700 years.
(Skjellerup)

1921 II   Discovered: March 14   Discovery Magnitude: 9.0
     Reid (Cape Town, South Africa) discovered this comet near Beta Capricorni
and described it as nebulous in appearance. On April 4, A. D. Dubiago (Kazan,
Russia) independently discovered the comet with a 16-cm comet-seeker and,
shortly thereafter, the comet became widely observed. By mid-April, the total
magnitude had increased to 6.5 and observers estimated the coma as 4 arcmin
across and the tail as 10 arcmin long. By the end of the month, the brightness had increased to 5.5.

     Perigee came on May 1 (0.6 AU), and the comet was then moving at a rate
of 3 degrees per day and was described as near magnitude 5, with a coma diameter of 6 arcmin. Two days later, van Biesbroeck (Yerkes Observatory) estimated
the total magnitude as near 4.5. The comet faded thereafter, despite a decreasing distance from the sun. On May 9, it passed only 4.5 degrees from
the north celestial pole and, the next day, it arrived at perihelion (r= 1.01

AU). Around this time, the tail reached its greatest length of 17 degrees on photographs.

Fading was rapid thereafter, with the total magnitude declining from 7 to 8.5 between May 15 and June 4. It appeared as a faint nebula near the horizon on June 30, and was lost after July 3. After conjunction with the sun, observers in Bergedorf recovered the comet on October 1. On November 8, the total magnitude was estimated as near magnitude 15.5 and the comet was last detected on the 26th. The orbit is hyperbolic with an eccentricity of 1.0004.
(Reid)

1921 V   Discovered: January 20, 1922   Discovery Magnitude: 9.5

During a sweep for comets, Reid (Cape Town, South Africa) spotted a very bright nebulous spot in the constellation Antlia. He stated that the comet possessed a star-like nucleus. Having passed perihelion on October 29, 1921 (r= 1.63 AU), the comet slowly faded after discovery as it headed away from both the sun and Earth. By early February, it had dimmed to near 10.5 magnitude and one month later, it was near 11.5. Due to the comet's increasing southern declination, Northern Hemisphere observers soon lost sight of the comet; however, Southern Hemisphere observers continued to follow it until April 25, when it had faded to near 12.5. After the orbit had been calculated, it was discovered that the comet could probably have been seen in the Northern Hemisphere long before Reid had found it, but no observations could be identified. The orbit is elliptical with a period near 1,400 years.
(Reid)

1922 II   Discovered: October 19   Discovery Magnitude: 10.5

This comet was discovered on photographs exposed on the Cygnus region by Walter Baade (Hamburg Observatory) and was described as nearly 2 arcmin across with a motion to the southeast. Photographs exposed during the next few days revealed a short tail extending away from the sun; however, on October 25, a photo taken at Yerkes Observatory revealed a 20-arcmin-long tail pointing almost directly towards the sun. Perihelion came on October 26 (r= 2.26 AU), and the comet faded very slowly thereafter as it moved away from both the sun and Earth. By early December, the total magnitude was near 11 and a tail extended nearly 5 arcmin, and by mid-March 1923, it had declined to 12th magnitude. The comet was lost in the sun's glare shortly afterward as it neared conjunction with the sun.

The comet was recovered by van Biesbroeck (Yerkes Observatory) in August, with a total magnitude near 14 and a tail 4 arcmin long. By December 10, Baade photographed the comet and estimated a total magnitude of 15.3. In addition, he described a double tail--the longest being 17 arcmin in length and the shortest being 2 arcmin long. The comet was last seen on January 28, 1924, at a magnitude near 16.5. A slight tail was still visible despite the comet being located over 5.2 AU from the sun. The orbit is hyperbolic with an eccentricity of 1.0008.
(Baade)

1923 I   Discovered: November 26, 1922   Discovery Magnitude: 7.0

Reid and Skjellerup (Cape of Good Hope, South Africa) both discovered this comet in Crater and described it as resembling a bright nebula with a diameter of 3 arcmin. Although approaching both the sun and Earth, the comet faded between November 28, when the discovery magnitude was confirmed by observers at Harvard Observatory, and December 2, when observers at Harvard and Yerkes estimated the total magnitude as 8. The fading continued and in late December, the comet arrived at perigee (0.7 AU). Perihelion came on January 4, 1923 (r= 0.92 AU), and despite a predicted magnitude near 5, the actual brightness was near 9. Observations continued until February 25, when the brightness was estimated as near 12.5. The orbit is elliptical with a period near 1,800 years. (Skjellerup)

1923 III   Discovered: October 12   Discovery Magnitude: 8.0

This comet was first discovered by Arturo Bernard (Madrid, Spain) in Monoceros and was moving at a rate of 4 degrees per day. On the 14th, an independent discovery was made by Dubiago (Kazan, Russia), who described it as 8 arcmin in diameter with a movement of 5 degrees per day. The comet was then near perigee (0.4 AU). Although a few observations were acquired at other Northern Hemisphere observatories, the comet's southwest motion soon ended northern observations. Widespread publicity on the comet was somewhat delayed in reaching Southern Hemisphere observatories until late October, but the comet was widely observed at the beginning of November. These observations indicated a rapid brightening as the comet neared perihelion with magnitude estimates near 5.5. On November 6, F. J. Morshead (New Plymouth, New Zealand) independently discovered the comet near Gamma Trianguli Australis. Perihelion came on November 18 (r= 0.78 AU), and the comet faded thereafter. By month's end, it had faded to magnitude 10 and by December 8, it had dropped to nearly 11, with a coma 2.5 arcmin across. The comet was lost in twilight thereafter, but upon exiting the sun's glare in February 1924, photographic searches by van Biesbroeck and Baade failed to locate it. Although parabolic orbits are generally given for this comet, there is a possibility that the orbit is hyperbolic. (Dubiago-Barnard)

1924 I   Discovered: March 25   Discovery Magnitude: 10.0

Sweeping through Fornax during a comet-hunting session, Reid (Cape Town, South Africa) discovered a faint nebulous "star" which soon displayed a northeast movement. With perihelion having occurred on March 13 (r= 1.76 AU), the comet faded very slowly. At the beginning of April, the magnitude was still near 10 and, by the end of May, it was near 11. The comet was lost in twilight after May 30, but was recovered in the Northern Hemisphere after September 26, while moving through Cancer. On the 27th, Baade photographed it near magnitude 16, and, on the 30th, van Biesbroeck estimated it as 0.2 arcmin in diameter. The comet was last detected on October 7, 1924. The orbit is elliptical with a period near 44,300 years. (Reid)

1924 II   Discovered: September 15   Discovery Magnitude: 4.0

Discovered with the aid of binoculars near 42 Comae Berenices by Dr. P. Finsler (Bonn, Germany), this comet was described as near magnitude 4 with a tail 4 degrees long. On the 17th, Finsler estimated a total magnitude near 5 and widespread observations after the 19th indicated a magnitude near 6. With a slower fading thereafter, some researchers believe the comet was discovered shortly after an outburst in brightness. Photographs at Yerkes Observatory on September 22 and 23 showed a bright coma 6 arcmin across and a fan-shaped group of tails. The main tail contained a bend 46 arcmin from the nucleus on the first day, which had receded to 52 arcmin by the 23rd.

The comet moved rapidly to the south after discovery and was located in Libra at the beginning of October. The magnitude was then slightly fainter than 7. After the 6th, northern observers found it difficult to see the comet, partly due to low altitude, but mostly due to a rapid fading which had again set in. By the 8th, the brightness was near 8.5, and photos at Yerkes on the 13th and 14th showed the comet as a diffuse object with a total magnitude near 12. The predicted brightness for the latter date, based on magnitude estimates made in late September, was 8.5. The comet was last seen on October 19, in Johannesburg. Perihelion had occurred on September 4 (r= 0.41 AU), and some mathematicians have remarked on the resemblance between this orbit and the orbit of the comet of 770, though the latter is only approximately known. (Finsler)

1925 I   Discovered: April 3   Discovery Magnitude: 9.0

This comet was discovered by L. Orkisz (Mount Lysin, Poland), who described it as somewhat bright in a telescope with a motion to the northeast. When worldwide observations began a few days later, observers consistently described the comet as 5 arcmin across, with a total magnitude slightly brighter than 8, a nuclear magnitude near 12 and a faint trace of tail. Upon passing perigee in early May (1.4 AU), the comet reached its maximum brightness of 7th magnitude. The coma was then described as 6 arcmin across visually and nearly 12 arcmin across on photographs, while the multiple tails extended as far as 1 degree. On June 1, the comet was located only 8 degrees from the north celestial pole and had then faded to magnitude 8.5. By the end of June, it had dropped another magnitude and the coma was estimated as 2 to 3 arcmin across. By mid-July, the magnitude was near 11 and, shortly thereafter, the comet was lost in the sun's glare. Van Biesbroeck recovered it in November and described it as near magnitude 14.0, with a diameter of 20 arcsec. He continued observations until May 12, 1926, when the total magnitude had declined to 17. The comet had passed perihelion on April 1, 1925 (r= 1.11 AU). It travels in a hyperbolic orbit with an eccentricity of 1.0006. (Orkisz)

1925 III   Discovered: March 24   Discovery Magnitude: 8.0

This comet was discovered 10 degrees south of Spica by Reid (Newlands, South Africa), who described it as "fairly large, very compressed in the center, with a distinct nucleus." The comet was then approaching both the sun and Earth, and, one month after discovery, it had brightened to a magnitude of 7.5 and possessed a tail 10 arcmin long. During May, the comet entered Centaurus and, by the end of the month, it had become visible to only Southern Hemisphere observers at a magnitude slightly brighter than 7. Also, in late May, the comet passed perigee (1.1 AU), but continued to brighten as it neared the sun.

Although descriptions of the comet from the Southern Hemisphere are hard to come by, Reid stated that it "almost reached naked-eye visibility" (magnitude near 6) during June and July. Perihelion came on July 29 (r= 1.63 AU), and the comet began to fade rapidly during August. Observations in the Southern Hemisphere came to an end in late December, but the comet had been recovered in the Northern Hemisphere on December 4, at a magnitude near 11. Observations continued throughout 1926, and the comet was last seen on December 31, when van Biesbroeck described it as small with a magnitude near 17. The orbit is elliptical with a period near 6,100 years. (Reid)

1925 VI   Discovered: March 22   Discovery Magnitude: 11.0
    This comet was discovered by G. A. Shajn on his first trial exposure with
the double astrograph at Simeis Observatory. It was then described as faint
and small, with a daily motion of 30 arcmin. The next evening, Professor J.
Comas Sola (Fabry Observatory, Spain) independently discovered the comet on a
photograph made in his regular search for asteroids. Then located 4.4 AU and
3.4 AU from the sun and Earth, respectively, the comet changed very little
from its initial brightness of 11 during the next few months as it approached
perihelion. At the same time, the coma remained near 1 arcmin in diameter
and no tail was observed. Observations continued until mid-June, when the
comet was lost in the sun's glare. Perihelion came on September 7 (r= 4.2 AU),
and the comet was recovered in October after conjunction with the sun. During
the remainder of the year, the total magnitude remained near 12.5. In January
1926, the total magnitude was near 13 and the coma was estimated as 1.2 arcmin
in diameter. By April, it had faded to 14, and shortly thereafter, the comet
was lost in twilight as it neared its second conjunction with the sun. Observers recovered it in late September, at a magnitude near 15[*]. The brightness
changed little during the final months of 1926, and estimates of the coma diameter usually indicated a value near 18 arcsec. The comet was last observed on
March 4, 1927, when van Biesbroeck (Yerkes Observatory) estimated a total magnitude near 16. The comet was then 6 AU from both the sun and Earth. The
orbit is hyperbolic with an eccentricity of 1.0024.
(Shajn-Comas Sola)

1925 VII   Discovered: November 17   Discovery Magnitude: 8.0
    Having passed perihelion on October 3 (r= 1.57 AU), this comet was dis-
covered by van Biesbroeck (Yerkes Observatory) while using a 10-cm finder to
set his telescope on the ephemeris position of comet 1925 I. The comet was
described as bright, with a 9th-magnitude nucleus and a tail 35 arcmin long.
The comet may then have been undergoing an outburst in brightness, since observations by van Biesbroeck and other observers during the next few days indicated a fading much more rapid than predicted. On November 21, van Biesbroeck
estimated a total magnitude of 9, and noted that the tail had decreased to
15 arcmin in length. Fading was slower thereafter, and one month later, the
total magnitude was near 9.5 and the tail extended 5 arcmin. Van Biesbroeck's
observations were the most extensive of any observer's and he continued to view
the comet into the new year. On January 19, 1926, he estimated a total magnitude of 10.2 and measured the tail length as 8 arcmin. By mid-February, the
magnitude had declined to between 10.5 and 11, while a photographic plate exposed with the 61-cm reflector revealed a broad tail 30 arcmin long. The magnitude had decreased to 13.5 on May 5, with a tail only 2 arcmin long, and on
June 10, the brightness of the then tailless nebulosity had dropped to magnitude 14.5. The comet was not observed thereafter. The orbit is hyperbolic
with an eccentricity of 1.0004.
(van Biesbroeck)

1925 XI   Discovered: November 14   Discovery Magnitude: 8.0
    This comet was discovered only four days prior to its closest approach to
Earth (0.58 AU). It was first detected by Leslie C. Peltier (Delphos, Ohio)
near Nu Boötis and was described as fairly bright with a rapid motion to the
southeast. Searches at Yerkes Observatory and Harvard College Observatory on
the 17th, failed to locate the comet; however, a few days later, word arrived

that A. Wilk (Krakow, Poland) had independently discovered the comet and
indicated a motion much more rapid than earlier believed. Later examination
of Harvard plates revealed an image of the comet at the very edge of a plate
exposed on the 17th, and a very faint image on a plate exposed on the 18th.
The comet was observed by van Biesbroeck (Yerkes) on November 21, and was
described as round and diffuse, with a total magnitude near 7.5, a coma diameter of 2 arcmin and a tail 15 arcmin long.

Perihelion came on December 7 (r= 0.76 AU), and the comet was then at
its brightest with a magnitude of 6.7. On the 10th, van Biesbroeck photographed a tail 1 degree long and a coma 3 arcmin across. From the round coma,
he described a "wide bundle of streamers, the axis of which is marked by a
straight threadlike streamer..." By the 18th, the total magnitude had faded
to 9.5 and, at low altitude on the 26th, the brightness was near 10. The
comet was last seen on December 31, when van Biesbroeck detected it at a magnitude of 10.5 when located only 5 degrees above the horizon. The orbit is
hyperbolic with an eccentricity of 1.0005.
(Wilk-Peltier)

1926 I   Discovered: January 16   Discovery Magnitude: 9.5
T. B. Blathwayt (Bloemfontein, South Africa) discovered this comet just
below Corvus and immediately reported his finding to the Union Observatory
(Johannesburg) where it was observed only a few hours later. The comet's
motion was then to the northwest at a rate of 2 degrees per day.

Perihelion had occurred on January 2 (r= 1.35 AU), and, after discovery,
the comet brightened very slowly as it neared Earth. Perigee came on February
4 (0.45 AU), and observers then described it as near magnitude 8.5, with a
coma 6 arcmin across. Fading was rapid thereafter, with the total magnitude
being near 10 on February 9, and 11 by the 16th. On March 10, the total magnitude was near 13.5, and, when last detected on April 9, van Biesbroeck (Yerkes
Observatory) found a magnitude near 14. No tail was ever observed during this
comet's apparition, as the coma always displayed a circular appearance. The
orbit is elliptical with a period near 2,300 years.
(Blathwayt)

1926 III   Discovered: December 13, 1925   Discovery Magnitude: 8.0
G. E. Ensor (Pretoria, South Africa) discovered this comet in Reticulum
and described it as fairly bright, with a rapid motion to the southwest. The
tail was estimated as 15 arcmin long. After reaching Tucana in late December,
the comet began moving northwestward and in late January, Reid described it
as "a very beautiful object, with a very distinct nucleus surrounded by a
large coma, and having a bright tail over a degree in length." Soon afterward,
the comet was lost in evening twilight.

Perihelion came on February 12 (r= 0.32 AU), and the comet was recovered
in Leningrad on February 23, by the Mirovedenie Society of Russia. One member
of this society, S. M. Selivanov, told how several members ascended in a fast-
ened hot-air balloon to a height of over 2,000 feet. Selivanov found the comet
near Epsilon Equulei after 20 minutes of searching with a pair of binoculars.
He described it as an indistinct spot 8 to 10 arcmin in diameter and possessing
a magnitude of 5. When the moon had set somewhat later, he noted a tail extending 20 arcmin from the coma.

As the comet further cleared morning twilight, observers were hoping for
a possible naked-eye object. On March 5, Dr. W. H. Steavenson searched for the

comet with a telescope, but found no cometary object brighter than 9th magnitude within 30 arcmin of the predicted position. Similarly, on March 6, Selivanov "explored the region of the sky within 100 square degrees around the place"
using a 17.5-cm refractor, but failed to find the comet. Other searches also
failed within the next several days.

The comet was finally recovered on March 10, by B. M. Peck (Herne Bay)
on a photograph and was described as a large, faint nebulous blur very near the
ephemeris position. On March 16 and 20, A. Schwassmann and J. Stobbe (Hamburg
Observatory) took 2.5-hour exposures which revealed a narrow, fan-shaped tail
near 30 arcmin long and extending at right angles to the line from the sun to
the comet. On the first photo the magnitude was near 12, while the second
showed it near 13 and neither showed any trace of a nucleus. The comet was
last detected on April 3 and 13, by Professor Schorr (Hamburg Observatory), who
described it as only a tail-like structure near 30 arcmin long and 5 arcmin
wide, with no indication of a nucleus or coma. The total magnitude equalled
17 on the latter date, although the predicted magnitude, based on pre-perihelion
observations, should have been 12. Although a parabolic orbit is generally giv-
en for this comet, there is a possibility that the orbit is a hyperbola.
(Ensor)

1926 VII   Discovered: January 25, 1927   Discovery Magnitude: 8.0

Having passed perihelion on December 30, 1926 (r= 0.75 AU), this comet was
discovered by Reid (Cape Town, South Africa) in Tucana and was described as large,
bright and circular. Then located at a declination of -57 degrees, the comet
was invisible to observers in the Northern Hemisphere, but was moving slowly
northeastward. Reid reported that the comet varied little in brightness throughout January and into February, and observations seemed to have ceased by February
20, when the brightness should have been between 8.5 and 9.* Thereafter, the
comet was lost in the sun's glare, and should have been visible in the morning
sky in the Northern Hemisphere in early March; however, photos by van Biesbroeck failed to locate it, thus, making it much fainter than magnitude 10.

Interestingly, Reid later reported that the comet was detected in early
March at a magnitude near 9.5 and was followed by him until early April. He
stated that "before disappearing (which it did very rapidly) it swelled out to
very large dimensions, at the same time becoming faint, and losing all signs of
the central condensation." Today, these apparent observations are considered
somewhat doubtful. Although the orbit is generally listed as a parabola, it
may possibly be a hyperbola with an eccentricity near 1.009.
(Reid)

1927 II   Discovered: January 11   Discovery Magnitude: 9.0

Blathwayt (Bloemfontein, South Africa) discovered this comet in Libra, almost one year after he had discovered comet 1926 I. He then described it as
fairly large, bright and round, with a strongly condensed center. The comet
moved rapidly southward, due to the comet's orbit being nearly perpendicular
to the ecliptic, and by late February, it was located less than 20 degrees from
the south celestial pole. Perihelion had occurred on February 14 (r= 1.04 AU),
but the comet had brightened little more than the discovery magnitude since
it always remained further than 1.1 AU from Earth. The final observation was
made on April 6, when the total magnitude was near 12. When the comet had
finally reached a declination accessible to Northern Hemisphere observers, it
was too faint for photographic observation.
(Blathwayt)

1927 IV   Discovered: March 10   Discovery Magnitude: 10.0
     Dr. C. L. Stearns (Van Vleck Observatory, Connecticut) discovered this
comet visually with the 51-cm refractor in the course of his regular work on
stellar parallaxes. It was then located 2 degrees north of Beta Librae and
was slowly moving to the northwest. Stearns described the comet as well-defined, with a tail 10 arcmin long and a nearly stellar nucleus. Upon announcement of this comet's discovery, numerous photographs were taken at observatories around the world. These photos indicated a magnitude closer to 9, as
well as two tails: one bright and stubby, and the other faint and narrow.
The coma was consistently measured as between 1 and 2 arcmin in diameter. The
comet passed perihelion on March 22 (r= 3.68 AU), but fading was slow thereafter, with the magnitude being estimated as 9.5 in late April, and 12 at the
beginning of July. During the latter time, the comet still possessed a short
tail and the coma diameter had decreased to less than 50 arcsec. At the end
of the year, the total magnitude had declined to 13.
     The comet's northern motion made it circumpolar by March 1928, when ob-
servers variously estimated the magnitude as between 12 and 13, and a photo
by van Biesbroeck revealed a tail 1.3 arcmin long. In early August, the comet
passed only 14 degrees from the north celestial pole and by the end of the
year, it had dimmed to magnitude 15.5 and possessed a tail 48 arcsec long.
     By April 1929, van Biesbroeck had become the most prolific observer of
this comet. Using the 61-cm reflector at Yerkes Observatory on the 14th, he
described the comet as a well-defined image near magnitude 16. No tail was
photographed during the year, as the comet always showed a circular form on
photos. The coma shrank from a diameter of 20 arcsec on July 5, to only 12
arcsec on October 1. At the year's end, the total magnitude had dropped to
16.5.
     Van Biesbroeck noted little variation in the appearance of the comet in
1930. The coma diameter seemed to remain near 15 arcsec, but the condensation
at its center became more diffuse. His estimated magnitudes declined from
16.5 to 17 during the year--in close accordance to the estimates of H. Struve
and Baade, who were the comet's only other observers. Van Biesbroeck last
observed the comet on March 12, 1931, when his photographs revealed a very
small, circular object, possibly 5 arcsec across, near magnitude 17.5. The
comet was then located at a distance of 11.5 AU from the sun. The orbit is
elliptical with a period near 2,000 years.
(Stearns)

1927 IX   Discovered: November 27   Discovery Magnitude: 3.0
     This comet was independently discovered in the morning sky by at least
10 observers. The first to officially announce its discovery was Skjellerup
(Melbourne, Australia), who described it as a 3rd-magnitude object with a 3-
degree-long tail on December 3. On December 6, Maristany (La Plata, Argentina)
independently announced the discovery. Other observations going back to November 27, were belatedly reported.
     The comet passed perihelion on December 18 (r= 0.18 AU), and was then
observed in broad daylight, only 5 degrees from the sun, at a magnitude near
-6.* On this date, also, two independent discoveries were made. Fading was
rapid after perihelion with magnitude estimates being near 1 on the 19th, and
3 on the 21st. As the comet moved out of twilight, the tail became an impressive sight, with estimates as large as 40 degrees between December 29 and

January 2. Thereafter, the tail shrank rapidly and observations could only
be made by observers in the Southern Hemisphere. In early February 1928, the
magnitude had reached 9, and by month's end, it was near 10, with a coma
diameter of 1 arcmin. On March 28, observers at La Plata estimated a magnitude
of 12 and at Johannesburg on April 28, observers detected the comet for the
final time at a magnitude near 14. The orbit is elliptical with a period near
36,500 years.
(Skjellerup-Maristany)

1930 I   Discovered: February 18   Discovery Magnitude: 10.0
     This comet was discovered on photographic plates by A. Schwassmann and
A. A. Wachmann (Hamburg Observatory) on February 18.9. It then appeared as an
abnormally large trail with an indicated daily motion of 5 degrees. On February 20, Peltier (Delphos, Ohio) independently discovered the comet.
     Perihelion had occurred on January 16 (r= 1.09 AU), and the comet passed
perigee on February 14 (0.21 AU). Subsequently, the comet faded rapidly after
discovery and, by February 24, it was estimated as near 12th magnitude. On
this same date, a one-hour exposure by H. Struve showed a coma 4 arcmin in
diameter with two faint rays and a tail extending to the west. On March 3,
the total magnitude had declined to 13.5 and when last detected on the 19th,
van Biesbroeck (Yerkes Observatory) described it "as a very faint nebulosity
of magnitude 16." The orbit is elliptical with a period between 14,000 and
176,000 years.
(Peltier-Schwassmann-Wachmann)

1930 II   Discovered: December 20, 1929   Discovery Magnitude: 7.0
     Wilk (Krakow, Poland) discovered this comet near Kappa Lyrae and described
it as a bright telescopic object with a coma diameter of 3 arcmin. The comet
moved southeast as it approached perihelion and conjunction with the sun; however, it changed little in brightness since its distance from Earth was increasing.
     The comet was widely photographed after discovery and by the end of the
month, the tail was 2 degrees long, while the coma was 5 arcmin across. Occasionally, observers detected a second, shorter tail. The comet seemed brightest during the first week of January 1930, as it was reported to be as bright
as magnitude 6.5. Thereafter, observations became more difficult as the comet
moved into twilight. Perihelion came on January 22 (r= 0.67 AU), and the comet
was then near magnitude 7.5. The final observation came on the 29th, when the
magnitude was near 8. The comet should have been brighter than magnitude 11
when it exited twilight in March, but observers in the Southern Hemisphere
could not detect it. The orbit is elliptical with a period near 18,000 years.
(Wilk)

1930 III   Discovered: March 21   Discovery Magnitude: 6.0
     Wilk (Krakow, Poland) discovered this comet near Eta Piscium and described
it as a bright telescopic object moving to the southwest. Although Wilk speci-
fically gave the total magnitude as 7.0, other observers within the next few
days indicated a value slightly brighter than 6.
     Perihelion came on March 29 (r= 0.48 AU), and the comet was then visible
to the naked eye at a magnitude near 5.5. During the first week of April, it
brightened to 5 and then faded as the comet moved quickly away from the sun and
slowly toward Earth. On the 1st of April, photographs showed a tail 4 degrees
long, which, by the 24th, had decreased to only 3.75 degrees. The total magni-

tude on the latter date was near 7.5.

The comet continued to approach Earth during May, and the fading also continued. When near perigee (0.6 AU) at the end of the month, the total magnitude had dropped to 9.5. A more rapid fading thereafter, brought magnitude estimates of 14 by the end of June, and 15 when last observed on July 3. The orbit is elliptical with a period near 485 years.
(Wilk)

1930 IV   Discovered: October 9, 1929   Discovery Magnitude: 14.0
On March 4, 1930, Max Beyer (Gummelt Observatory, Germany) discovered a diffuse trail 0.8 arcmin long on a plate he had exposed on February 26, using the 15-cm astrograph. Unfavorable weather prevented Beyer from again observing the comet until March 11, when a photographic plate exposed on the extrapolated position revealed the diffuse object. Later orbital calculations allowed the comet to be found on over a dozen photographs taken at various observatories between October 9, 1929, and March 3, 1930.

On March 11, the comet was located 2.11 AU and 1.71 AU from the sun and Earth, respectively. Although nearing perihelion, the comet's increasing distance from Earth caused it to brighten only slightly during the next few weeks. Van Biesbroeck (Yerkes Observatory) made the most extensive observations of the comet, which began on March 15. On the 17th, he estimated a total magnitude of 10 and by the 21st, he found it to be 9.5. On both dates, the coma diameter was 2 arcmin and a sharp nucleus was detected. Also, on the latter date, Peltier (Delphos, Ohio) made an independent discovery of the comet. Thereafter, the comet faded, since it was receding from Earth at a very rapid pace and by April 4, it had dropped to magnitude 11. Perihelion came on April 18 (r= 2.08 AU), and shortly thereafter, van Biesbroeck noted a round coma 1 arcmin across with a sharp nucleus.

Fading was steady but slow during the remainder of the year. On May 23, the total magnitude was 11.5 and the coma measured 48 arcsec across. By September 17, van Biesbroeck estimated the total magnitude and nuclear magnitude to be 13 and 13.5, respectively, with a coma diameter still near 48 arcsec. At the end of December, the total magnitude had faded to 15. Throughout January, February and March, 1931, the total magnitude was estimated as 15.5 and by April 24, it had declined to 16, with a coma 30 arcsec across. Observations continued up to August 13, when van Biesbroeck described it as diffuse with a magnitude of 17.5. The comet was not seen afterwards. The orbit is hyperbolic with an eccentricity of 1.0005.
(Beyer)

1930 V   Discovered: May 29   Discovery Magnitude: 9.0
A. F. I. Forbes (Cape Town, South Africa) discovered this comet in Sculptor and described it as somewhat faint with a motion to the north. Observers during the next few days indicated a slight brightening to about 8.5 after the first week of June. Perihelion had occurred on May 10 (r= 1.15 AU), and, although the comet approached Earth during June, it suddenly began to fade more rapid than expected during the second week of June. Perigee came on June 21 (0.38 AU), and van Biesbroeck then estimated the total magnitude as near 11 and measured a broad tail 3 arcmin long. Thereafter, fading was even more rapid, with total magnitude estimates reaching 13 by the end of June and 15 on July 17. The comet was last seen on July 21. Observations indicate the coma was at least 10 arcmin across when at perigee and only 1 arcmin across

when last detected. Although a parabolic orbit is generally given, there is a possibility that the orbit is elliptical.
(Forbes)

1931 III   Discovered: July 16   Discovery Magnitude: 7.0

Masani Nagata (Brawley, California) discovered this comet while observing Neptune shortly after sunset. This region of the sky was soon lost near the horizon, but the next evening, the comet was observed 1 degree east of its previous position. Due to its brightness, Nagata doubted that he had been alone in this discovery and he made a modest inquiry to Mt. Wilson Observatory that evening, where the comet was confirmed by photography and announced as new.

Perihelion had occurred on June 11 (r= 1.05 AU), and, after discovery, the comet rapidly faded as it moved away from both the sun and Earth. Interestingly, the comet is a candidate for a prediscovery outburst in brightness, with E. M. Pittich (1969) showing that if the discovery brightness was not due to an outburst, it should have been discovered at least 72 days earlier.

On August 5, the comet had faded to 9th magnitude and possessed a wide tail 15 arcmin long. In mid-September, the total magnitude had declined to 10 and by October 5, it had dropped to 12.5. Between the latter date and October 6, the comet began to brighten unexpectedly, with brightness estimates being near 9.5 on the 6th, and 8 over the next few days. On October 20, the total magnitude had dropped to 10 and after the first week of November, it was back to 12. As November progressed, the comet neared evening twilight and at a low altitude on the 27th, the comet was estimated as near magnitude 13, with a circular coma 18 arcsec across. The comet was lost shortly thereafter.

The comet was recovered in the morning sky in early February, at a magnitude near 15.5. One month later, the total magnitude had declined to 16 and when last detected on April 15, 1932, it was described as a tiny diffuse spot of magnitude 17.5. The orbit is elliptical with a period near 357 years.
(Nagata)

1931 IV   Discovered: August 10   Discovery Magnitude: 5.0

This comet was discovered in the morning sky by P. M. Ryves (Zaragoza, Spain) only 2 degrees from U Geminorum--a variable star which was being observed extensively by Ryves during the year. The comet was then described as bright, with a motion towards morning twilight. The comet brightened rapidly as it neared both the sun and Earth, and was estimated as near magnitude 4 on August 14. It then possessed a tail 1 degree long and a coma 30 arcsec across. Observations continued until August 20, when the comet was seen deep in twilight. Descriptions of it then gave a total magnitude between 2 and 3 and a tail length of 2 degrees. Observations were impossible thereafter.

Perihelion came on August 26 (r= 0.08 AU), and the comet remained lost in the sun's glare until October 9, when van Biesbroeck (Yerkes Observatory) found it in the morning sky as a very diffuse 9th-magnitude object 3 to 4 arcmin in diameter. The comet faded fairly rapidly thereafter, and on October 17, van Biesbroeck estimated a total magnitude of 9.5 and photographed a sunward tail extending 45 arcmin. On the 22nd, the total magnitude was near 10.5 and after the first week of November, it was near 12.5, with a nuclear magnitude near 15 and a coma diameter of 2 arcmin. Observations continued into December, with the comet being seen for the final time on the 16th, at a magnitude near 14.5. The orbit is elliptical with a period of 289 years. Calculations

reveal the comet passed as close as 0.14 AU to Jupiter in October 1930   and
must have undergone considerable perturbations.
(Ryves)

1931 V   Discovered: April 5, 1932   Discovery Magnitude: 12.0
     Having passed perihelion on November 30 (r= 2.33 AU), this comet was not
discovered until 5 months later, when R. Carrasco (Madrid, Spain) found it on
April 22, on photographs exposed in Coma Berenices. Later calculations helped
to reveal prediscovery images on plates exposed at Yerkes Observatory on April
5 and 6, and on Harvard plates exposed on April 13.
     Observations soon after discovery confirmed Carrasco's magnitude estimate
of 12, but also indicated a coma diameter of 1 arcmin, a short, wide tail and
a stellar nucleus. At the end of May, the total magnitude had dropped to 13
and the nucleus was near 14. By the end of June, the total and nuclear magnitudes were 14 and 15, respectively. The comet was last detected on July 5.
Although a parabolic orbit is generally given for this comet, there is a chance
that the orbit is slightly hyperbolic.
(Carrasco)

1932 I   Discovered: April 1   Discovery Magnitude: 9.0
     Having passed perihelion on February 28 (r= 1.25 AU), this comet was dis-
covered by H. E. Houghton (Cape Town, South Africa) when only 15 degrees from
the south celestial pole. The next evening, Ensor (Pretoria, South Africa)
made an independent discovery. Both observers had been studying the variable
star T Apodis at discovery and described the comet as faint, with a rapid motion
to the north.
     The comet brightened slightly as it neared Earth, and reached 8th magnitude
between April 7 and 10, with a coma diameter of 5 arcmin. Perigee came in mid-
April (0.45 AU), when the total magnitude was near 9, and the comet faded rapidly thereafter. Observations began in the Northern Hemisphere on April 23,
when the comet was low over the horizon, and on the 27th, the total magnitude
was near 10, with a coma diameter between 6 and 8 arcmin. By the end of May,
the total magnitude was near 13.5, while estimates of the coma diameter were
near 2 arcmin. The comet was last seen on June 5, at a magnitude near 15. The
orbit is elliptical with a period near 302 years.
(Houghton-Ensor)

1932 V   Discovered: July 13   Discovery Magnitude: 9.0
     This comet was discovered near the Perseus-Aries border by Peltier (Delphos,
Ohio) on August 8, and was described as bright. The next morning, Fred L.
Whipple (Harvard College Observatory) independently discovered the comet on a
photographic plate exposed on August 6, and on the evening of August 9, H. T.
Sase (California) also, discovered the comet. Later calculations helped to
locate prediscovery images of the comet on plates exposed on July 13.
     After discovery, the comet brightened as it neared the sun. On August 10,
van Biesbroeck (Yerkes Observatory) estimated the total and nuclear magnitudes
as 7 and 8, respectively. He also measured the coma as 5 arcmin across and the
tail as 1 degree long. On August 31, observers estimated a total magnitude near
6.8 and, the next night, the comet was located within 10 degrees of the north
celestial pole. The comet was at perihelion (1.04 AU) on September 2, and
reached a maximum brightness of 6.5 a few days later. Thereafter, the comet
changed rapidly with the total magnitude dropping to 8.5 and the tail shrinking
to 30 arcmin on September 20, and reaching values of 9.5 and 7 arcmin, respec-

tively, by the end of the month. After the first week of October, the total magnitude had faded to 11 and by the end of the month, it was near 14.5. During November, the comet faded to invisibility. It was estimated as near 15th magnitude at the beginning of the month and was photographed for the last time on the 30th, by van Biesbroeck, who described it as "a vague small coma not brighter than a star of magnitude 17." The orbit is elliptical with a period near 291 years.
(Peltier-Whipple)

1932 VI   Discovered: August 14, 1931   Discovery Magnitude: 8.5

This comet was discovered less than 6 degrees from the south celestial pole by Murray Geddes (Otago, New Zealand) on June 22, 1932. Later, after the orbit had become well established, Whipple (Harvard Observatory) found a prediscovery image on a photo exposed on August 14, 1931, when the total magnitude was near 13.

After the initial discovery, the comet slowly faded as the increasing distance from Earth countered the decreasing distance from the sun. By the end of June, it was near magnitude 9 and, by mid-August, it was near 10. Due to the comet's northward motion, it was lost in the sun's glare after August. Perihelion came on September 21 (r= 2.31 AU), and the comet was recovered after conjunction with the sun by Northern Hemisphere observatories during December 1932. The total magnitude was then near 11.5. As the new year began, the comet slowly brightened as it drew closer to Earth and by February, was estimated as near 10.5. Observers then estimated the coma diameter as 3 arcmin and the tail length as 30 arcmin. The brightness slowly decreased thereafter, with estimates of 12.5 in early June and 13.5 by the end of July. The comet was followed throughout the remainder of 1933, primarily by van Biesbroeck (Yerkes Observatory), who described it as a small, round coma of magnitude 14.5 on September 21. Between October and December, he reported the same physical appearance, although the total magnitude was then closer to 14. On February 12, 1934, the magnitude was near 15 and on April 13, it was near 16. The comet was followed until July 19, 1934, when observers at Lick Observatory estimated the total magnitude as 17.5. The orbit is hyperbolic with an eccentricity of 1.0014.
(Geddes)

1932 VII   Discovered: June 1   Discovery Magnitude: 13.0

K. A. Newman (Lowell Observatory, Arizona) discovered this comet on a plate exposed on June 20. He then described it as near magnitude 12.5, with a centrally condensed coma 40 arcsec across. Shortly thereafter, Newman found prediscovery images on plates exposed on June 1 and 7.

During the next couple of months, the comet changed very little in brightness as it approached perihelion, partly due to its great distance from the sun and partly due to its passing perigee in early July (1.25 AU). Of primary interest, however, is the possible discovery of a companion by A. Schmitt (Athens) on June 25. It was located one-half degree from comet Newman and was at first presumed to be a different comet--thus gaining the preliminary designation of 1932 h; however, Schmitt said it had "at least approximately the same motion" as Newman's comet. Schmitt again detected it, both visually and photographically, on June 29, but failed to detect it on July 1, 3 and 4. A similar object was detected by Steavenson (Norwood) on July 1, and by E. Delporte (Uccle, Belgium) on July 5 and 6; however, attempts to compute an orbit based on the available positions proved to be impossible. The identity and existence of the object or objects detected remains undecided.

Perihelion came on September 24 (r= 1.65 AU), and the comet began to noticeably fade thereafter. By mid-November, it was near magnitude 13.5 and in mid-December, it had dropped to magnitude 14. Observations continued up until January 20,1933, when van Biesbroeck reported a total magnitude of 15. The orbit is hyperbolic with an eccentricity of 1.0006.
(Newman)

1932 X   Discovered: December 15   Discovery Magnitude: 8.0
This comet was first discovered by Forbes (Hermanus, South Africa) near Alpha Piscis Austrinus, but his announcement was delayed due to a public holiday. Word of the comet's existence actually spread as a result of a second discoverer, G. F. Dodwell (Adelaide, Australia), who first detected it on the 17th. Both men reported the comet as bright, with a round coma 3 arcmin across. Observers during the remainder of December confirmed the comet's initial appearance, although a stellar nucleus of magnitude 14 was occasionally noted.

Perihelion came on December 30 (r= 1.13 AU), and by January 17, van Biesbroeck reported that it had faded to magnitude 9 and possessed a tail 1 degree long on photographs. Thereafter, fading was more rapid, with the comet reaching a total magnitude of 12.5 by late February, and 15 by the end of March. Observations continued until April 22, when observers at Lick Observatory estimated a total magnitude of 17.5. The orbit is elliptical with a period of 262 years.
(Dodwell-Forbes)

1933 I   Discovered: February 16   Discovery Magnitude: 8.0
Peltier (Delphos, Ohio) discovered this comet between Delta and Iota Cephei, while observing variable stars. He described it as bright, with a rapid southeastern motion. Having passed perihelion on February 6 (r= 1.00 AU), the comet faded after discovery and reached a total magnitude of 9.5 by the end of February. The coma diameter was then near 3 arcmin. By the end of March, observers estimated a total magnitude of 12.5 and a coma diameter of 2 arcmin. Observations continued only until April 14, when H. M. Jeffers (Lick Observatory) estimated the photographic magnitude as 16.
(Peltier)

1934 II   Discovered: June 3, 1935   Discovery Magnitude: 13.0
This small, faint comet was discovered by Cyril Jackson (Union Observatory, South Africa) on photographs taken during his systematic study of asteroids. It was then described as diffuse, without a tail and was again photographed by Jackson on June 8, 11 and 19, before being officially announced. Perihelion had occurred on September 7, 1934 (r= 3.49 AU), and the comet rapidly faded after discovery as it moved away from both the sun and Earth. Harvard and Yerkes observatories both reported a total magnitude of 15 in late June, and by July 21, van Biesbroeck described it as near magnitude 16, with a round, diffuse coma 10 arcsec across. The comet was last detected on August 5, at a total magnitude of 16.
(Jackson)

1935 I   Discovered: January 7   Discovery Magnitude: 10.0
E. L. Johnson (Union Observatory, South Africa) photographically discovered this comet near Pi Phoenicis and described it as very faint and diffuse, with a rapid northerly motion. After discovery, the comet slowly brightened as

it headed towards both the sun and Earth.

Northern Hemisphere observatories began observing this comet on January 30, when it was near magnitude 9.5 and possessed a coma 3 to 4 arcmin across. A maximum brightness of 8.5 was reached on February 7, and, since perigee came a few days later (0.8 AU), the total magnitude changed little through the remainder of the month as the comet continued to approach the sun. Perihelion came on the 26th (r= 0.81 AU), and the next evening, the coma was described as nearly 3 arcmin across with a 35-arcmin-long tail visible on photographs. On February 28, van Biesbroeck reported a diffuse sunward tail visible on his photographs. Thereafter, the comet faded and by the end of March, it was near magnitude 10. During mid-April, the comet reached magnitude 12 and by the end of the month, it was near 15. The comet reached its maximum northern declination of 81 degrees on May 10, and was last detected on the 24th, when the total magnitude had declined to 17.5. The orbit is elliptical with a period near 900 years.
(Johnson)

1936 I   Discovered: July 3, 1935   Discovery Magnitude: 14.5

This comet was discovered due to a peculiar chain of events. During July, van Biesbroeck had been following the asteroid 1143 Odysseus when he found an unidentified asteroid. He decided to follow this object to secure positions for an accurate orbit and on August 21, his photographs revealed a 14th-magnitude comet only 8 arcmin from the asteroid. He visually confirmed the comet's existence the next evening and described it as round, with a coma 20 arcsec across and a 15th-magnitude nucleus.

Then located 4.62 AU and 3.74 AU from the sun and Earth, respectively, the comet faded very slowly during the following months with the increasing distance from Earth overshadowing the decreasing distance from the sun. By mid-September, the total magnitude was 14.5 and by late November, it was near 15. The comet was last detected on December 21, with observations thereafter being affected by twilight. Also, during December, H. E. Wood (Union Observatory) found prediscovery images of the comet on plates exposed by Johnson on July 3, 22, 29 and August 5. On the latter date, the comet was actually measured shortly after the plate had been exposed, but its faintness did not allow its cometary character to be revealed and it was thought to be an asteroid.

The comet was recovered on February 28, 1936, as an object of magnitude 14.5, with a coma 15 arcsec across and a motion to the north; however, instead of brightening, the comet seems to have abruptly faded. Observations at the end of March and throughout April indicated a total magnitude of 16--nearly three magnitudes fainter than expected. Perihelion came on May 12 (r= 4.04 AU), and the comet was then described as near magnitude 16.5, with a coma 15 arcsec across. The total magnitude was estimated as 16 throughout June, July and August, and during the remainder of the year estimates were close to 16.5. Van Biesbroeck again observed the comet on January 13, 1937, as a 17th-magnitude object with a coma 6 arcsec across. With the comet located in the north circumpolar region, it was followed all year and as it approached its October opposition, it brightened to magnitude 16. By mid-November, the comet had faded to 16.5 and was located only 4 degrees from the north celestial pole. Observations continued only until January 26, 1938. The orbit is hyperbolic with an eccentricity of 1.0020.
(van Beisbroeck)

1936 II   Discovered: May 15   Discovery Magnitude: 9.0
     Peltier (Delphos, Ohio) discovered this comet in Cepheus and described it
as rather bright with a very slow motion to the southeast. The comet brightened
and increased in size during the next couple of months as it neared both the
sun and Earth. During mid-June, the total magnitude was near 8 and a tail was
measured as about 20 arcmin long. By the end of June, the comet had brightened to 7 and the tail was extending over one-half degree. Perihelion came on
July 9 (r= 1.10 AU), and observers described the comet as near magnitude 6.5,
with a tail slightly over one degree long. Visual observations by van Biesbroeck indicated a fan-shaped streamer was extending toward the sun from the
stellar nucleus. Although the comet moved away from the sun thereafter, its
distance from Earth began to decrease considerably and the comet continued to
brighten. The total magnitude reached 5 on July 23, and by the 26th, it had
increased to 4. Observers on the latter date indicated a visual tail 1 degree
long and a photographic tail over 2 degrees long. The coma was then about 9
arcmin across. On August 1, the total magnitude was near 3 and the coma was
measured on photographs as near 20 arcmin across. Perigee came on August 4
(0.13 AU), and the total magnitude was then near 2. The tail reached a maximum length of 6 degrees on the 6th. Fading was rapid thereafter, with estimates of 5 by the middle of August and 7 by month's end. During September,
the comet seemed to fade faster than predicted as, shortly after mid-month, it
reached magnitude 11. Soon afterward, the comet passed below the horizon for
observers in the Northern Hemisphere, however, observers in the Southern Hemisphere continued to photograph it into October. It was last detected on the
22nd, at a declination near -70 degrees, so that during its time of visibility
it traveled from one pole to the other. The orbit is elliptical with a period
near 1,550 years.
(Peltier)

1936 III   Discovered: July 17   Discovery Magnitude: 5.5
     This comet was first discovered by Kaho (Sapporo, Japan) shortly after
sunset and was described as diffuse, with a tail less than 1 degree long. Independent discoveries were made by Lis (Mt. Lubormirs, Poland) a few hours after Kaho on the 17th, and by Kozik (Ashkhabad, Russia) on the 19th. Lis did
not immediately announce his discovery until he could confirm the comet's
existence a few days later.
     Perihelion had occurred on July 16 (r= 0.52 AU), and, thereafter, the
comet rapidly faded as it moved away from both the sun and Earth. On July 18,
observers described the tail as 2 degrees long and estimated the total magnitude as 5.5. By the end of July, the total magnitude was near 6.5 and a short
tail was visible in twilight. The comet was observed until August 3, with further observations being temporarily halted due to twilight. The comet was recovered after conjunction with the sun on the morning of August 29, when van
Biesbroeck estimated the total magnitude as near 12. This observer managed
to follow the comet with the aid of photography until October 24, when the
total magnitude was near 15.5 and the coma was measured as 42 arcsec across.
The orbit is elliptical with a period near 890 years.
(Kaho-Kozik-Lis)

1936 V   Discovered: August 4, 1937   Discovery Magnitude: 13.0
     Edwin P. Hubble (Mt. Wilson Observatory, California) discovered this comet
photographically and described it as faint, with a coma diameter of 30 arcsec.

The comet was then located about 8 degrees north of Alpha Piscis Austrini and was moving to the southwest.

Perihelion had occurred on November 14, 1936 (r= 1.95 AU), and, at discovery, the comet was located 3.55 AU and 2.62 AU from the sun and Earth, respectively, with both distances increasing. Fading was subsequently slow, with van Biesbroeck estimating a total magnitude of 14 on September 14 and observers at Lick Observatory finding a magnitude of 15 when the comet was last detected on October 28. The orbit is elliptical with a period near 599 years. (Hubble)

1937 II   Discovered: February 27   Discovery Magnitude: 7.0

This fairly bright comet was found shortly after sunset by Wilk (Krakow, Poland) and, a few hours later, by Peltier (Delphos, Ohio). It was then described as 2 arcmin across, with a fuzzy nucleus and a well-defined tail. Having passed perihelion on February 22 (r= 0.62 AU), the comet faded as it slowly approached Earth. By mid-March, it was estimated as near magnitude 8, with a coma 6 arcmin across and, by the end of the month, it had faded to magnitude 9.5. Perigee came on April 1 (0.58 AU), and, afterwards, the comet faded rapidly. On April 9, the total magnitude was 11 and by the 12th, it was near 11.5. Interestingly, both van Biesbroeck (Yerkes Observatory) and Beyer (Hamburg Observatory) noted that the comet was 1 magnitude brighter than expected on April 11, after which it was back to normal 24 hours later. The comet was last detected on May 12, at a total magnitude of 15.5. The orbit is elliptical with a period near 589 years. (Wilk)

1937 IV   Discovered: February 4   Discovery Magnitude: 12.0

On February 14, Fred L. Whipple (Harvard College Observatory) was examining a plate he had exposed on the 7th when he discovered a faint comet with a nucleus and a short tail. A short time later, he found a prediscovery image on a plate exposed on February 4.

At discovery, the comet was located in Canes Venatici and was moving to the northeast. During the next few months the comet slowly brightened as it approached both the sun and Earth with magnitude estimates of 10 at the end of February, 9.5 in the middle of April and 9 at the end of May. Perihelion came on June 20 (r= 1.73 AU), and perigee occurred about the same time (1.27 AU). The comet was then about magnitude 9 and slowly faded thereafter. By mid-August, it was near magnitude 9.5. Curiously, between February and August, the comet seemed to almost constantly fluctuate in brightness. Beyer indicated four major peaks--May 13, June 5, June 17 and July 15--and van Biesbroeck independently gave magnitude estimates averaging near one-half magnitude brighter than normal near the same dates. During this same time, the nucleus remained near magnitude 12, the coma was estimated as between 3 and 4 arcmin across and the tail reached a maximum length of 20 arcmin. Observations continued after August, and the comet was finally detected for the final time on October 28, when the total magnitude was near 14. The orbit is hyperbolic with an eccentricity of 1.0001. (Whipple)

1937 V   Discovered: July 4   Discovery Magnitude: 7.0

Professor Finsler (University of Zurich, Switzerland) discovered this comet in the morning sky near Rho Persei and described it as bright and diffuse, with no indication of a tail. With it approaching both the sun and Earth, the bright-

ness rapidly increased and observers reported magnitudes near 6.5 on July 9,
6 on July 15 and 5 on July 27. Photographs beginning on July 16 revealed a
complex multiple tail over 1 degree long and, according to van Biesbroeck,
this "complex bundle of streamers...changed considerably from day to day."
The total magnitude reached a maximum of 4.2 on August 4 and 5, and, thereafter, the comet faded very slowly for the next couple of weeks. Photographs
then indicated a tail 6 to 8 degrees long, although wide-field photos taken
by Quenisset (Juvisy, France) on August 3 and 6 gave a length of nearly 20
degrees. Perigee occurred during the first half of August (0.6 AU), and
perihelion came on the 16th (r= 0.86 AU). Thereafter, the comet faded and,
in early September, was estimated as near magnitude 6. It was last observed
on September 23, with observations thereafter being temporarily halted due to
evening twilight. After conjunction with the sun, the comet was detected for
a final time on December 30. The orbit is elliptical with a period between
37,800 years and 13.8 million years.
(Finsler)

1939 I   Discovered: January 17   Discovery Magnitude: 8.0
     This comet was independently discovered by Kozik (Tashkent Observatory,
Russia) and, 2 days later, by Peltier (Delphos, Ohio). The latter observer
was using a 15-cm refractor and described the comet as fairly bright with a
nucleus and a tail length of over 1 degree. The comet was then located near
the meeting of Cygnus, Pegasus and Vulpecula and was moving to the southeast.
     The comet brightened rapidly as it neared both the sun and Earth, with
total magnitude estimates being near 7 on January 25 and 6 on January 31. On
the latter date, the coma was estimated as 3 arcmin across. Perihelion came
on February 7 (r= 0.72 AU), and the comet was described as near magnitude 6,
with a photographic tail extending over 3 degrees. Perigee came on February
12 (0.54 AU), and the comet was then near magnitude 6.2 and possessed a faint,
narrow sunward tail extending 3 to 5 arcmin on photographs. Fading was rapid
thereafter, with the comet nearly reaching magnitude 7.5 on February 20; however, on the 21st, observers reported a total magnitude near 5.5 as the comet
underwent an outburst. By March 1, the comet had faded to near 7th magnitude,
but 5 days later, a second outburst--this time increasing by near 1.5 magni-
tudes--occurred. No further activity was noted during March, and, by the
end of the month, the total magnitude was down to 8.5. Observations during
March were made exclusively by observers in the Southern Hemisphere and they
kept the comet under observation until April 21, when it was located in Pictor.
The orbit is elliptical with a period near 1,770 years.
(Kozik-Peltier)

1939 III   Discovered: April 15   Discovery Magnitude: 3.0
     Having passed perihelion on April 10 (r= 0.53 AU), this comet first ap-
peared as a naked-eye object of magnitude 3, with a tail 4 degrees long. Several
independent discoveries were made within the first five days of visibility. The
first two were made by Jurlof (Votkinsk, Russia) and Achmarof (Balezino, Russia)
almost simultaneously. They were followed 10 hours later by Lewis V. Smith
(Edmonton, Canada) and, after an additional 8 hours, by Hassel (Oslo, Norway).
E. W. Barlow (England) and E. Buchar (Prague, Czechoslovakia) discovered the
comet on the 18th, and C. L. Friend (Escondido, California) and Kozik (Ashkhabad,
Russia) found it on the 19th. Several more independent discoveries followed
during the next few days.

The comet slowly faded after discovery as it moved away from both the sun and Earth. Observers estimated the total magnitude as near 4 on April 20, 5 on April 25, and 6 on April 30. On the 20th and 21st, the tail was visually estimated as 6 degrees long, while photographs revealed it closer to 20 degrees in length. Curiously, the comet seems to have undergone a great change between the 21st and 22nd. This was best indicated by a series of photos taken between April 19 and 25, by Dr. R. L. Waterfield (Headley, England), which showed a tail extending 9 degrees on the 21st, and less than 1 degree on the 22nd. Other observers confirmed the change visually and van Biesbroeck noted "rapid changes in the inner part of the tail" between photos taken on the 20th and 23rd. The comet faded to a magnitude of 7 by May 12, and 8 on the 21st. Thereafter, observations became more difficult as the comet moved back into evening twilight and it was last detected on May 27. The orbit is elliptical with a period near 6,500 years.
(Jurlof-Achmarof-Hassel)

1940 IX   Discovered: November 1   Discovery Magnitude: 9.0
Clarence L. Friend (Escondido, California) discovered this comet in the evening sky just above Zeta Herculi. He described it as a faint, round object low in the southwestern sky. Perihelion came on November 5 (r= 0.95 AU), and, due to Friend's delayed announcement, observations at other localities did not begin until November 6, when van Biesbroeck (Yerkes Observatory) described it as 3 arcmin across with a total magnitude of 9.2. Due to the comet's decreasing distance from Earth after discovery, the total magnitude changed little during the next two weeks as it moved away from the sun. Observations consistently placed the total magnitude as near 9 and indicated a coma diameter of 4 arcmin. In addition, several observers detected a star-like nucleus between magnitude 12 and 13. Perigee came on November 20 (0.75 AU), and, thereafter, the comet began to fade more rapidly. During the first days of December, it had dropped to 10 and van Biesbroeck commented that the coma had "become more diffuse and less condensed in a central nucleus." He gave the coma diameter as 6 arcmin on the 9th. By December 31, the total magnitude had declined to 13 and, when last detected on January 9, 1940, van Biesbroeck described it as "a vague nebulosity not brighter than 15th magnitude."
(Friend)

1940 III   Discovered: September 30   Discovery Magnitude: 9.0
Sigeki Okabayasi (Kurasiki, Japan) discovered this comet in the morning sky, while searching for comets with a 7.6-cm telescope. A rapidly approaching sunrise prevented him from detecting any motion and cloudy weather prevented him from again seeing the comet until October 3. On that morning, he found the comet to have moved to the north, and described it as faint, with a nucleus and a trace of tail. Also, on October 3, Minoru Honda (Seto, Japan) made an independent discovery.

Having passed perihelion on August 16 (r= 1.06 AU), the comet faded quickly after discovery, although it was slowly approaching Earth. On October 4, the total magnitude was near 11 and by the end of the month, it was near 13. Observations continued until November 8, when observers estimated the total magnitude as between 13.5 and 14, with a coma 1 arcmin across. Further observations during November and early December were fruitless, partly because of moonlight, but mostly due to the comet's abrupt fading as it passed perigee soon after mid-November. Van Biesbroeck (Yerkes) finally managed to recover

it on December 21, when he described it as a diffuse 17th-magnitude object on
photographs. Jeffers (Lick Observatory) photographed the comet on January 2,
1941, and described it as very small and faint. The comet was last detected
on January 3, when van Biesbroeck estimated the total magnitude as 17.5. The
orbit is hyperbolic with an eccentricity of 1.0015. The comet may be related
to a meteor shower which was observed in late January of 1932 and 1935.
(Okabayasi-Honda)

1940 IV   Discovered: July 28   Discovery Magnitude: 10.5

During September, Whipple (Harvard College Observatory) discovered a comet
on a Harvard plate exposed on the Aquila region on August 8. He described it
as near magnitude 10.5, with a round coma and a movement to the southwest.
Shortly thereafter, he detected further images on plates exposed on July 29
and August 5. On September 30, he announced the discovery and included a
detailed ephemeris, which indicated the comet had moved too far south for observation in the Northern Hemisphere shortly after mid-August. Upon receiving
the announcement, J. Bobone (Cordoba, Argentina) and B. Dawson (La Plata, Argentina) obtained accurate positions on October 4. The comet arrived at perihelion on October 8 (r= 1.08 AU), and was then independently discovered by Dr.
J. S. Paraskevopoulos (Bloemfontein, South Africa) as a 10th-magnitude object.
Later, after receiving Whipple's ephemeris, he also detected the comet on a
plate he had exposed on August 25. With a more accurate orbit at hand, observers at Sonneberg, Germany, detected faint images on plates exposed on July 28,
August 1 and August 4. Dawson was the only observer to follow the comet as it
faded and he last detected it on January 1, 1941. Although he did not provide
magnitude estimates, the comet must have been near magnitude 15 when last seen.
The orbit is elliptical with a period near 425 years.
(Whipple-Paraskevopoulos)

1941 I   Discovered: August 25, 1940   Discovery Magnitude: 13.0

Leland E. Cunningham (Oak Ridge station of the Harvard College Observatory)
discovered this comet on a patrol plate exposed on September 5, 1940, in the
Cygnus region. He described it as near magnitude 13, with a motion to the
south. Later, he found further images on 11 patrol plates taken at Oak Ridge
and Cambridge between August 25 and September 15, 1940.

The comet brightened rapidly as it approached both the sun and Earth, and
by the end of October, it had reached a magnitude of 8 and possessed a coma 3
arcmin across. A 20-minute exposure on November 1 by van Biesbroeck (McDonald
Observatory, Texas) with the 208-cm reflector revealed a "broad and somewhat
dissymmetrical tail...visible over a length of 8 arcmin." The comet's perfor-
mance up to this point sparked excitement in the astronomical community that
the comet would be a spectacular object at the end of the year and this fore-
cast received much print in the daily press; however, in mid-November, the
brightening abruptly slowed and by the end of the month, the total magnitude
was near 6, with a coma 8 arcmin across. By mid-December, the comet had almost
brightened to magnitude 5 and possessed a coma between 12 and 19 arcmin across.
Photographs then revealed a double tail with lengths of 6 degrees and 42 arcmin.
On January 2, 1941, van Biesbroeck estimated the total magnitude as 3.5--nearly
3 magnitudes fainter than predictions based on the pre-November observations--
and his photographs revealed a tail 15 to 20 degrees long and a coma diameter
near 13 arcmin. Observing conditions rapidly deteriorated thereafter, and the
comet was last seen in evening twilight on January 5.

The comet was in conjunction with the sun thereafter, and passed perihelion on January 16 (r= 0.37 AU). At about the same time, the comet was nearest
Earth (0.6 AU). The comet was recovered on January 21, by Bobone (Cordoba),
who described the total magnitude as 3. Observations were possible from only
Southern Hemisphere observatories and the comet was kept under observation for
the next few months as it faded away. Observers after May 28, reported the
comet to have become more and more diffuse, and observations continued until
June 17, when the total magnitude had dropped to nearly magnitude 14.* The
orbit is hyperbolic with an eccentricity of 1.0005.
(Cunningham)

1941 II   Discovered: December 31, 1940   Discovery Magnitude: 10.0
    Friend (Escondido, California) discovered this rather inconspicuous object
in Lacerta on January 17, 1941, while using a 12.7-cm refractor. Independent
discoveries were later made by Reese (Uniontown, Pennsylvania) on the 18th, and
Honda (Tanokami, Japan) on the 21st. Orbital calculations later helped to locate a prediscovery image of the comet on a plate exposed by Harvard College
Observatory on December 31, 1940.
    Perihelion came on January 20 (r= 0.94 AU), and the comet brightened dur-
ing the next month as it approached Earth. On January 20, the comet was described as near total magnitude 9.5, with a coma diameter of 3 arcmin and a
short, stubby tail. At the beginning of February, the comet was slightly brighter than magnitude 9 and possessed a coma diameter of 4 arcmin. Perigee came
on February 18 (0.15 AU), and observers then described the total magnitude as
near 7, with a coma diameter of nearly 10 arcmin. Fading was rapid thereafter,
and the comet was last detected on March 3, when the total magnitude had dropped
to 12. Moonlight interfered for the next several days and photographs taken
after mid-March showed no trace of the comet. The orbit is elliptical with
a period near 355 years.
(Friend-Reese-Honda)

1941 IV   Discovered: January 15   Discovery Magnitude: 5.8
    On the morning of January 15, R. P. de Kock (Paarl, South Africa) was ob-
serving the variable star R Lupi when he discovered this comet at a total magnitude of 5.8. The comet brightened rapidly during the following days and, beginning on January 23, brought a flurry of independent discoveries. Paraskevopoulos (Bloemfontein, South Africa) found it on the 23rd, and described it as
near magnitude 3.5 with a tail 5 degrees long. That same morning, R. Grandon
(Santiago, Chile) also found it. On January 24, at least six more observers
accidentally found the comet and reported the total magnitude as near 3.
    Perihelion came on January 28 (r= 0.79 AU), and the comet was at perigee
on the 30th (0.26 AU). On the latter date, the total magnitude was near 2 and
the tail extended more than 6 degrees. The comet traveled rapidly northward
and moved into the evening sky for observers in the Northern Hemisphere during
the first days of February. On the 2nd, Friend (Escondido, California) estimated the tail as 20 degrees long and on the 4th, van Biesbroeck (Yerkes) estimated the total magnitude as near 3. After the full moon on February 12, the
comet had faded, but was still well observed. By the 19th, it had dropped to
magnitude 5 and by March 16, it was near 8. The comet was last detected on
March 29, before it became lost in evening twilight. It was recovered on July
4 by van Biesbroeck at a magnitude near 15 and was followed until September 17,
when Jeffers (Lick Observatory) estimated the magnitude as 17. The orbit is

elliptical with a period near 880 years.
(de Kock-Paraskevopoulos)

1941 VIII   Discovered: May 27   Discovery Magnitude: 11.0
     H. van Gent (Union Observatory, South Africa) discovered this comet in
Corona Australis and described it as faint, with a central condensation and a
short tail. Wartime conditions delayed word of the discovery and the comet was
independently discovered on June 16, by Giovanni Bernasconi (Bologna, Italy) and
Hobart (Newcastle, Australia), both of whom described the comet as near magnitude 9.
     After discovery the comet moved to the northwest and brightened as it
approached both the sun and Earth. By the end of June, observers reported a
total magnitude near 8.5 and the comet was also at perigee (0.6 AU). Thereafter, the comet still brightened rapidly to magnitude 8 on July 12, and 7 by
July 24. At that time, and for the next two months, the increasing distance
from Earth perfectly countered the decreasing distance from the sun and the
comet remained near magnitude 7, with a coma 3 to 4 arcmin across and a tail
near 1 degree long. Perihelion came on September 3 (r= 0.87 AU), and the total
magnitude slowly decreased as the comet-Earth distance again began to decline.
At the end of September, the total magnitude was near 8 and by the end of
November it had finally reached 9. Thereafter, with both the sun and Earth
distances increasing, the comet faded rapidly. On December 18, van Biesbroeck
estimated the total magnitude as 11 and by January 21, 1942, it had dropped to
13. The comet was last observed on February 18, and van Biesbroeck then estimated the magnitude as 16.5 and measured the coma as 12 arcsec across. The
orbit is hyperbolic with an eccentricity of 1.0002.
(van Gent)

1942 IV   Discovered: December 28, 1941   Discovery Magnitude: 10.0
     Whipple (Harvard College Observatory) discovered this comet on patrol plates
exposed on January 25, 1942. Upon announcing his discovery, he described it as
near magnitude 10, with a central condensation and a short tail. Whipple soon
detected prediscovery images on plates exposed on December 28, 1941, and January 17, 1942. Located in Coma Berenices at discovery, the comet moved to the
southwest and rapidly brightened as it neared both the sun and Earth. Independent discoveries were subsequently made where wartime conditions had prevented
the distribution of Whipple's announcement. Bernasconi (Como, Italy) and G.
Kulin (Budapest Observatory, Hungary) found it on February 11, and Antonin
Becvar (Skalnate Pleso Observatory, Czechoslovakia) discovered it on February
18. On the former date, the comet's magnitude was near 8 and by the 18th, it
had brightened to 7.
     Perigee came on February 28 (0.49 AU), and throughout March, the comet
remained at 6th magnitude as the increasing distance from Earth countered the
effects of a decreasing distance from the sun. The comet was subsequently discovered on the 17th by D. du Toit (Bloemfontein, South Africa). Also during
March, the comet displayed two tails on photographs. The brighter one, which
was also visible to visual observers, nearly attained a length of 1 degree and
changed little in appearance; however, the fainter, photographic tail underwent several changes and, at times, even disappeared. The comet was observed
until April 19, with twilight interfering thereafter. It had then faded to a
magnitude of 8, as its distance from Earth began increasing more rapidly.
     Perihelion came on April 30 (r= 1.45 AU), and the comet was recovered in

the Southern Hemisphere in late May. It was kept under observation throughout
the remainder of the year and was last detected on January 8, 1943. The orbit
is hyperbolic with an eccentricity of 1.0009.
(Whipple-Bernasconi-Kulin)

1942 VIII   Discovered: February 12   Discovery Magnitude: 15.0
    Dr. Liisi Oterma (Turku, Finland) discovered this comet on a photograph
exposed on the Leo region. She described it as faint and diffuse, without
condensation. Traveling in a retrograde orbit of small inclination, the comet
slowly retrogressed among the ecliptic constellations. Observations elsewhere
were few, due to the temperamental wartime telegraph, however, Dr. Hirose
(Tokyo Observatory, Japan) was able to detect it at magnitude 14 on February
20. Observations continued at Turku until April 20, when the comet was located
in Cancer. Perihelion came on September 27 (r= 4.11 AU), and word of the comet
finally reached the United States in late October. Dr. Whipple then computed
a rough ephemeris, but although the comet should have then been bright enough
for large telescopes, it was not located due to the uncertain position.
(Oterma)

1943 I   Discovered: November 5, 1942   Discovery Magnitude: 10.0
    Whipple (Harvard College Observatory) discovered this comet on a plate
exposed on December 8. Located near the Gemini-Cancer border, the comet was
described as faint and diffuse, with a motion to the east-northeast. Independent discoveries were made by Fedtke (Königsberg) on December 11, and Tevzadze
(Abastumani, Russia) on photographs exposed on December 15. Both observers
estimated the total magnitude as 8. Whipple later detected prediscovery images
of the comet on 20 patrol plates exposed between November 5 and December 8.
    The comet developed rapidly as it approached both the sun and Earth. Van
Biesbroeck (Yerkes) described it on December 13 as having a total magnitude of
7.6, a coma diameter of 7 arcmin and a narrow tail 42 arcmin long. By January
12, he found the magnitude to have increased to 5.4, the coma to have grown to
12 arcmin and the tail to have lengthened to over 1 degree. At the end of
January, the comet arrived at perigee (0.4 AU) and also attained its expected
maximum brightness of 4. Perihelion came on February 7 (r= 1.35 AU), and the
comet faded as it moved away from both the sun and Earth. On February 8, it
had reached magnitude 5 and possessed a tail at least 6 degrees long. About
mid-February, the comet suddenly brightened, and van Biesbroeck reported the
total magnitude as slightly brighter than 4 between February 16 and 28. During
this same time, photographs taken at Sonneberg (Germany) indicated a tail length
of 15 degrees and a coma diameter of nearly 30 arcmin. By March 8, the comet
had faded to 4.3 and continued to decline until March 25, when observers reported another outburst which amounted to over 1 magnitude. By April 9, van
Biesbroeck reported it to have declined to magnitude 5.3 and estimated the
coma as 6 arcmin across. Van Biesbroeck was then visiting McDonald Observatory
(Texas) and on April 3, he reported a secondary nucleus on photographs obtained
with the 208-cm reflector. Later, an examination of the plates revealed the
secondary nucleus on photos taken on March 31, April 1, 2, 3 and 9. During
this time it increased in separation from 9.8 arcsec to 16.2 arcsec and also
grew more diffuse. This data indicate a probable split date of March 9.
    The comet underwent two more 1-magnitude outbursts in brightness on April
23 and May 23, but, beginning in June, it faded as expected with no further
notable events. At the end of June, the total magnitude was near 11 and by the

end of July, it was near 15. The comet was last detected on August 2, when van
Biesbroeck described it "as a vague nebulosity of magnitude 16." The orbit is
elliptical with a period near 2,300 years.
(Whipple-Fedtke-Tevzadze)

1943 II   Discovered: September 3   Discovery Magnitude: 8.0
     This comet had already passed perihelion on August 21 (r= 0.76 AU) when
discovered by Daimaca (Bucharest, Rumania) on September 3. It was then described as bright, with a tail and a rapid motion northward. Delays due to the
war prevented the announcement from being sent to the United States until after
September 10, and then, due to relays via Copenhagen and Zurich, it did not
arrive until September 17. By that time the rapid daily motion of more than
3 degrees per day had carried it far from the September 10 position given in
the radio message and searches revealed nothing. On September 19, Peltier
(Delphos, Ohio) independently discovered the comet near Iota Draconis and gave
the rate of motion as nearly 5 degrees per day. His estimated magnitude of
10.5 indicated a very rapid fading. Peltier remained the only observer in the
United States and his observations ended on September 23, when the comet had
moved to a point near Epsilon Herculis and had faded to a magnitude of 13. One
month later, a rough orbit based on European observations was received by van
Biesbroeck (McDonald Observatory), but a photograph exposed on October 24, which
showed stars as faint as magnitude 18, failed to show the comet. The failure
was attributed to the rapid fading.
(Daimaca)

1944 I   Discovered: November 27, 1943   Discovery Magnitude: 9.0
     Van Gent (Union Observatory, South Africa) discovered this comet in Puppis
and described it as diffuse and somewhat faint, with a rapid motion to the southwest. Upon receiving word of the discovery, Henry L. Giclas (Lowell Observatory, Arizona) observed the comet on December 3, at a low altitude in southern
Puppis. He indicated a rate of motion faster than van Gent's estimate and
could not detect the comet the next evening, since it was then below the horizon.
After the comet reached a declination of -52 degrees on December 8, it began to
move northward. Perigee came on the 10th (0.25 AU), but no reports of observation have ever been made. Around mid-December, the comet was near magnitude
6 and had moved far enough north to allow several independent discoveries.
Daimaca (Bucharest, Rumania) was the first to find it. His observation was
made on December 16, when the comet was located in Aquarius. Peltier (Delphos,
Ohio) discovered it on the 17th, in Pegasus, Kellaway (England) found it on the
19th, and Finsler (Switzerland) detected it on the 24th. On the latter date,
van Biesbroeck (Yerkes) described it as near magnitude 8.5, with a coma 5 arcmin
across and a slender tail 1 degree long. By January 5, 1944, the brightness had
declined to 9. Perihelion came on the 12th (r= 0.87 AU), and van Biesbroeck
obtained the final observation on January 24, when the comet appeared as a
vague nebulosity of magnitude 15.5. The orbit is generally given as a parabola, but it may be hyperbolic with an eccentricity near 1.003.
(van Gent-Peltier-Daimaca)

1944 IV   Discovered: May 23   Discovery Magnitude: 12.0
     Van Gent (Union Observatory) discovered this comet in Vela and described
it as faint and diffuse, with a motion to the northeast. Though the announcement was delayed, it did arrive in time for observers at Lick Observatory (California) to obtain several observations in July, before the comet became lost

in the sun's glare. They reported a brightness near magnitude 11. Perihelion
came on July 18 (r= 2.23 AU), and the comet remained in the sun's glare until
late December, when it appeared in the morning sky as a very faint object located near the Virgo-Boötes border. Observations continued into 1945, with the
comet slowly fading from a magnitude of 15.5 on January 12 to 16.0 by March
8, as its distance from Earth slowly decreased from 3.2 AU to 3.1 AU. Thereafter, it faded more rapidly as the distances from both the sun and Earth increased, and the comet was last detected on August 11, at a magnitude near 19.
The orbit is hyperbolic with an eccentricity of 1.0021.
(van Gent)

1945 I   Discovered: April 18, 1944   Discovery Magnitude: 14.5
    Y. Väisälä (Turku, Finland) discovered this comet during a regular pro-
gram of studying asteroids. He described it as faint with a coma diameter of
12 arcsec. Then located in Virgo at distances of 3.60 AU and 4.40 AU from the
sun and Earth, respectively, the comet moved slowly southward on its way to a
1945 perihelion. On May 22, Hoffmeister (Sonneberg Observatory, Germany) estimated the total magnitude as 14. The discovery announcement was lost on its
way to the United States, but on May 20, a short ephemeris of a "new Väisälä
comet" was received which gave the predicted positions for May 14, 22 and 30.
On June 4 and 13, G. H. Herbig (Lick Observatory) searched near the extrapolated positions, but detected nothing of a cometary nature. A later revised
ephemeris allowed van Biesbroeck (Yerkes) to locate the comet on August 7, as
a faint diffuse spot of magnitude 13.5. Then at low altitude, and nearing the
sun's glare, the comet was last observed on August 15, as Herbig described it
as near magnitude 14, with a coma 15 arcsec across.
    The comet remained lost in the sun's glare for the remainder of the year.
Perihelion came on January 4, 1945 (r= 2.41 AU), and, thereafter, the comet
began to move slowly northward. After several futile attempts, the comet was
finally recovered in August, when van Biesbroeck photographed it on the 7th,
8th and 9th, as a diffuse spot near magnitude 15.5. He added that the comet
was not well condensed and was located nearly 1 degree from the predicted
positions. Van Biesbroeck remained the only observer of the comet from then
on. During the first week of September, he described it as very diffuse with
a magnitude near 15.5 and on October 10, he described it as round and diffuse,
with a coma diameter of 15 arcsec and a total magnitude near 16. Afterwards,
van Biesbroeck traveled to McDonald Observatory (Texas) where he continued
observations with the 208-cm reflector. Between November 30 and December 4,
he described it as 12 arcsec across with a magnitude of 16.5 and when last
observed on January 1 and 2, 1946, it was 10 arcsec across and near magnitude
17. The orbit is elliptical with a period near 7,100 years.
(Väisälä)

1945 III   Discovered: June 11   Discovery Magnitude: 10.0
    Having passed perihelion on May 17 (r= 1.00 AU), this comet was discovered
by du Toit (Bloemfontein) on a photograph exposed on June 11. Then located
just east of Beta Ceti, the comet moved southwestward and was described as
faint. The comet approached Earth to within 0.3 AU on June 30, but seems not
to have brightened much over the discovery magnitude. On that date, it also
reached its maximum southern declination of -69 degrees and thereafter began
moving northward. Observations ended on July 15, and the comet was never detected by Northern Hemisphere observers.
(du Toit)

1945 VI   Discovered: November 22   Discovery Magnitude: 7.0
     Friend (Escondido, California) discovered this comet in the evening sky
near Zeta Herculi and, two days later, Peltier (Delphos, Ohio) made an independent discovery. Both observers described the comet as bright, with a nucleus and a tail. Evening twilight made observations difficult as the comet
neared the sun, although Giclas (Lowell Observatory) managed to keep it under
observation from November 27 until December 7, when it had brightened to magnitude 6. Observations thereafter were impossible due to the comet's nearness
to the sun. Perihelion came on December 17 (r= 0.19 AU), and the comet remained in the sun's glare until May 1946. By that time, the comet had faded
considerably and, since its orbit was not known with great accuracy, its
exact position was impossible to find.
(Friend-Peltier)

1945 VII   Discovered: December 11   Discovery Magnitude: 7.0
     Du Toit (Bloemfontein) discovered this comet on a photograph exposed near
the Circinus-Triangulum Australe border and described it as bright and diffuse.
Further photographs were obtained on the following 4 days, but, thereafter,
the comet had become a difficult object as it headed rapidly towards the sun.
Perihelion came on December 28 (r= 0.0075 AU), and although the comet could
have been observed during the first half of January, no observations were made.
This comet is a member of the sungrazing family of comets.
(du Toit)

1946 I   Discovered: January 23   Discovery Magnitude: 9.0
     Matthew Timmers (Vatican Observatory, Italy) discovered this comet on the
edge of a photographic plate he had exposed on February 2. The comet was then
situated in Ursa Major and was described as somewhat bright with a well-marked
central condensation in a coma 2.5 arcmin across. A short tail was also visible. Soon afterward, Whipple (Harvard College Observatory) found prediscovery images on patrol plates exposed on January 23, 24, 28 and 29.
     The comet moved northwest after discovery and although approaching the
sun, it faded very slowly due to an increasing distance from Earth. The most
precise estimates of the comet's brightness came from van Biesbroeck and S. K.
Vsekhsvyatsky (Kiev Observatory, Russia). Their estimates during the remainder
of February indicate virtually no change in brightness, but as the fading began
in March, both men indicated several distinct variations in brightness. The
first came on March 2 and 3, when the comet was found to be nearly one-half
magnitude brighter than the previous days' estimates of magnitude 9. A similar
event occurred on the 18th, but a 1-magnitude flare occurred on the 27th, just
as the comet had reached magnitude 10. On this latter date, van Biesbroeck
photographed a complex tail structure which consisted of three distinct tails
extending up to 20 arcmin from the nucleus. By April 4, he reported "nothing
left of the complex tail except a short stub 3 arcmin long." Perihelion came
on April 13 (r= 1.72 AU), and on the 21st, a final outburst of at least one
magnitude occurred. Thereafter, the comet faded without further events. At
the end of May, when passing within 5 degrees of the north celestial pole, observers indicated a total magnitude near 10.5. This brightness declined to
12 by mid-August, 13.5 by early September, and 15.5 by the end of October. The
comet was lost in the sun's glare thereafter, but in 1947, Jeffers (Lick Observatory) managed to photograph it on June 24, June 25, July 13 and August 9, when
near opposition. The total magnitude was then near 19.4. The orbit is hyper-

bolic with an eccentricity of 1.0012.
(Timmers)

1946 II   Discovered: May 30   Discovery Magnitude: 6.0
     Having passed perihelion on May 11 (r= 1.02 AU), this comet rapidly bright-
ened thereafter, as it neared a very close approach to Earth. The first to
detect it was Miss Ludmilla Pajdusakova (Skalnate Pleso Observatory, Czechoslovakia) on May 30. She was followed 6 hours later by David Rotbart (Washington D. C.) and 24 hours later by Anton Weber (Berlin, Germany). Although
Pajdusakova reported the comet as an 8th-magnitude object, the latter two
observers estimated it as near magnitude 6. Other observers during the next
few days confirmed the estimate of magnitude 6 and added observations of a
10th-magnitude nucleus and a 1-degree long tail.
     Perigee came on June 2 (0.16 AU), and the comet's retrograde motion then
carried it across the sky at a rate of 14 degrees per day. During the next
few days it remained near magnitude 6 and estimates of the tail length varied
from 1 to 7 degrees; however, fading was rapid thereafter, with estimates of
7 on June 7, 9.5 on June 18 and 12.5 on July 2. Observations continued until
July 29, when Jeffers (Lick Observatory) photographed it as a diffuse object
near magnitude 17.
(Pajdusakova-Rotbart-Weber)

1946 VI   Discovered: August 6   Discovery Magnitude: 9.0
     Albert F. Jones (Timaru, New Zealand) discovered this comet while locat-
ing the variable star U Puppis. He described it as somewhat bright, with a
centrally condensed coma 1 arcmin across. Then located 1.69 AU and 2.42 AU
from the sun and Earth, respectively, the comet brightened as both distances
decreased; however, the southeast motion allowed only observations from Southern Hemisphere observers. The comet was observed until October 3, before it
temporarily vanished in the sun's glare. It was then near magnitude 7.
     Perihelion came on October 26 (r= 1.14 AU), and the comet was recovered
on January 3, 1947, by a remarkable observation made by Giclas (Lowell Observatory). With the total magnitude near 10, Giclas located it at a very low altitude when only 23 degrees from the sun. Fading was slow as the comet moved
away from both the sun and Earth and it was well observed as it moved northeastward. From late March until the end of the year, the comet was well observed as it faded from a total magnitude of 10.5 to 17 and shrank from a coma
diameter of 1.5 arcmin to 0.5 arcmin. One distinguishing feature of this comet
was the persistence of its tail, which by the end of 1947 was still extending
about 1 arcmin. Further observations were scarce until the summer of 1948, when
van Biesbroeck and Jeffers both made observations when the comet was near opposition. During July, both observers indicated a total magnitude near 17.5
and both detected a fan-shaped tail nearly 1 arcmin long. Van Biesbroeck last
observed the tail during October, when the comet was located 7.7 AU from the
sun. The comet was last detected by Jeffers on November 23, 1948, when the
total magnitude was estimated as 19.3. The orbit is hyperbolic with an eccentricity of 1.0008.
(Jones)

1947 I   Discovered: October 31, 1946   Discovery Magnitude: 10.0
     Michiel Johann Bester (Bloemfontein, South Africa) discovered this comet on
a plate taken with the 7.6-cm Ross-Fecker camera. He described the comet as
faint and round, with a coma diameter of 2 arcmin. Located 2.64 AU and 2.06 AU

from the sun and Earth, respectively, the comet slowly brightened until it reached its maximum brightness of 9 in December. At the same time, it developed a tail 6 arcmin long. Thereafter, as the distance from Earth increased, the comet slowly faded.

Perihelion came on February 7 (r= 2.41 AU), and observers indicated a total magnitude between 10 and 10.5 and a tail 3 arcmin long. From then until late July, the comet rapidly faded to magnitude 15 as the distances from both the sun and Earth increased; however, in July, the comet-Earth distance began to decrease as opposition approached and for the next 2 months the brightness remained virtually unchanged. The comet was again fading in December, and by January 2, 1948, it was estimated as near 16. Few observations were made thereafter, with Jeffers (Lick Observatory) estimating a magnitude near 18.7 on July 27, 1948, and van Biesbroeck (McDonald Observatory) finding a total magnitude near 18.5 when he last observed the comet on October 2. On the latter date, the coma was measured as near 20 arcsec across. The orbit is hyperbolic with an eccentricity of 1.0009.
(Bester)

1947 III   Discovered: March 27   Discovery Magnitude: 9.0

During a routine search for comets using 20x100 binoculars, Antonin Becvar (Skalnate Pleso Observatory, Czechoslovakia) discovered this comet only 8 degrees from the north celestial pole. He described it as diffuse and 4 arcmin across, with little condensation.

After discovery, the comet continued northward and passed very near the north pole on March 31--the date of perigee (0.65 AU). Thereafter, the comet moved southwards and, although approaching the sun, it slowly faded. Perihelion came on May 4 (r= 0.96 AU), and the comet was approaching evening twilight, which prevented further observations after May 8. The magnitude on the latter date was nearly 12.
(Becvar)

1947 IV   Discovered: March 24   Discovery Magnitude: 10.5

Esteban Rondanina (Montevideo Observatory, Uruguay) was examining photographic plates on March 26 when he found a comet on a plate of the Kappa Crucis region taken two days earlier. He described it as near magnitude 11, with no visible nucleus. That evening, Bester (Bloemfontein) exposed a patrol plate which acquired him an independent discovery. He estimated the magnitude as near 10.

After discovery, the comet moved northward and brightened rapidly as it approached both the sun and Earth. Shortly before mid-April, it reached magnitude 6 and, thereafter, brightened more slowly as the comet-Earth distance began to increase. Perihelion came on May 21 (r= 0.56 AU), and observers then described it as near magnitude 4.5, with a bright, diffuse central condensation and a slender tail several degrees long. In addition, during the solar eclipse on May 20, van Biesbroeck (then on an eclipse expedition to Brazil) saw the comet with the naked eye. The comet began to fade soon afterward, and by the end of June, it was near magnitude 8. Observations finally ended on September 21, when Jeffers (Lick Observatory) estimated the total magnitude as near 18. The orbit is elliptical with a period near 3,200 years.
(Rondanina-Bester)

1947 V   Discovered: May 18   Discovery Magnitude: 11.0

Bester (Bloemfontein) discovered this comet on May 19, while examining a

photographic plate taken of the Lupus region the night before. He described
it as faint and diffuse, with no apparent central condensation and a rapid
movement to the north.

Then located 1.42 AU and 0.47 AU from the sun and Earth, respectively, the
comet brightened slightly as it passed both perigee (0.4 AU) and perihelion (r=
1.40 AU) by the end of May. Thereafter, the comet faded and after the first
week of June, observers estimated the total magnitude as near 12. Around this
time van Biesbroeck (Yerkes) detected a tail 2 arcmin long, while using the
61-cm reflector, and W. H. Steavenson (England) described the comet as unusually diffuse without any marked boundary, while using a 76-cm reflector. By July
7, van Biesbroeck described the comet as a "small condensation in a very faint
coma; the total light corresponded to a stellar magnitude of 15." Observations
continued until August 7, when Jeffers (Lick Observatory) estimated the total
magnitude as 17.3. This may have been the last observation of the comet as it
entered the sun's glare; however, a photo exposed by Jeffers on the predicted
position of the comet on November 25 showed a poorly outlined object whose
identity with 1947 V is still in question.
(Bester)

1947 VI   Discovered: July 18   Discovery Magnitude: 12.0
    Carl A. Wirtanen (Lick Observatory) found this comet on July 23, as a trail
on a plate exposed on July 18 for the special proper motion program. He described it as small and diffuse, with a nearly stellar condensation and a tail
3 arcmin long. Observers during the next few days generally agreed with Wirtanen's initial descriptions as they estimated the total magnitude as 12 to 13,
the tail length as 3 arcmin long and the coma diameter of 12 arcsec.

    Perihelion had occurred on July 18 (r= 2.83 AU), and the comet faded with
unusual rapidity as it moved away from both the sun and Earth. On September
14, van Biesbroeck (Yerkes) estimated the magnitude as 17 and, on November 7,
Jeffers (Lick) found it to be 17.5. Both estimates were nearly 2 magnitudes
fainter than expected. The comet was lost near the sun after November, but
was recovered by Jeffers on May 14, 1948, when a nearly stellar object of magnitude 18.5. When last observed by van Biesbroeck (McDonald Observatory) with
the 208-cm reflector on October 2, it was described as a diffuse coma 4 arcsec
across with a total magnitude near 20. The orbit is hyperbolic with an eccentricity of 1.0011.
(Wirtanen)

1947 VIII   Discovered: October 7, 1948   Discovery Magnitude: 14.0
    Having passed perihelion on September 4, 1947 (r= 3.26 AU), this comet was
discovered 13 months later by Wirtanen (Lick) on a plate exposed during the
proper motion program. It was then described as diffuse, with a central condensation and a tail 1 arcmin long.

    Then located in Aquarius at distances of 4.89 AU and 4.19 AU from the sun
and Earth, respectively, the comet slowly faded and by December 2, it was near
15th magnitude with a coma 24 arcsec across. Shortly thereafter, the comet was
lost in the sun's glare, but was recovered at opposition on August 2, 1949, when
van Biesbroeck described it as near magnitude 16, with a circular coma 21 arcsec
across. On September 24, Jeffers estimated the total magnitude as 18 and the
coma diameter as 5 arcsec. The comet was again detected at opposition during
the summer of 1950, and was last photographed on September 11, near magnitude
18. It was then 9.5 AU from the sun. The orbit is hyperbolic with an eccen-

tricity of 1.0023.
(Wirtanen)

1947 X   Discovered: November 13   Discovery Magnitude: 9.0

Honda (Kurashiki, Japan) discovered this comet as a result of systematic searching with a 38-cm reflector. He described the comet as a slightly condensed object near Epsilon Corvi, which moved southward. After again observing the comet on November 14, Honda cabled news of the discovery to the Harvard College Observatory, but by the time it arrived on the 18th, the comet was too far south for observation in the Northern Hemisphere. Honda last detected it on the 17th, when the brightness had increased to 8.

Perihelion came on November 18 (r= 0.75 AU), and the comet was recovered by Johnson (Union Observatory, South Africa) in twilight on November 28. He then estimated the magnitude as 8, but a rapid fading coupled with the slow movement out of twilight ended observations on December 19, when the magnitude was near 10. Efforts to locate the comet in dark skies during January 1948 proved fruitless.
(Honda)

1947 XII   Discovered: December 7   Discovery Magnitude: 0.0

Having passed perihelion on December 3 (r= 0.11 AU), this comet rapidly moved away from the sun and suddenly appeared in Southern Hemisphere skies soon after sunset on December 7 and 8. Estimates of the total brightness varied considerably, but averaged magnitude 0, while the nucleus was estimated as near magnitude 1. The head was distinctly orange in color and the tail stretched vertically upwards for 20 or 30 degrees. The comet faded rapidly as it quickly moved away from both the sun and Earth, and, when observations began in the Northern Hemisphere on December 15, the magnitude had declined to near 3. At this time, the tail was being widely photographed and observers were noting a much more complex structure than was revealed visually, with up to 5 components being visible to lengths of slightly more than 4 degrees. As the month progressed the fading continued with magnitude estimates being near 4.5 on the 20th and 7.5 on the 31st. The tail also diminished rapidly, and by the end of the month it extended only slightly more than 1 degree. Observations continued during January 1948, as the comet moved northeastward. Unfortunately, its movement slowed to such a pace that its elongation from the sun began to slowly decrease. Observations were impossible after January 20, with van Biesbroeck then estimating the magnitude as near 10.5 and describing the comet as an extremely diffuse spot at low altitude.

The most interesting feature of this comet was the discovery of a secondary nucleus on December 10. W. H. van den Bos (Union Observatory) was the first to detect it when the separation was only 6.3 arcsec, but many other observers studied it and when last seen on January 20, it was located 33 arcsec from the main nucleus. In terms of brightness, the secondary nucleus was nearly 4 magnitudes fainter than the primary when first discovered, but the difference slowly declined and during December 15 and 16, it was considered slightly brighter than the primary. Thereafter, it again faded relative to the primary and when last seen was nearly 1 magnitude fainter. Investigations into the splitting by various researchers indicate November 30 as the most likely date of the split. Elliptical orbits have been calculated for both nuclei, with the primary having a period of 3,800 years and the secondary having a period of 5,100 years.
(Southern Comet)

1948 I   Discovered: September 25, 1947 Discovery Magnitude: 11.0

Bester (Bloemfontein) discovered this comet on a sky-patrol plate taken with a 7.6-cm Ross camera. He described it as a small, diffuse, centrally condensed object moving southwestward in Eridanus. The comet was then located 2.50 AU and 2.01 AU from the sun and Earth, respectively, with both distances decreasing.

During October, van Biesbroeck (Yerkes) noted the comet was brightening fairly rapidly from a magnitude of 9.8 on the 9th to 8.8 by the 23rd. He also detected a broad tail extending 10 arcmin. By the end of October, the comet was out of reach of northern observers and the brightening slowed as the comet-Earth distance increased. Observations by Johnson (Union Observatory) between November 9 and December 2, indicated a brightening of only 1 magnitude, although the tail had increased to a length of 15 arcmin on the latter date. After reaching a declination of -55 degrees on November 22, the comet moved basically westward and observations ended after January 12, 1948, as it was lost in evening twilight.

Perihelion came on February 16 (r= 0.75 AU), and a rapid northward motion allowed a recovery on March 3, when the magnitude was near 7. By mid-month, naked-eye visibility was attained and the brightness changed little from that time until the first days of April. Perigee came on April 6 (0.74 AU), and observers then described the comet as near magnitude 6, with a coma 5 arcmin across and a tail nearly 4 degrees long. The comet faded thereafter, and by the end of April, it was near 9.5. Observations continued throughout the year as the comet faded to 11 by May 28, 14 by July 3, and 17.7 by November 11. It was last observed on February 6, 1949, when described as star-like, with a magnitude of 17.5. The orbit is hyperbolic with an eccentricity of 1.0004. (Bester)

1948 II   Discovered: December 20, 1947 Discovery Magnitude: 9.5

Antonin Mrkos (Skalnate Pleso Observatory) discovered this comet during a systematic search for comets on December 20, 1947. Then located near the Serpens-Libra border, the comet was described as diffuse, without a condensation. Twilight and subsequent bad weather prevented the object from being confirmed and it was not announced. On January 18, 1948, Mrkos succeeded in recovering the comet near Kappa Ophiuchi and described it as near 10th magnitude, with a condensation and a trace of tail. Upon announcing the discovery, B. Protitch (Belgrade, Yugoslavia) succeeded in locating the comet on a photo exposed on January 10.

Although approaching both the sun and Earth, the comet remained near magnitude 10.5 until mid-February, as observers described the coma as 2 arcmin across, with a faint nucleus and a photographic tail 7 arcmin long. Perihelion came on February 17 (r= 1.50 AU), and with perigee coming at about the same time (1.65 AU), the comet slowly faded thereafter. By March, it had declined to magnitude 13, and by July, it was near 14.5. The comet was last detected on November 30, as van Biesbroeck estimated the magnitude as 15 and the coma diameter as 24 arcsec. Observations were reported by Mrkos on December 5, 1948, and January 28, 1949, with indicated magnitudes of 16 and 17.5, respectively, but these were never confirmed. The orbit is hyperbolic with an eccentricity of 1.0011. (Mrkos)

1948 III   Discovered: September 1   Discovery Magnitude: 13.0
     Ernest Leonard Johnson (Union Observatory) discovered this comet about 2
degrees from Beta Sculptoris on a plate exposed with the Franklin-Adams camera
for the regular minor planet program. He described it as faint and uncondensed.
     Perihelion had occurred on April 9 (r= 4.71 AU), and, after discovery,
the comet slowly faded as it moved away from both the sun and Earth. At the
beginning of October, van Biesbroeck (McDonald Observatory) photographed the
comet with the 208-cm reflector and reported a magnitude of 13.7 and a tail
length of 10 arcmin. By November 21, it had faded to magnitude 15 and observations ended after December 3.
(Johnson)

1948 IV   Discovered: June 3   Discovery Magnitude: 3.5
     This bright comet was discovered with the naked eye by Honda (Kurashiki,
Japan), who then confirmed it with his 15-cm reflector. The next evening, it
was independently found by Bernasconi (Como, Italy) and on the 5th, it was
discovered by Tosikazu Higasi. The latter observer was on board a ship returning to Japan from the eclipse expedition of May 9, and could not announce
his discovery until several days later. The combined initial observations of
these three observers indicated a magnitude near 3.5, a coma diameter near
8 arcmin and a tail length of several degrees.
     Perihelion had occurred on May 16 (r= 0.21 AU), and the comet remained a
faint naked-eye object until mid-June. Between the 12th and 14th, the comet
experienced an outburst in brightness that took the magnitude from 5.5 to 4.
Thereafter, it faded rapidly to magnitude 6 by the 17th. With the distances
from both the sun and Earth increasing, the comet faded to 8.7 by July 4, and
11.5 by the beginning of August. A rapid fading then set in, which ended observations after September 3. On this last date, van Biesbroeck estimated the
total magnitude as 19 (nearly 6 magnitudes fainter than predicted), the coma
diameter as 7.8 arcsec and the tail length as 18 to 24 arcsec. The orbit is
elliptical with a period near 67,700 years.
(Honda-Bernacsoni)

1948 V   Discovered: February 15   Discovery Magnitude: 10.0
     This comet was discovered by Pajdusakova (Skalnate Pleso Observatory) on
March 13, during a routine search for comets with 25x100 binoculars; however,
she wrongly identified it with the faint galaxy NGC6615, which had been shown
in the Skalnate Pleso star atlas. Her colleague, Mrkos, realized that this
galaxy should have been too faint (magnitude 15) to be seen in these binoculars
and, later that night, he was successful in locating this new comet. Shortly
thereafter, van Biesbroeck (Yerkes) located prediscovery images on plates exposed on February 15 and March 5, when the magnitude was near 11.
     Located 2.24 AU and 2.20 AU from the sun and Earth, respectively, this
comet slowly brightened as both distances decreased. Upon reaching perihelion
on May 17 (r= 2.11 AU), it was described as between magnitude 9 and 9.5, with
a coma 3 arcmin across and a tail nearly 10 arcmin long. On May 3, the comet
had possessed a short sunward tail. By early June, the total magnitude was
still near 9.5, but by early September, it had faded to 11.5. The comet passed 2 degrees from the north celestial pole during November, and was then described as near magnitude 12.
     Observations continued into 1949, with observers reporting a total magni-

tude of 14 on January 18 and 19, and 16 by April 2 and 3. After April, the
comet was lost in twilight, but in early October, Mrkos photographed it near
opposition at a magnitude of 17 and reported a short tail. Jeffers (Lick)
made the final observations of this comet. On November 18, he reported a
total magnitude near 18.5 and estimated the coma as 12 arcsec across, while
on February 9, 1950, these values had declined to 19.2 and 3 arcsec, respectively. The orbit is hyperbolic with an eccentricity of 1.0008.
(Pajdusakova-Mrkos)

1948 X   Discovered: November 24   Discovery Magnitude: 7.5
     Having passed perihelion on October 23 (r= 1.27 AU), this comet was dis-
covered in Volans by Bester (Bloemfontein) on a patrol plate exposed for 2
hours with the 7.6-cm Ross camera. It was described as bright, with a small
head, a bright nucleus and a tail 3 arcmin long.
     Moving away from both the sun and Earth, the comet faded as it moved to
the northwest. By the end of November, the magnitude had dropped to 9 and
the coma contained a nucleus near magnitude 10. Van Biesbroeck (Yerkes) became the first Northern Hemisphere observer when he detected this comet on
January 3, 1949. It was then described as between magnitude 11 and 12, with
a very diffuse coma 42 arcsec across and a wide tail. By January 23, the
magnitude was near 12.5 and on February 16, it was near 15. The comet was
last detected by van Biesbroeck on February 26, when at magnitude 16. The
orbit is elliptical with a period near 11,500 years.
(Bester)

1948 XI   Discovered: November 1   Discovery Magnitude: -3.0
     Having passed perihelion on October 27 (r= 0.14 AU), this comet was first
detected during the total solar eclipse of November 1, 1948. Then located
105 arcmin from the sun's center, the comet was described as very bright, with
a long tail which stretched towards the horizon. A number of photographs
were then taken.
     The first person to detect the comet after the eclipse was Captain
Frank McGann of Pan American-Grace Airways, who accidentally discovered it on
November 4, while flying over Kingston, Jamaica. During the next three days,
the comet was found by many other Southern Hemisphere observers, who reported
a total magnitude near 1 and a tail length of 20 degrees. Thereafter, although
moving away from the sun, the comet faded slowly as it moved closer to Earth.
On November 12, it was near magnitude 2, with a tail 15 degrees long, and, when
near perigee on November 26 (0.55 AU), it was estimated as near magnitude 3.5,
with a tail at least 5 degrees long.
     The comet continued to be an easy object for observation during December.
On the 7th, van Biesbroeck (Yerkes) and Giclas (Lowell) independently estimated
the total magnitude as near 5 and by the 20th, A. Jones (Timaru, New Zealand)
described the comet as slightly fainter than magnitude 6, with a coma 5 arcmin
across, a nucleus of magnitude 10.5 and a tail length of 2 degrees. Thereafter,
fading became a little more rapid as the comet declined to 8 by December 31,
9 by January 18, 1949, and 11 by February 16. Observations continued to April
3, when the comet had faded to magnitude 17, according to van Biesbroeck. The
orbit is elliptical with a period near 95,000 years.
(Eclipse Comet)

1949 I   Discovered: July 15, 1948   Discovery Magnitude: 15.5
     Wirtanen (Lick Observatory) discovered this comet on a plate taken with

the 51-cm Carnegie astrograph during the proper motion program. Then located
in Equuleus, the comet was described as faint and diffuse, with a coma 6 arcsec
across and a tail 2 arcmin long.

The comet was 10 months away from perihelion at discovery and brightened
very slowly during the following months. By October 6, the magnitude had increased by only one and, thereafter, a slight fading set in as the comet approached conjunction with the sun. Observations ended in December, and the
comet moved southwestward until it was recovered in the Southern Hemisphere
on March 1, 1949, when near magnitude 13. On April 29, A. Jones (Timaru, New
Zealand) estimated the total magnitude as 13.3 and described the coma as 42
arcsec across. Perihelion came on May 1 (r= 2.52 AU), and the comet's motion
then shifted to almost directly south. During May and June, it was brightest,
with a total magnitude of 12, and was also well observed as it passed within
5 degrees of the south celestial pole at the end of May. Thereafter, a slow
fading set in which brought the total magnitude to nearly 14 by the end of
the year. Observations continued into 1950, as the comet slowly faded to magnitude 16.5 by May 15, when van Biesbroeck (McDonald Observatory) estimated
the coma as 12 arcsec across. After conjunction with the sun, the comet was
recovered in November, by L. E. Cunningham (Mount Wilson), who estimated
the nuclear magnitude as near 19. He continued observations until March 4,
1951, when the total magnitude had declined to 19. The orbit is elliptical
with a period near 242,000 years.
(Wirtanen)

1949 III   Discovered: November 19   Discovery Magnitude: 16.0
This comet was discovered by Albert G. Wilson and Robert G. Harrington
on a series of photographic plates taken with the 122-cm Schmidt camera between
November 19 and 25. The plates had been taken for the National Geographic
Society-Palomar Sky Survey, and the comet's images were described as strong
and entirely asteroidal, except for a short, faint tail, which appeared on
both a red-light and blue-light photograph exposed on November 19. Preliminary
orbital calculations by L. E. Cunningham indicated that the comet was probably
of short period with a period as small as 2.31 years, although this figure is
uncertain by more than 2 years. Since no further observations were made, the
orbit is generally listed as parabolic, with a perihelion date of October 12,
1949. The orbit also indicates a very close approach to Earth (0.16 AU) on
November 19.
(Wilson-Harrington)

1949 IV   Discovered: June 29   Discovery Magnitude: 13.0
The discovery of this comet is considered an effort by three astronomers
at Harvard's Oak Ridge Observatory. It began on July 2, when Vaina Bappu, a
graduate student, made a 60-minute exposure of the Cygnus region for a special
program being supervised by Bart J. Bok. The next day, Gordon A. Newkirk, Jr.,
an undergraduate student, was invited to look at the excellent quality of the
plate, when the trail of this comet was discovered. It was then described as
faint and diffuse, with a central condensation, and shortly thereafter, a prediscovery image was found on a patrol plate exposed on June 29.

After discovery, the comet brightened slowly as the increasing distance
from Earth nearly countered the decreasing distance from the sun. By the end
of July, magnitude estimates ranged from 12 to 12.5, and, by the end of September, it was between 11.5 and 12. During this same time, the coma was nearly 3

arcmin across, a tail was occasionally extending up to 5 arcmin and a 14th-mag-
nitude nucleus was observed. Perihelion came on October 26 (r= 2.06 AU), and,
thereafter, a slow fading began. By the end of the year, the comet had declined to magnitude 12.5. During February 1950, the comet dropped to magnitude
13 and by mid-April, it was described as near 15.5, with a coma 24 arcsec
across. The comet was followed until May, but observations resumed in late
November, and by December 16, the nuclear magnitude was near 19. The final
observation came on March 4, 1951, when still near magnitude 19. The orbit is
elliptical with a period near 1,500 years.
(Bappu-Bok-Newkirk)

1950 I   Discovered: May 20, 1949   Discovery Magnitude: 13.0
     Johnson (Union Observatory) discovered this comet on plates exposed with
the 25.4-cm Franklin-Adams camera during routine work on minor planets. Then
located near Epsilon Lupi, the comet was described as small and diffuse, with
a central condensation, but no tail. It was situated 3.61 AU and 2.67 AU from
the sun and Earth, respectively, and moved slowly northwest; however, its
brightness changed little during the next few months as the increasing distance
from Earth countered the effects of a decreasing distance from the sun. The
comet was lost in the sun's glare after August, but was recovered in mid-December, at a magnitude near 12.5.
     Perihelion came on January 19, 1950 (r= 2.55 AU), and observers then des-
cribed the comet as near magnitude 12, with a nuclear magnitude near 13.5, a
coma 2 arcmin across and a tail 3 arcmin long. Thereafter, the comet brightened slightly during February, as it neared Earth, and reached a maximum brightness of 11.5 during the first week of March. The tail was then 6 arcmin long
and the coma diameter was 2.5 arcmin. The comet faded to magnitude 13 by May,
and was next observed at magnitude 17 by Cunningham (Mount Wilson) in November.
Observations continued until March 8, 1951, when the magnitude had faded to 18.
On November 3, 1951, van Biesbroeck (McDonald Observatory) took one photographic
plate with the 208-cm reflector, which showed a small, diffuse image near magnitude 20; however, the reality of this image, which is at the very limit of
the plate, is still in question. The orbit is hyperbolic with an eccentricity
of 1.0007.
(Johnson)

1951 I   Discovered: May 19, 1950   Discovery Magnitude: 11.0
     Rudolph Minkowski (Palomar Observatory, California) discovered this comet
on a plate taken with the 122-cm Schmidt camera during the National Geographic
Society-Palomar Sky Survey. Then in Ophiuchus, the comet was described as 25
arcsec in diameter, with considerable condensation and a tail 50 arcsec long.
The original magnitude estimate of 8 was apparently an overestimate, since
observers during the next few days found the visual magnitude closer to 11.
     The comet brightened slightly during the next couple of months as it neared
both the sun and Earth. By July, it reached magnitude 10.5 and possessed a
coma 2 arcmin across. A faint tail then extended nearly 4 arcmin. Thereafter,
as the comet-Earth distance increased, the comet faded, and when last seen on
October 10, it was near magnitude 12. After conjunction with the sun, the
comet was recovered on December 17 by van Biesbroeck, who described the magnitude as near 10.
     Perihelion came on January 15, 1951 (r= 2.57 AU), and the comet continued
to brighten for the next 2 months as it neared Earth. Its increasing southern

declination made it a difficult object for Northern Hemisphere observers and they even reported the comet as fading slightly by March; however, in the Southern Hemisphere, the comet could be well observed high in the sky, and they reported it to brighten as expected. On March 15, A. Jones (Timaru, New Zealand) described it as magnitude 9.8, with a coma 2 arcmin across, a nuclear magnitude of 11.7 and a tail nearly 4 arcmin long. Thereafter, the comet faded, and by the end of June, it was near magnitude 12.

After a second conjunction with the sun, the comet was recovered at the end of October, and on November 4, van Biesbroeck described it as near magnitude 14, with a coma 7.8 arcsec across and a faint, wide tail 1 arcmin long. Observations continued throughout 1952, with magnitude estimates of 17 at the end of April, and 19 in December. The comet was last seen on January 19, 1953, when van Biesbroeck (McDonald Observatory) described it as near magnitude 18, with a coma diameter of 20 arcsec. The orbit is hyperbolic with an eccentricity of 1.0012.
(Minkowski)

1951 II   Discovered: February 4   Discovery Magnitude: 8.5
Pajdusakova (Skalnate Pleso Observatory) discovered this comet during a routine search for comets with 25x100 binoculars. She described it as bright, with a very condensed coma 1 arcmin across and a slender tail 6 arcmin long. Having passed perihelion on January 30 (r= 0.72 AU), the comet traveled northeast and slowly brightened as its distance from Earth decreased. After reaching magnitude 8 shortly after mid-month, the comet slowly faded as it moved away from both the sun and Earth, and by March 2, it was slightly fainter than 9. An outburst of 1 magnitude occurred by the next evening, after which the comet faded back to normal by the 8th. At the end of April, observers estimated the total magnitude as between 13 and 14, and the comet was last detected on May 8, when near magnitude 16. An attempt by van Biesbroeck to locate the comet on a plate exposed on May 25 with a 25.4-cm camera failed, and his report of a 19th-magnitude image on a plate taken with the 208-cm reflector on November 4 was found to be erroneous.
(Pajdusakova)

1952 I   Discovered: August 6, 1951   Discovery Magnitude: 15.0
Wilson and Harrington (Palomar) discovered this comet on routine sky survey plates. Then located in Ophiuchus, the comet was described as 30 arcsec across, with a near-stellar nucleus of magnitude 16 and a tail 2 arcmin long. As with 1951 I, the magnitude was overestimated, this time being called 10, while observers during the following days found it closer to 15. The comet brightened slightly during the next month as it approached the sun, but observations ended by mid-September, as the comet neared conjunction.

The comet was recovered as strictly a Southern Hemisphere object at the end of December, with the magnitude being estimated as near 8.5. The coma was then 1.3 arcmin across and the tail extended 8 arcmin. On January 4, 1952, Jacobus A. Bruwer (Johannesburg, South Africa) estimated the magnitude as 7.3. Perihelion came on January 13 (r= 0.74 AU), but the comet continued to brighten as it neared Earth and when at perigee (0.38 AU) at the end of January, it was near magnitude 6, with at least 3 observers reporting naked-eye visibility. The coma was then 3 arcmin in diameter and the tail extended more than 1 degree. During February, the comet faded as it moved northwestward, and by the end of the month, it was near 9th magnitude. Thereafter, the comet moved into even-

ing twilight and was last observed on March 16, when R. Weitbrecht (Yerkes
Observatory) reported a magnitude of 12. Attempts to recover the comet at
Lick Observatory during September and October failed. The orbit is elliptical
with a period near 151,000 years.
(Wilson-Harrington)

1952 V   Discovered: April 27   Discovery Magnitude: 10.0
     This comet was discovered 10 degrees west of the Andromeda Galaxy by
Mrkos (Skalnate Pleso) during his observatory's routine program of comet
hunting with 25x100 binoculars. He described it as diffuse, with condensation, but the comet's slow motion, plus clouds, prevented him from confirming
the cometary character. Clouds and moonlight prevented further observations
until May 14, when Mrkos found it southwest of the previous position and then
reported the discovery.
     Observers during the next few days confirmed the initial magnitude esti-
mate of 10 and the comet brightened as it approached both the sun and Earth.
Perihelion came on June 9 (r= 1.28 AU), and the comet was then near magnitude
9, but the brightening continued as its distance from Earth continued to decrease. During July, the comet arrived at perigee (0.42 AU on July 19) and
then moved at a rate of over three degrees per day. Van Biesbroeck noted an
increase in the diameter of the coma to 6 arcmin by mid-July, while several
other observers estimated the total magnitude as 8 and the nuclear magnitude
at near 12. Thereafter, the comet began to rapidly move out of northern skies
and into the southern--fading along the way. As both the sun and Earth distances increased, the comet became a difficult object during August. On the
13th, Bruwer (Johannesburg) estimated the total magnitude as 13.5 and the
comet was last detected on the 25th, when Jorge Bobone (Cordoba Observatory)
estimated the magnitude as near 14.5. Further attempts to locate the comet
were hampered by a waxing moon, and a final effort by Martin Dartayet (Bosque
Alegre, Argentina) on September 17 failed with the 152-cm reflector despite
dark skies, thus indicating a magnitude fainter than 17.5. The orbit is
elliptical with a period of 590 years.
(Mrkos)

1952 VI   Discovered: June 20   Discovery Magnitude: 10.0
     Peltier (Delphos, Ohio) discovered this comet with a 15-cm refractor 3
degrees north of Alpha Draconis and described it as faint and diffuse, without
a tail. Thereafter, the comet brightened as it approached both the sun and
Earth, and reached its maximum magnitude of 9 shortly after mid-July. It was
then described as 4 arcmin in diameter, with a nuclear magnitude of 14. Perihelion had occurred on July 15 (r= 1.20 AU), but the comet faded slowly as it
continued to near Earth. Beyer (Hamburg Observatory) reported outbursts in
brightness which amounted to 1 magnitude on July 23 and August 14. Perigee
came at the beginning of September (0.7 AU), and the comet was then described
as large and diffuse, with a magnitude near 10.5. Thereafter, fading was
rapid, and observations continued only until November 22, when Cunningham
(Mount Wilson) estimated the magnitude as 18. The orbit is elliptical with
a period near 1.7 million years.
(Peltier)

1953 I   Discovered: August 18, 1952   Discovery Magnitude: 13.0
     Harrington (Palomar) discovered this comet on a plate taken with the 122-
cm Schmidt camera during the sky survey program. He described it as diffuse,

with a nucleus; however, his magnitude estimate of 15 seems to have been an underestimate, since Steavenson (England) visually estimated the magnitude as 13 on August 23, and estimated the coma diameter as 1 arcmin. On August 21, Elizabeth Roemer (Lick Observatory) estimated the nuclear magnitude as 16.5

The comet brightened during the next several months as it approached both the sun and Earth. By mid-November, when at perigee (1.0 AU), it was near magnitude 11, with a coma 2 arcmin across, and by the end of December, it was near magnitude 8.5, with a coma 3 arcmin across. Perihelion came on January 5, 1953 (r= 1.66 AU), and the comet faded very slowly during January, as its distances from both the sun and Earth slowly increased. By mid-month, it was near magnitude 9 and at the end of the month it was near 9.5. An outburst in brightness seems to have occurred at the beginning of February, when the magnitude reached 8 on the 2nd and 3rd, but, by the 8th, it was back to normal at magnitude 9.5. At the beginning of March, the comet had faded to magnitude 10 and, two months later, it had dropped to 12. Observations continued until June 29, when observers at Cordoba Observatory estimated the total magnitude as 13.5. The orbit is elliptical with a period near 8,200 years.
(Harrington)

1953 II   Discovered: November 28, 1952   Discovery Magnitude: 10.0
Mrkos (Skalnate Pleso) discovered this comet near Alpha Virginis during a routine comet search with 25x100 binoculars. As with his previous discoveries, the comet could not be immediately confirmed due to bad weather, but it was rediscovered by him on December 9. On this latter date, the comet was described as diffuse with some condensation and photos by other observers in the next few days revealed a short tail.

The comet headed south after discovery and brightened as it approached the sun. On December 15, Roemer (Lick Observatory) described the comet as near magnitude 9 with a strong condensation. Shortly thereafter, the comet was lost to Northern Hemisphere observers, but continued to be followed by southern observers as it continued to brighten. On December 24, Bruwer (Johannesburg) estimated the magnitude as near 8.5. Perihelion came on January 25 (r= 0.78 AU), and observers at that time estimated the magnitude as between 7 and 8 and the coma diameter as near 4 arcmin. The comet slowly faded afterwards, and was lost in the sun's glare after mid-February. Following conjunction with the sun, the comet was recovered at McDonald Observatory and Lick Observatory on July 5 and 9, respectively. The comet was then described as between magnitude 17.5 and 18, with a coma mainly extending to the northwest. Observations continued until September 5, 1953, when Cunningham (Lick) estimated the total magnitude as near 19. The orbit is hyperbolic with an eccentricity of 1.0003.
(Mrkos)

1953 III   Discovered: April 12   Discovery Magnitude: 9.0
This comet was discovered by Mrkos (Skalnate Pleso Observatory) and Honda (Kurashiki, Japan) within 17.5 hours of each other. Both were involved in routine searches for comets with 100-mm binoculars and both described the comet as large and diffuse. On April 14, van Biesbroeck (Yerkes) described the comet as 3 arcmin across and near magnitude 8.7, and on the 15th, S. Vasilevskis (Lick) estimated the nuclear magnitude as 15.

The comet brightened to near magnitude 8 by mid-May, and then possessed

a stellar nucleus of magnitude 13 and a coma diameter of 6 arcmin. Perihelion
came on the 26th (r= 1.02 AU), and the comet faded thereafter, with van Biesbroeck (McDonald) describing it as 30 arcsec in diameter and near magnitude
13 on July 4. A short time later, the comet entered evening twilight, but was
recovered by van Biesbroeck on the morning of September 13, when 50 arcsec
across and near magnitude 15.5. Observations continued up to January 1, 1954,
when Roemer (Lick) estimated the magnitude as near 19.3. The orbit is elliptical with a period near 7,800 years.
(Mrkos-Honda)

1954 I   Discovered: June 24   Discovery Magnitude: 19.0
       Having passed perihelion on January 18, 1954 (r= 2.06 AU), this comet was
discovered by Harrington (Palomar) during the sky survey with the 122-cm
Schmidt camera. It was described as very faint, with a slight tail, and further Palomar observations were obtained on June 25, 26 and 28. Thereafter,
due to the comet's faintness, it was tracked by only Lick Observatory. On
July 28 and 31, Roemer described the comet as near magnitude 18.5, with a
not quite stellar nucleus. The comet was last observed on August 25 and 31,
when Roemer described it as near magnitude 19, with a coma 0.2 arcmin across.
(Harrington)

1954 II   Discovered: November 7, 1953   Discovery Magnitude: 11.0
       Pajdusakova (Skalnate Pleso) discovered this comet on December 3 in
Cetus during a routine comet-hunting session with 25x100 binoculars. She described it as diffuse, with a central condensation and a tail over 3 arcmin
long. A photo by van Biesbroeck (Yerkes) on December 5, confirmed Pajdusakova's magnitude estimate of 11, and gave the diameter of the coma as 20 arcsec
and the length of the tail as 10 arcmin. A while later, a prediscovery image
was found on a photograph exposed on November 7.
       The comet was expected to brighten rapidly as it neared the sun, but, in-
stead, observers indicated a very slow magnitude increase which reached 10.5
by December 25. On that date, van Biesbroeck estimated the tail length as 20
arcmin long and determined the nuclear magnitude as 12. During the next few
days, the comet faded rapidly as it grew large and diffuse--despite a decreasing
distance from the sun--and it was last detected on January 5, 1954, when Vasilevskis (Lick Observatory) photographed it as a faint object measuring 7 arcmin
by 2 arcmin. Later that same day, Beyer (Hamburg Observatory) visually observed the comet as near magnitude 10. Attempts to locate the comet during the
following days failed. Perihelion came on January 25 (r= 0.07 AU).
(Pajdusakova)

1954 V   Discovered: April 13, 1955   Discovery Magnitude: 15.0
       George O. Abell (Palomar) discovered this comet with the 122-cm Schmidt
camera during the sky survey. He described it as diffuse, with a central condensation and a short tail pointing southeast. Perihelion had occurred on
March 24, 1954 (r= 4.50 AU), so, after discovery, the comet got steadily fainter and observations were only made with the large telescopes at Lick, Yerkes
and McDonald observatories. Between April 21 and 25, van Biesbroeck (McDonald)
photographed it with the 208-cm reflector and described the image as near magnitude 16, with a tail extension of 1 arcmin. On May 10, Roemer (Lick) estimated the magnitude as 17 and by June 23, Jeffers (Lick) estimated it as 17.5.
The comet was last detected before conjunction with the sun by van Biesbroeck
(Yerkes) at magnitude 18.

After conjunction, Roemer detected the comet on November 25 and 26, while it was moving slowly northeast through Coma Berenices. She described the nucleus as well condensed with a magnitude of 17.5 and a fan-shaped tail 1 arcmin long. On February 15, 1956, the comet was described as 6 arcsec in diameter with a magnitude of 18 and a faint tail. Observations continued until April 30, when it appeared as a moderately sharp nucleus of magnitude 19.3, with a faint tail 0.8 arcmin long. The orbit is hyperbolic with an eccentricity of 1.0028.
(Abell)

1954 VIII   Discovered: July 28   Discovery Magnitude: 9.0
Margaret Vozarova (Skalnate Pleso) discovered this comet as part of her observatory's regular program of comet hunting. She described it as fairly bright in 25x100 binoculars, with no visible tail or nucleus.

Perihelion had occurred on June 2 (r= 0.68 AU), and, after discovery, the comet faded fairly rapidly. At the beginning of August, it was near magnitude 9.5 and a slender sunward tail was photographed at Skalnate Pleso, Sonneberg and Yerkes observatories. This tail was measured as 10 to 15 arcmin long and the regular tail was then only a short stub. By the end of September, observers at Lick described the comet as 12 arcsec in diameter, with a faint fan-shaped tail 42 arcsec long and a nucleus near magnitude 16. On October 27, van Biesbroeck (McDonald) photographed the comet with the 208-cm reflector and described it as near magnitude 16.5, with a trace of tail. The comet was last detected on December 18, when Jeffers (Lick) described it as nearly stellar, with a magnitude of 18.5. The orbit is hyperbolic with an eccentricity of 1.0003.
(Vozarova)

1954 X   Discovered: October 15, 1953   Discovery Magnitude: 15.0
Once again a comet was found on sky survey plates exposed with the 122-cm Schmidt camera. This time, Abell was the discoverer and he described it as 0.6 arcmin across, with a nearly stellar nucleus and a tail 1.5 arcmin long.

Then located in Camelopardalis, the comet was moving northward and slowly brightened as it approached the sun. By November 14 and 16, van Biesbroeck's photos with the 61-cm Yerkes reflector indicated a brightening to magnitude 14 and an increase in the diameter of the coma to 1 arcmin. On the 20th, the comet passed 1 degree from the north celestial pole. At the end of December, observers estimated the total magnitude as 13 and by the end of January 1954, it was near 12. During the next few months, observers reported the following magnitude estimates: 11 on March 1, 9.5 on April 18, 8.5 on May 21 and 8 on May 30. By the end of June, the comet reached 7th magnitude and was described as 3 arcmin in diameter, with a tail 10 arcmin long and a nuclear magnitude near 10.

Perihelion came on July 7 (r= 0.97 AU), and the comet faded rapidly thereafter, as its distances from both the sun and Earth increased. Observations were then possible only in the Southern Hemisphere and continued only until September 15, 1954. The comet returned to northern skies in mid-1955, and Roemer (Lick) obtained a 43-minute exposure on July 20, and a 70-minute exposure on the 21st, using the Crossley astrograph, which revealed no trace of the comet. She concluded that it was then fainter than magnitude 18. The orbit is hyperbolic with an eccentricity of 1.0006.
(Abell)

1954 XII   Discovered: June 26   Discovery Magnitude: 10.0
     L. Kresak (Skalnate Pleso) discovered this comet in Boötes in the course
of systematic searches with 25x100 binoculars. Three days later, an independent discovery was made by Peltier (Delphos, Ohio). The comet was then described as diffuse, without a condensation.
     The comet traveled rapidly southwestward and was described as very large
and diffuse during the first days of July. Observers then estimated the coma
as 10 to 15 arcmin in diameter and the total magnitude as near 9. Subsequently,
observers with large telescopes actually had a more difficult time observing
the comet than did observers with small telescopes. Also, as July progressed,
the diffuse nature made observations very difficult as the comet neared the
horizon for northern observers and after July 31, it could only be seen in the
Southern Hemisphere. Perihelion came on August 30 (r= 0.75 AU), and the comet
faded rapidly thereafter, as it headed northward. On October 9, Roemer (Lick)
described the comet as 1 arcmin across with a total magnitude of 13, and on
October 24, the comet was last seen by observers at the Skalnate Pleso Observatory. Photos exposed on October 29, at Skalnate Pleso, and October 30, at
Greenwich, revealed no trace of the comet. Further attempts were made at Lick
on November 25 and 29, 1954, and January 15, 1955, but these also failed. The
orbit is hyperbolic with an eccentricity of 1.0002.
(Kresak-Peltier)

1955 III   Discovered: June 12   Discovery Magnitude: 3.0
     Having passed perihelion on June 4 (r= 0.53 AU), this comet was discovered
in the morning sky by Mrkos (Skalnate Pleso) near Capella during his routine
searches with 25x100 binoculars. Mrkos described it as bright, with a tail
over 1 degree long.
     As the comet moved away from the sun after discovery, it rapidly faded.
Observations by Roemer (Lick) began on June 14, when she determined the total
magnitude as 4 with binoculars and described the tail as 2 degrees long. Two
days later, she found the tail to be 1 degree in length and the coma was 4
arcmin in diameter. Beyer (Hamburg Observatory) reported a decline from magnitude 4.8 on June 17, to 6.9 by July 9, and on the latter date he determined the
coma diameter as 6 arcmin and the nuclear magnitude as 11. By July 19, the
tail was less than 30 arcmin long visually, but photos taken at Kiev (Russia)
revealed a wide, curved tail 50 arcmin long and a straight jet over 1 degree
long. By mid-August, Roemer estimated the coma as 2 arcmin across and the
nucleus as near magnitude 14, while van Biesbroeck (Yerkes) estimated the total
magnitude as near 10. The comet was last detected before entering the sun's
glare on September 8, and Roemer then estimated the nuclear magnitude as near
18. The comet was recovered after conjunction on April 17, 1956, when Roemer
estimated the total magnitude to be 17.5. No further observations were made.
The orbit is very similar to that of the comet of 1707. It is also elliptical
with a period of 356 years.
(Mrkos)

1955 IV   Discovered: July 13   Discovery Magnitude: 8.0
     A. M. Bakharev (Stalinabad Observatory, Russia), a meteor specialist, had
been observing meteors with binoculars when he discovered this bright, diffuse
object in Pegasus. An independent discovery was made 13 hours later by Lewis
Macfarlane and Karl Krienke (Seattle, Washington), who were both observing different celestial objects with the latter's father. The comet was then moving

at a rate of 2 degrees per day to the northwest.

Perihelion had occurred on July 11 (r= 1.43 AU), but the comet brightened slightly after discovery as its distance from Earth decreased. Descriptions during the latter half of July gave the total magnitude as 7.5, the nuclear magnitude as 12 to 13, the coma diameter as 5 arcmin and the tail length as 15 to 20 arcmin. During August, the comet slowly faded and at mid-month, it was estimated as near magnitude 8.5, with a nuclear magnitude of 13 to 14, a coma diameter of 4 arcmin and a tail 1 degree long. By mid-September, the comet had faded to 11 and one month later, it was near 13, with a diameter of 2 arcmin. The comet was last detected on November 25, when Roemer estimated the nucleus as about magnitude 17. The sun's glare prevented observations thereafter, and photographs exposed at Yerkes and Lick in March 1956, after the comet had passed conjunction with the sun, failed to show the comet. The orbit is elliptical with a period near 3,800 years.
(Bakharev-Macfarlane-Krienke)

1955 V   Discovered: July 29   Discovery Magnitude: 8.0
Honda (Kurashiki, Japan) discovered this comet near the Orion-Eridanus border and described it as diffuse, with a central condensation and a rapid northward motion. The comet brightened rapidly as it neared both the sun and Earth, and between August 2 and 4, observers at Lick described it as near magnitude 6, with a coma 1 arcmin across.

Perihelion came on August 4 (r= 0.88 AU), and for the next two weeks, the brightness changed little as the comet neared Earth, although the coma and tail grew. With perigee coming on August 19 (0.27 AU), observers on August 22, described the comet as between magnitude 6 and 6.5, with a coma 10 arcmin across and a tail 2 to 3 degrees long. Also on the 22nd the comet passed only 10 degrees from the north celestial pole. Fading was rapid thereafter, as the comet moved away from both the sun and Earth. On the 30th, it was near magnitude 7.5 and by September 2, it had dropped to 8.

Up to September 4, the comet was considered an average comet since it had behaved more or less as expected; however, beginning on the 4th, the comet began to provide a number of surprises. On that date, observers found a magnitude of 5--indicating an outburst within the previous 24 hours. This brightness was sustained through the 5th, whereupon the comet then began to fade. By the 7th, the comet had dimmed to magnitude 6 and by the 20th, it was near 8.5. During this same time interval, Roemer (Lick) photographed variations in the coma and in the structure of the tail. Another surprise occurred on the 21st, when Roemer photographed a double nucleus, with both components being of equal brightness and separated by 5 arcsec. Observations of the components continued until October 16. Researchers are unsure of the exact date of splitting for the comet. For a long time, it was believed that the division occurred at the time of the comet's outburst, but the discovery of several photos exposed by van Biesbroeck between October 5 and 16, which showed the secondary nucleus, indicated a much slower separation rate--pointing to a splitting date closer to July 1953. The problem with this date is that when the comet was closest to Earth in late August, the secondary nucleus should have been located 45 arcsec from the main nucleus; however, nothing was found on photographs taken on this date.

Observations continued throughout October, as the comet continued to fade without further unexpected events. By mid-November, observers at Lick and

McDonald observatories described the comet as near magnitude 16, with a faint
tail nearly 2 arcmin long. Observations continued until November 20. The
orbit is hyperbolic with an eccentricity of 1.0009.
(Honda)

1955 VI   Discovered: July 31, 1954   Discovery Magnitude: 15.0
    Walter Baade (Palomar) discovered this comet on two 122-cm Schmidt camera
plates exposed by Abell. The comet was then in Ursa Minor and was located
5.04 AU and 5.22 AU from the sun and Earth, respectively. It was described as
faint and diffuse, with a short tail.
    At discovery, the comet was still one year from perihelion and it slowly
brightened as it approached the sun. At the end of December 1954, it was estimated as near magnitude 12.5, and by mid-March 1955, it was near 12. On the
latter date, observers described the comet as 1 arcmin in diameter, with a tail
4 arcmin long. Also, the comet was then passing less than 10 degrees from the
north celestial pole. Perihelion came on August 13 (r= 3.87 AU), and, with
the comet-Earth distance having been increasing for the past four months, the
comet had faded to magnitude 13, but the tail was still near 3 arcmin across.
As 1955 drew to a close, the comet-Earth distance decreased and the comet
brightened to nearly 12th magnitude by the end of December. Observations continued throughout 1956, as the comet slowly faded from 12.5 during the first
week of January to 13 by April 7. The tail continued to extend to a length of
nearly 5 arcmin during this time and the nucleus was estimated as near magnitude
16.5. After mid-April, the comet was lost in the sun's glare, but it was recovered in September, as a very diffuse object near magnitude 17. Opposition
came during February 1957, and the comet was then photographed at Lick and
described as 20 arcsec across with a nuclear magnitude of 16.5. After a second
conjunction with the sun during the summer months, the comet was observed for
the last time during September and November 1957. During the latter month, it
was described as a somewhat diffuse object with a magnitude of 19.5. It was
last seen on November 26. The orbit is hyperbolic with an eccentricity of
1.0005.
(Baade)

1956 I   Discovered: December 18, 1954   Discovery Magnitude: 16.0
    During a study of flare stars in the dark clouds of Taurus, this comet
was discovered on a Schmidt camera plate by Guillermo Haro and Enrique Chavira
(Tonantzintla Observatory, Mexico). Although at first reported as an asteroid,
photos taken in mid-January 1955 at the Lick and Yerkes observatories proved
the object was a comet. Later examination of the discovery plate by van
Biesbroeck revealed a faint coma 15 arcsec in diameter and a faint tail nearly
30 arcsec long.
    At discovery, the comet was still 13 months away from perihelion and it
slowly brightened during the first half of 1955, as it neared both the sun and
Earth. Van Biesbroeck consistently photographed a faint wide tail during this
time and by August 14, he estimated the total magnitude as 14.5. During the
latter half of the year, the magnitude changed little as the distance from
Earth increased fast enough to counter the effects of a decreasing distance
from the sun and the comet slowly neared the north celestial pole. Perihelion
came on January 26, 1956 (r= 4.08 AU), and although the comet moved slowly
away from the sun thereafter, it remained near magnitude 14 through the end of
May, as its distance from Earth slowly decreased. Beginning in June, the comet

faded slowly, with total magnitude estimates of 15.5 coming in August, and
estimates of 16 coming in September. During the former month, a faint narrow
tail extended 10 arcmin long on some photographs. At the end of the year, the
comet was described as 25 arcsec in diameter, with a short tail. During 1957,
the comet was observed until September, when it then became lost in the sun's
glare. Between March and July, the total magnitude was near 17 and a tail
extended 24 arcsec, but by September, it had faded to 19. The comet was recovered on February 7, 1958, and was followed until May 15. On the latter
date, it was located 7.7 AU from the sun and possessed a nearly stellar nucleus of magnitude 20.9, with a very faint coma 30 arcsec across. The orbit
is hyperbolic with an eccentricity of 1.0047.
(Haro-Chavira)

1956 III   Discovered: March 12   Discovery Magnitude: 9.0
    Mrkos (Skalnate Pleso) discovered this comet in the Milky Way region of
Sagittarius during a routine search with 25x100 binoculars. He described it
as diffuse, without central condensation, and added that it was moving rapidly
to the northeast. The rapid motion amounted to about 3.5 degrees per day and
this made it difficult to locate for early observers. Van Biesbroeck (Yerkes)
did, however, obtain one of the first photographs of the comet on March 13,
when he found it far from the center of a plate taken with the Ross 7.6-cm
camera. He then described it as 3 arcmin in diameter, with a total magnitude
of 9.
    The comet was one month from perihelion at discovery, but was very close
to perigee and therefore brightened rapidly shortly after it was found. On
March 21, van Biesbroeck described it as near magnitude 8.3, with a narrow
tail 5 arcmin long. Perigee came on March 22 (0.35 AU), and thereafter, the
comet faded rapidly, despite a decreasing distance from the sun, as it moved
quickly away from Earth. On March 28, Beyer (Hamburg Observatory) estimated
the brightness as 8.8, and by April 10, it had dropped to about 10.5. Perihelion came on April 14 (r= 0.84 AU), and the comet then faded even more
rapidly, as well as becoming more diffuse. By April 28, a photograph exposed
at Lick revealed a very diffuse image near magnitude 17.5 and the comet was
last detected on May 5, when van Biesbroeck (McDonald) used the 208-cm reflector
to photograph the comet. He described it as only vaguely suspected and hardly
measurable with a magnitude near 18.
(Mrkos)

1957 III   Discovered: September 14, 1956 Discovery Magnitude: 10.0
    Silvio Arend and G. Roland (Uccle Observatory, Belgium) discovered this
comet on two astrograph plates taken on November 8, during routine asteroid
observations. Due to a delay in the examination of these two photos, the discovery did not occur until over a week after the photographs had been taken
and the announcement was not made until November 19. Prediscovery images were
subsequently found at various observatories--the oldest of which was obtained
on September 11. By late December, the comet had only increased to magnitude
9, and possessed a tail 8 arcmin long; however, orbital calculations were then
indicating the comet would become very bright during April and May 1957.
    Although approaching the sun at the beginning of 1957, the comet brighten-
ed very slowly as its distance from Earth increased, and on February 22 and 23,
observers indicated a total magnitude near 8.5, and a tail length of nearly one
degree. Observations continued during the next few days, but were impossible

after the 27th, as the comet became lost in the sun's glare. After conjunction with the sun, the comet was recovered on April 2, by J. G. Gow (Tapui, New Zealand) in the morning sky and was described as brighter than Beta Ceti; hence, near magnitude 2. A couple of days later, observers were estimating the tail length as between 3 and 5 degrees.

Perihelion came on April 8 (r= 0.32 AU), but the comet continued to brighten as it neared Earth. During the first half of April, observations were only possible in the Southern Hemisphere and observers reported the brightness to increase to nearly first magnitude by the 14th. Interestingly, a photograph secured by K. Gottlieb and A. Przybylski (Mount Stromlo Observatory, Australia) on the morning of April 11, with the 66-cm Yale-Columbia refractor, suggested the nucleus was double with a separation of about 9 arcsec; however, no further observations were reported.

Perigee came on April 21 (0.57 AU), and the comet was then being well observed in the Northern Hemisphere as an object between magnitude 1 and 2. During the next few days, the comet exhibited a very unusual feature as a sunward tail appeared that was as bright as the main tail. Observers then estimated the main tail length as 25 to 30 degrees and the anti-tail length as 15 degrees. The anti-tail remained visible until May 2, as it slowly swung eastward around the comet's nucleus and faded. Thereafter, the comet faded with magnitude estimates of 6 on May 18 and 7.8 on June 3.

On August 25, van Biesbroeck (McDonald) estimated the nuclear magnitude as near 16 and by December 3, observers at Lick found the nucleus near 18. Observations continued into 1958, with van Biesbroeck determining the nuclear magnitude as 20 on January 25, and Roemer estimating the nuclear magnitude as 21.0 on April 11. The comet was not observed after the latter date. The orbit is hyperbolic with an eccentricity of 1.0002.
(Arend-Roland)

1957 V   Discovered: July 29   Discovery Magnitude: 1.0
While measuring night-sky glow at his observatory on Lomnicky Stit (Czechoslovakia), Mrkos discovered this comet on August 2, with the naked eye. When first seen, only the tail was visible, but shortly before sunrise, the comet's head was observed. Mrkos immediately announced his discovery. Numerous independent discoveries were also made, of which two were earlier than Mrkos'. Sukehiro Kuragano (Yokohama, Japan) found the comet on July 29, but his announcement was not sent from Japan until August 13. Peter Cherbak, an American Airlines pilot, detected the comet on the morning of July 31, while flying between Denver and Omaha. He confirmed his discovery the next morning and informed the Griffith Planetarium (Los Angeles, California). A telegram announcing Cherbak's discovery was delayed until August 4, to allow confirmation at the California observatories, which was delayed due to bad weather.

Perihelion had occurred on August 1 (r= 0.35 AU), and beginning on the 4th, the comet was well seen. Observers consistently estimated the total magnitude as near 1 and the tail length as 4 to 5 degrees. As the comet faded, the tail grew, and by August 13, observers estimated the total magnitude as near 2.5 and the tail length as near 13 degrees. A straight, narrow tail, which had first appeared on August 10, was then extending 9.5 degrees. The comet was closest to Earth on August 13 (1.07 AU), and, thereafter, began to fade more rapidly. By September 2, the total magnitude was 4.5 and by mid-month, it was too faint for naked-eye observations. The original tail also faded, and by August 21, it

had greatly decreased in length; however, on the same date, the straight tail
had brightened considerably and displayed several knots of material along its
length. By September 30, the total magnitude had declined to 6.5 and on October 5, it was near 7. With the comet approaching conjunction with the sun,
observations became impossible after October 11.

The comet was recovered after conjunction on February 17, 1958, when
Roemer (U. S. Naval Observatory, Arizona) photographed a moderately condensed
nucleus of magnitude 16.8. During April and May, the nucleus was near magnitude 18 and the coma was 12 arcsec across. Observations continued until July
9, 1958, when Roemer found the nucleus near magnitude 19.0. The orbit is
elliptical with a period near 13,200 years.
(Mrkos)

1957 VI   Discovered: March 16, 1956   Discovery Magnitude: 15.0

Wirtanen (Lick) discovered this comet on a plate taken with a 51-cm astrograph. He described it as faint and diffuse, with a central condensation and a
short tail. Then located 6.14 AU and 5.28 AU from the sun and Earth, respectively, the comet was still 18 months from its perihelion passage and changed
little during the remainder of its visibility in 1956. This period of visibility ended after only June 2, due to an approaching conjunction with the sun.

The comet was accidentally recovered on April 30, 1957, by Bruwer and T.
Gehrels (Johannesburg, South Africa) and was reported as a new comet. The magnitude was estimated as near 10. Van Biesbroeck independently recovered the
comet on May 1, and described it as 15 arcsec in diameter with a tail 4 arcmin
long. His estimate of the total magnitude was 13 and he added: "There is a
second nucleus at 8 arcsec from the main nucleus and 2.5 magnitudes fainter."
Numerous other observations were made during the summer months. The total magnitude was 10 in May and June, but slowly faded thereafter as the distance from
Earth increased. The secondary nucleus also remained visible throughout this
period and was usually 2 to 3 magnitudes fainter and separated by 8 or 9 arcsec.
Perihelion came on September 2 (r= 4.45 AU), and the comet was observed until
September 19, when observations ended due to a second conjunction with the sun.

The comet was recovered on January 25, 1958, when van Biesbroeck (McDonald)
described the comet as 25 arcsec across. The main nucleus was near magnitude
16 and there was a vague suspicion of the second nucleus shining as magnitude
18 and located 15 arcsec away. Observations of both nuclei continued throughout the year and on December 11, Roemer estimated the magnitudes as 18.5 and
19.0. After a third conjunction with the sun, the comet was found on May 10,
1959. Roemer described the main nucleus as stellar and of magnitude 19.2,
while the second nucleus was slightly diffuse and of magnitude 20.5. The separation was then 30 arcsec. Observations continued into September, and the
comet was soon lost as it neared its fourth conjunction. It was not seen
again until it arrived at opposition and Roemer then photographed it on September 25, 1960. Only one nucleus was then visible and the total magnitude
was near 20. Researchers have established that the nucleus divided in 1955.
The orbit is hyperbolic with an eccentricity of 1.0027.
(Wirtanen)

1957 IX   Discovered: October 2   Discovery Magnitude: 8.0

This comet was first observed by Latyshev (Ashkhabad Observatory, Russia)
on October 16, while observing the short-period variable X Arietis with 7x50
binoculars. The comet was described as diffuse, with a motion of 10 degrees

per day to the southwest. Independent discoveries were made by Paul Wild
(Berne Observatory, Switzerland) on October 18,   and Robert Burnham, Jr.,
(Prescott, Arizona) on October 19.

The comet's rapid motion was due to an approaching perigee, which came on
October 21 (0.13 AU). The southwestward movement took the comet into an in-
convenient position with respect to the sun, and the final observation was
made on October 25. Shortly thereafter, prediscovery images were found on
sky patrol plates taken at Sonneberg on October 2, 3 and 5. An orbital calculation based on these positions and the few accurate positions obtained later
in the month revealed a perihelion date of December 5 (r= 0.54 AU). After
conjunction with the sun, the comet was searched for in January 1958 by
observers at Flagstaff, McDonald and Johannesburg observatories, but no
cometary object was found.
(Laytshev-Wild-Burnham)

1958 III   Discovered: February 10   Discovery Magnitude: 11.0

Burnham had been working at Lowell Observatory (Arizona) on February 21,
but when he arrived home, he quickly ate and then went outside to try his new
20-cm reflector. After only 15 minutes, he spotted this diffuse 9th-magnitude
object near Alpha Orionis. Burnham then proceeded to drive the 90 miles back
to Lowell and reported his find to acting director Earl C. Slipher. A telephone call was made to the U. S. Naval Observatory's Flagstaff station where
Elizabeth Roemer promptly made a photographic confirmation with the 102-cm
reflector. She described the comet as diffuse with a central condensation, and,
the next evening, she estimated the nuclear magnitude as 15.5. Upon the
announcement of the comet's discovery in Europe, astronomers at Sonneberg
Observatory located two prediscovery photographs taken on February 10 and
16. The total magnitude was estimated as about 11--two magnitudes fainter
than when discovered a short time later.

As early observations began coming in, two facts quickly surfaced about
this comet. First, it was two months from perihelion and, second, observers
consistently found the magnitude to be near 10. The change in magnitude was,
at first, attributed by some observers as an overestimation due to Burnham,
but later observations indicated the comet was experiencing minor variations
in brightness. These variations continued during March, April and May. One
of the largest occurred in mid-March, when van Biesbroeck (Yerkes) found a
magnitude of 10.2 on March 10, and 8.6 on the 12th. A few days later it had
dropped back to about 9.5.

Perihelion came on April 16 (r= 1.32 AU), and the comet slowly faded
thereafter. As May dawned, another large brightness variation occurred. On
May 6, van Biesbroeck estimated the magnitude as 9.0, on the 7th, it had brightened to 8.2, but on the 10th, it had dropped to 10.0. The tail was longest
in May, as Roemer observed it on the 10th as 25 arcmin long. By June 20, van
Biesbroeck recorded it as only a slight extension and a similar feature was
photographed on August 12. The coma varied little in size during the comet's
apparition. Roemer estimated the diameter as 2.5 arcmin on March 10, and both
she and van Biesbroeck found it to reach a maximum diameter of 4 arcmin between
May 6 and 10. It shrank rapidly after May, as the comet moved away from both
the sun and Earth, and by June 23, Roemer estimated it as 0.3 arcmin. Afterwards, observers noted only a trace. As the coma shrank, the brightness faded,
with magnitude estimates of 9.5 on June 20, 15 on July 21 and 17 on August 20.

The comet was last observed on September 14 and 15, when Roemer described it as a fairly well-condensed nucleus of magnitude 19, with little coma. The comet was then 2.8 AU and 2.5 AU from the Earth and sun, respectively. The orbit is elliptical with a period near 3.5 million years.
(Burnham)

1959 I   Discovered: September 7, 1958   Discovery Magnitude: 14.0
     Burnham and Charles D. Slaughter (Lowell Observatory) discovered this comet on a photo exposed with the 33-cm telescope during the proper motion survey. It was described as diffuse, with a central condensation and a fan-shaped tail.

     The comet changed little in brightness during the next few months, but by the end of December, it had brightened to magnitude 13.5. Although the tail remained visible during October to lengths of over 1 arcmin, photographs taken around November 4 and 10 by Roemer (U. S. Naval Observatory) showed no real tail--only an elongated coma measuring 0.7 arcmin wide and 1.3 arcmin long. The nucleus was then nearly stellar on the latter date, with a magnitude of 16.7. The tail was again visible during December, and was frequently measured as 30 arcsec long. During the first three months of 1959, the comet brightened and was considered to be at its maximum magnitude around mid-March. Beyer (Hamburg Observatory) then estimated the magnitude as 11.1, while van Biesbroeck (Yerkes) found a value closer to 13. The nucleus was then near magnitude 15.2 and the coma was estimated as 25 arcsec across. Perihelion came on March 11 (r= 1.63 AU), and the comet slowly faded thereafter. By June 2, it had faded to magnitude 14, and, at low altitude, the coma was measured as 8 arcsec across. The comet was not seen after June 4, as it neared conjunction with the sun.

     The comet was recovered on December 4, 1959, by Roemer and was described as well-condensed, with a nuclear magnitude near 19.5. Roemer continued observations for the next few months until April 21, 1960, when she described it as weak, but fairly sharply condensed, with a magnitude near 19.7. The orbit is elliptical with a period near 1.3 million years.
(Burnham-Slaughter)

1959 III   Discovered: June 2   Discovery Magnitude: 8.0
     During the first half of August 1959, M. J. Bester and C. Hoffmeister (Boyden Observatory, South Africa) were comparing photographs of the Sagittarius region taken on July 26 and 31 when they noticed the trailed images of an 8th-magnitude comet. The discovery was announced on August 15, and an approximate orbit was immediately calculated by J. Schubart (Sonneberg Observatory), which enabled the comet to be recovered at Boyden on August 20. Prediscovery images were subsequently found on Boyden photos taken between June 2 and 7 and between July 7 and 13.

     Perihelion had occurred on July 17 (r= 1.25 AU), and the comet had been closest to Earth on the 16th (0.23 AU). Subsequently, the comet faded after discovery as it traveled southeastward into Microscopium. The Boyden observers continued observations until September 11. The orbit is hyperbolic with an eccentricity of 1.0027.
(Bester-Hoffmeister)

1959 IV   Discovered: August 24   Discovery Magnitude: 10.0
     G. E. D. Alcock (Peterborough, England) discovered this comet in Corona Borealis while sweeping for comets with 25x105 binoculars. He described it as

faint and diffuse, with a rapid southeastward motion. Having passed perihelion on August 17 (r= 1.15 AU), the comet moved away from both the sun and Earth after discovery and faded rapidly. On September 1, van Biesbroeck described the comet as near magnitude 10, with a coma 1.3 arcmin across and a tail 4 arcmin long. By the 22nd, he described the comet as near magnitude 13, with a coma 1 arcmin across. Van Biesbroeck last detected the comet on October 21, when the coma had shrunk to 8 arcsec and the magnitude had faded to near 18. Roemer continued observing the comet until November 5, when she reported a magnitude near 19.5. The orbit is hyperbolic with an eccentricity of 1.0009.
(Alcock)

1959 VI   Discovered: August 30   Discovery Magnitude: 6.0
    One week after his discovery of 1959 IV, Alcock found this comet in the morning sky near Zeta Cancri. He described it as bright and sharply condensed, with a tail over 1 degree long. The comet moved into evening twilight there-after, as it neared perihelion. On September 1, H. L. Giclas (Lowell Observa-tory) photographed the comet and described the tail as narrow and slightly curved, with a length of 5 degrees. He also reported a straight jet 1 degree long to one side of the tail and a short spike on the other side of the tail. The comet brightened rapidly during the following days, with observers esti-mating the total magnitude as 5.7 on the 3rd, and 5.0 on the 4th. On the 6th, the comet was observed for the final time. It was then 22 degrees from the sun and Roemer indicated a magnitude near 4.5. The comet was closest to Earth on September 8 (0.86 AU), and perihelion came on the 16th (r= 0.17 AU). After conjunction with the sun, several observers searched for the comet dur-ing the latter half of October. On October 19, it should have been located 34 degrees from the sun and the total magnitude should have been near 10[*]; how-ever, photographs by van Biesbroeck (McDonald) on the 18th and 23rd, with a 25-cm camera, failed to detect the comet, although stars to magnitude 12 were recorded.
(Alcock)

1959 VII   Discovered: January 21, 1960   Discovery Magnitude: 14.0
    Burnham (Lowell Observatory) discovered this comet on a plate taken for the proper motion survey program. On confirmatory plates exposed at New Mexico State University that evening, the comet possessed a tail 8 arcmin long and was moving almost due north at a rate of nearly 30 arcmin per day. Having passed perihelion on September 28, 1959 (r= 1.17 AU), the comet faded rapidly after discovery. On February 19 and 20, van Biesbroeck (Yerkes) estimated the total magnitude as near 17, with a coma 10 arcsec across. By March 21, he estimated the magnitude as near 18, with a faint tail 1 arcmin long. On April 17, Roemer (U. S. Naval Observatory) obtained the final observation of the comet when she photographed "a weak cometary image of magnitude 19.8." No further observations were reported.
(Burnham)

1959 IX   Discovered: December 3   Discovery Magnitude: 8.0
    Having passed perihelion on November 13 (r= 1.25 AU), this comet was dis-covered three weeks later, by Mrkos (Skalnate Pleso Observatory) and was des-cribed as bright and diffuse, with a nucleus and a short tail. Numerous obser-vations were made during the remainder of December, as the comet faded from magnitude 8 to 10, but as 1960 began, the comet was only observed by Roemer (U. S. Naval Observatory) and van Biesbroeck (Yerkes). The latter observer

observed the comet with the 61-cm reflector until May 1, when he described the
comet as 10 arcsec in diameter, with a total magnitude near 17. Roemer continued observations for the next several months with a 101.6-cm reflector and
between April 21 and June 21, she noted a slight brightening of the nucleus
from magnitude 17.8 to 17.0 as the comet-Earth distance decreased. Thereafter,
fading was more rapid and when Roemer last observed the comet on September 26,
the nucleus was estimated as near magnitude 19.1. The orbit is elliptical with
a period near 351,000 years.
(Mrkos)

1959 X   Discovered: June 18, 1960   Discovery Magnitude: 17.0
    While conducting a photographic search for supernovae with the 122-cm
Schmidt camera at Mount Palomar Observatory, Milton L. Humason discovered this
faint comet in Hercules. Observations by Roemer (U. S. Naval Observatory)
within the next few days gave a nuclear magnitude of 18.2 and a tail length
of 5 arcmin.
    Perihelion had occurred on December 11, 1959 (r= 4.27 AU), and at discov-
ery, the comet was located 4.64 AU and 4.44 AU from the sun and Earth, respectively. In the following months the comet faded so slowly that by December
26, the nuclear magnitude had dropped by only 0.7 magnitude. Roemer continued
observations during the first half of 1961. On January 17, she described the
comet as 18 arcsec in diameter with a nuclear magnitude of 19.0 and when last
observed on June 7, she described it as possessing only a trace of coma, while
the essentially stellar nucleus was near 19.8. The orbit is hyperbolic with an
eccentricity of 1.0009.
(Humason)

1960 II   Discovered: December 30, 1959   Discovery Magnitude: 13.0
    Burnham (Lowell Observatory) discovered this comet on a photo of the Pisces
region, but was unable to immediately confirm it due to snowstorms. Instead,
Gibson and Wirtanen (Lick) photographically confirmed the comet the next evening.
Burnham's initial magnitude estimate had been 11, but observations by H. L.
Giclas (Lowell) on January 2, 1960, and van Biesbroeck (Yerkes) on January 5
indicated a total magnitude of 13. The latter observer then estimated a coma
diameter of 20 arcsec and a tail length of 1 arcmin. The comet brightened
during the next few weeks as it neared the sun. On January 28, Roemer (U. S.
Naval Observatory) estimated a total magnitude of 10.5 to 11 and photographed
two tails--one curved and 3 arcmin long, and the other straight and narrow and
15 arcmin long. The coma was then 1.3 arcmin across. By February 17, she
reported a total magnitude of 9.5 and a coma diameter of 0.7 arcmin. Shortly
thereafter, the comet moved into evening twilight.
    Perihelion came on March 21 (r= 0.50 AU), and the comet was recovered .on
March 25 by Spigl (Perth, Australia) in the morning sky. On the 26th, A. F.
Jones (Timaru, New Zealand) estimated the magnitude as 6.6 and, with the comet
rapidly approaching Earth, this amateur estimated the magnitude as 6 on April
8. Also around the 8th, the comet was again observed in the Northern Hemisphere
as it rapidly moved northward. On April 27, it was closest to Earth (0.20 AU)
and observers then estimated the total magnitude as between 4 and 4.5, with a
coma over 7 arcmin across and a narrow, straight tail which extended at least
11 degrees on photographs.
    After reaching a declination of +77 degrees on April 29, the comet moved
into the evening sky and began to fade. By May 12, van Biesbroeck estimated

the magnitude as 8.9 and by the 18th, it was near 10.5. On the latter date, the coma was 4 arcmin across and the tail was 28 arcmin long. On June 15, he estimated the total magnitude as 17. The comet was last detected on July 13, when Roemer photographed only a trace of the comet during a 60-minute exposure. The orbit is hyperbolic with an eccentricity of 1.0001.
(Burnham)

1961 II   Discovered: December 17, 1960   Discovery Magnitude: 8.0
    On December 26, 1960, M. P. Candy (Royal Observatory, England) was test-ing a new eyepiece on his 12.7-cm comet-seeker when he discovered this comet near the Cepheus-Draco border, less than 14 degrees from the north celestial pole. He described it as fairly bright and diffuse. A short time later, astronomers at Sonneberg Observatory (East Germany) found prediscovery images on sky patrol plates taken on December 17 and 24.
    The comet moved rapidly southward after discovery, but, although approach-ing perihelion, its increasing distance from Earth caused it to slowly fade. During the first half of January 1961, observers consistently estimated the total magnitude as near 8, but by the end of the month, it had dropped to 8.5. Perihelion came on February 9 (r= 1.06 AU), and by February 17, observers estimated the magnitude as close to 9. With the comet moving into evening twilight, it was last detected on February 21, as a fairly easy object, with a coma 4 arcmin across and a fairly sharp central nucleus. After conjunction with the sun, observations were resumed in the Southern Hemisphere. It was first detected at Perth on a photograph exposed on April 23. Observations continued until May 14. The orbit is elliptical with a period near 1,077 years.
(Candy)

1961 V   Discovered: July 23   Discovery Magnitude: 3.0
    Between July 23 and 25, this comet was independently discovered by numer-ous observers around the world. The first observers to report their sightings were A. Stewart Wilson, who spotted the comet near Tau Geminorum while navi-gating a Pan American 707 jet from Honolulu to Portland, Oregon, on July 23.48, and William B. Hubbard (McDonald Observatory), who discovered the comet on July 24.42. Later it was revealed that Anna Ras, a South African Airways stewardess, was the first observer of the comet when she detected it on July 23.11, while her plane was flying over Libya. Her description gave the tail length as 15 degrees, while Wilson and Hubbard gave magnitude estimates near 3.
    The comet had passed perihelion on July 17 (r= 0.04 AU), and after dis-covery, it moved northwestward and slowly faded. The comet was well observed during the remainder of July, with photographs revealing a tail between 21 and 25 degrees long on the 25th, and 12 degrees long on August 1. During the same time the total magnitude faded from 3.2 to 4.8, according to Alan McClure (Los Angeles, California). Also, on the 25th, McClure, van Biesbroeck and Hubbard all reported a sunward tail which extended 2 to 3 degrees. This anti-tail had vanished after only a couple of days. With the comet moving quickly away from both the sun and Earth, fading became more rapid during August, with the total magnitude dropping to between 6.5 and 7 by the 10th. On September 6, Roemer (U. S. Naval Observatory) estimated the nuclear magni-tude as 18.0 and by the 30th, she found it near 18.4. Roemer's last observa-tion came on October 12, when she photographed the nucleus at magnitude 19.5.

Roemer took two 120-minute exposures on November 3, but failed to detect any trace of the comet; however, Koichiro Tomita (Okayama, Japan) succeeded in securing two photographs of the comet on November 7 and 9, while using the new 188-cm reflector at that observatory. The orbit is elliptical with a period near 8,700 years.
(Wilson-Hubbard)

1961 VIII   Discovered: October 10   Discovery Magnitude: 8.0

Tsutomu Seki (Kochi Observatory, Japan) discovered this comet on the day of its perihelion passage (r= 0.68 AU). Then located near Beta Leonis, the comet was described as fairly bright, with a rapid motion to the south. Despite an increasing distance from the sun, the comet brightened in the following weeks as it neared Earth, with van Biesbroeck (Yerkes) estimating the total magnitude as 7 on October 17, and 6 by November 4. During this same time, observers described the tail as relatively faint and inconspicuous; however, photographs revealed it to be fairly long and narrow, with lengths of 4 degrees on October 14, and nearly 11 degrees on November 5.

Perigee came on November 15 (0.10 AU), with observers then estimating the total magnitude as near 4. Three days earlier, van Biesbroeck commented that the coma was nearly 1 degree in diameter and possessed no stellar nucleus. On the 11th, Roemer described the nucleus as weakly condensed, with a magnitude of 15.5. The comet faded rapidly thereafter, and was near magnitude 9 by December 6. Roemer obtained the final photograph of the comet on December 29, and described it as 1.5 arcmin across, with a nuclear magnitude of 19.0. The orbit is elliptical with a period near 759 years.
(Seki)

1962 III   Discovered: February 4   Discovery Magnitude: 8.5

This comet was discovered by Richard D. Lines and his wife (Phoenix, Arizona) with a 20-cm reflector. Nine hours later, Seki (Kochi Observatory) independently found the comet 3 degrees northeast of Zeta Puppis. These observers described the comet as between magnitude 8 and 9, with a coma 5 arcmin across and a suggestion of a tail.

The comet brightened after discovery, but was not an easy object for observation in the Northern Hemisphere as it moved southwestward from its discovery declination of -38 degrees. On February 10, observers estimated a total magnitude near 6.5 and gave a coma diameter of 6 arcmin and a tail length of 10 to 15 arcmin. During the latter half of February, the comet began to move northwestward, and it became well observed in the Northern Hemisphere. On the 26th, McClure (Hollywood, California) estimated the total magnitude as 5.5 in 12x70 binoculars, while his 20-cm reflector showed a tail 80 arcmin long. The next evening, he photographed the tail as 8 degrees long. By March 11, the comet had brightened to magnitude 4.7. Observations continued nearly until the end of March, although the comet's entrance in evening twilight caused many observational problems. During its last days of visibility before conjunction with the sun, the comet could only be detected in the Southern Hemisphere and the last observer was A. Marks (Sydney, Australia), who estimated the total magnitude as -1 on March 27.

Perihelion came on April 1.7 (r= 0.03 AU), and, although it should then have been located 2 degrees west of the sun at a predicted magnitude of -7.5, no daylight observations were reported. The comet was independently recovered by several amateur astronomers on April 3 and 4, the first of whom was K.

Hindley (York, England), who found it just after sunset on April 3, at a total
magnitude of -2.5. After moving further out of evening twilight, the comet
began to be well observed after April 7. On that date, Dennis Milon (Houston,
Texas) found the comet in Aries and estimated the tail length as 13 degrees
to the naked eye. On April 9, McClure photographed the tail as 15 degrees
long and described it as possessing a highly complex structure and on the 13th,
van Biesbroeck (Yerkes) estimated the tail length as 20 degrees long and the
total magnitude as 3.0. Thereafter, the comet rapidly became less conspicuous
as its tail shrank and the brightness faded. By April 24, Beyer (Hamburg
Observatory) estimated the total magnitude as 5.4, while the tail still extended a couple of degrees, and by May 7, van Biesbroeck found the magnitude near
7. At the beginning of June, the comet was lost in twilight as it neared
conjunction with the sun. It was recovered on October 27, when photographs
taken by Roemer revealed weak images of magnitude 20.2. Roemer continued to
photograph the comet until January 25, 1963, when she described it as weak, but
condensed, with a magnitude near 20.4. The orbit is hyperbolic with an eccentricity of 1.000.
(Seki-Lines)

1962 IV   Discovered: April 28   Discovery Magnitude: 8.0
      Honda (Kurashiki, Japan) discovered this fairly bright comet near Phi
Persei as it was moving at a rate of nearly 1 degree per day to the northeast.
Having passed perihelion on April 20 (r= 0.65 AU), the comet was consistently
described as diffuse without a nucleus after discovery, which apparently caused
great problems in determining the comet's total brightness. Calculations
indicated very little change in brightness throughout May; however, the comet
faded, with the rate of fading varying from one observer to another as the
comet became more diffuse. Van Biesbroeck (Yerkes) indicated a fading from
magnitude 8.2 on May 4, to 10 on May 24, 12 on May 30 and 14 on June 8. During
this same interval Beyer (Hamburg Observatory) found the total magnitude to
fade to only 8.9 by June 2, with an apparent rapid fading thereafter. Roemer
was closely monitoring the nuclear magnitude of this comet during May--as she
has done with so many other comets--and she indicated a fairly rapid fading
throughout May. On the 1st, the nucleus was described as a poorly condensed
spot of magnitude 13.9 and by the 22nd, it was described as weak and near
magnitude 15.5. After mid-June, Roemer was the only observer of the comet,
with her last observations coming on June 30 and July 2. On those dates, she
described the comet as a weak diffuse object nearly 1 arcmin in diameter and
without condensation.
(Honda)

1962 VIII   Discovered: September 1, 1961   Discovery Magnitude: 14.0
      Humason (Palomar Observatory) discovered this comet on a plate taken with
the 122-cm Schmidt camera and described it as faint, with a slow westward motion of 18 arcmin per day. Then located 5.26 AU and 4.60 AU from the sun and
Earth, respectively, this comet was still over a year away from passing perihelion and slowly brightened in the following months. Although most comets
appear as nearly stellar objects when located at the distance of this comet,
comet Humason soon surprised observers as it developed a tail several minutes
of arc in length. Of even greater interest was the unusual activity displayed
soon afterward as the comet appeared to be undergoing violent changes. The
first observer of this activity was Roemer, who on October 2 described the

coma as possessing a crab-like shape, with faint evidence of material some
5 arcmin to the south and southeast. Two days later, she said the coma had
extensions in all directions, as well as a contorted streamer at least 15
arcmin long. Before the end of 1961, the comet experienced two more episodes
of unusual activity and by the end of December, the total magnitude was near
11.

The comet was followed until mid-February 1962, when it had brightened
to magnitude 10, and was thereafter lost in the sun's glare as it neared
conjunction with the sun. Recovery came at the end of May, with the magnitude then being near 9. During the next two months, the comet continued to
brighten, though more rapidly than before, and by late August, it was at its
brightest. Magnitude estimates were then between 5.5 and 6. Also during the
summer months, the comet's tail displayed much turbulent motion and on 5
occasions between May 31 and August 28, the tail actually separated from the
comet's head. A slow fading set in thereafter, as the comet-Earth distance
slowly increased and observers commented on the surprising amount of activity
during September and October. Perihelion came on December 10 (r= 2.13 AU),
and the comet must then have been near magnitude 9*, although it was then lost
in the sun's glare. Observations during the first part of the year were only
possible in the Southern Hemisphere, but on May 24, 1963, Roemer again began
observations as the comet moved northward. Since her last observation on
September 9, 1962, the nucleus had faded from magnitude 14.8 to only 15.3.
Her photos taken during November and December 1963 gave nuclear magnitudes
of 16.6 and 16.8, as well as revealing further contortions in the comet's
tail.

On January 9, 1964, Roemer described the comet as possessing a nearly
stellar nucleus of magnitude 17.3 and by May 12, it had faded to magnitude
17.8. Her photos between June 2 and 14, revealed a further surprise as the
comet had increased drastically in brightness. Her estimates of the nuclear
magnitude were then near 14 but other observers estimated the total magnitude
as near 10.5. The comet was then situated about 6 AU from both the sun and
Earth. By the end of the year, Roemer found the nuclear magnitude back to
near 17. She continued her observations into 1965, and estimated a nuclear
magnitude of 17.7 on April 30, as well as a trace of tail extending to the
southeast for 0.7 arcmin. After conjunction with the sun, the comet was last
observed on October 31 and November 1, 1965, as a very weak and diffuse object.
The orbit is elliptical with a period near 2,925 years.
(Humason)

1963 I   Discovered: January 2   Discovery Magnitude: 12.0
Kaoru Ikeya (Maisaka, Japan) was searching for comets with his homemade
20-cm reflector when he discovered this comet near Pi Hydrae. He decribed it
as faint, with a fairly rapid southern motion. Observations by Northern Hemisphere observers continued until January 27, when Milon (Houston, Texas) used
6x30 binoculars to detect the 7.5-magnitude object. The coma was then 7 arcmin
in diameter and possessed a slight condensation. Southern observers intently
watched the comet as it brightened during February, and by February 11, the
total magnitude was near 4.2. Perigee came on the 16th (0.32 AU), and Ralph
L. Sangster (Brighton, South Australia) then described the comet as 20 arcmin
across, with "an intensely glowing nuclear condensation two arcmin in diameter."
Thereafter, the comet moved rapidly northward and was again visible in the

Northern Hemisphere by February 20. As perihelion neared, the comet continued
to brighten and reached a magnitude of 3 on March 2. During the remainder of
March, it slowly faded as the increasing distance from Earth countered the
decreasing distance from the sun, and when last detected on March 23, before
entering evening twilight, it had dropped to magnitude 4.7. The tail grew
as perihelion approached, and between March 14 and 20, photographs revealed a
length of about 19 degrees. Perihelion came on March 21 (r= 0.63 AU).

        After conjunction, the comet was recovered at a total magnitude near 9,
but instead of fading in the following months, it slowly brightened to near
magnitude 8 by late June. Thereafter, fading was fairly rapid, with the total
magnitude dropping to near 10.5 by July 18. The comet was followed until October 12, 1963, when Roemer (U. S. Naval Observatory) described the comet as a
weak, nearly stellar image of magnitude 19.2. The orbit is elliptical with a
period near 932 years.
(Ikeya)

1963 III   Discovered: March 1   Discovery Magnitude: 8.0
        Alcock (Peterborough, England) discovered this comet on March 19, after
more than a thousand hours of searching with 25x105 binoculars since his discovery of 1959 VI. The comet was then situated in Cygnus and was described as
fairly bright, with a strong condensation and a faint spine pointing away from
the sun. The comet brightened slowly in the following weeks as it approached
both the sun and Earth, and by May 1, it was described as near magnitude 7.
Perihelion came on the 6th (r= 1.54 AU), and the comet brightened an additional
one-half magnitude by the time it passed perigee around mid-month (0.73 AU).
By the 27th, observers found the comet near magnitude 7, but two days later, a
remarkable outburst in brightness had occurred and the comet attained a magnitude of 4. After a slow fading to magnitude 7 by June 23, the comet displayed
yet another surprise--this time showing a multiple nucleus. R. L. Waterfield
(Ascot, England) was the first to report his visual and photographic detection
of the 4 condensations on that night, but Gordon Solberg (Las Cruces, New
Mexico) independently found them the same evening. Despite these sightings,
researchers are still undecided as to the actual splitting of the comet, with
a photograph taken by Roemer on June 16 being used as evidence against the
split. She then described the nuclear condensation as sharp, and detected a
curved spine extending 1.2 arcmin to the east-southeast. The condensations
were also strung out in that direction on the 23rd. After July 16, observations were only possible in the Southern Hemisphere, and observers there followed the comet until August 8. The orbit is elliptical with a period near
22,300 years.
(Alcock)

1963 V   Discovered: September 14   Discovery Magnitude: 2.0
        This comet was discovered in Hydra by Zenon M. Pereyra (Cordoba Observa-
tory) as it moved southward away from the sun. Pereyra's initial magnitude
estimate of 2 may have been an overestimation, since observations by other observers between the 16th and 18th gave magnitudes between 6 and 6.5; however,
some researchers have suggested a possible outburst in brightness at discovery.
Pereyra also described the comet as diffuse, with a bright nucleus and a tail
12 degrees long. On the 16th, McClure (Hollywood, California) estimated the
tail length as 10.5 degrees.

        Perihelion had occurred on August 24 (r= 0.005 AU), and the comet was

identified as another member of the famous sungrazing group of comets. After
discovery, the comet faded rapidly with observers describing it as near magnitude 7 on September 23, with a nuclear magnitude of 13.2, and near magnitude
7.5 at the end of the month. Observations continued throughout October as the
comet faded to near 10th magnitude, but, thereafter, only Roemer (U. S. Naval
Observatory) and Tomita (Dodaira, Japan) were able to provide further observations. On November 9, the former observer described the nucleus as moderately
well condensed, with a magnitude of 17.2. She added that a possible secondary
nucleus was located 0.1 arcmin away. Though this was the only date the secondary nucleus was detected, researchers consider it to be real. Tomita photographed the comet on November 16 and 26, and Roemer photographed it on December 14 and 18. On the latter date, the nucleus was near magnitude 18.2. The
orbit is elliptical with a period near 900 years.
(Pereyra)

1963 IX   Discovered: November 22   Discovery Magnitude: 16.0
    In May 1967, Jean H. Anderson (Department of Astronomy, University of
Minnesota) reported her discovery of a comet on plates taken by Dr. W. J. Luy-
ten on four consecutive nights in November 1963. The photos were taken with
the 122-cm Schmidt camera at Mount Palomar Observatory and the comet was described as faint, with a wide tail 3 arcmin long. K. Aksnes and Brian Marsden
calculated two orbits that would satisfy the observations: one was parabolic
with a perihelion date of October 9, 1963 (r= 2.09 AU), and the other was elliptical, with a perihelion date of November 7, 1963 (r= 1.95 AU). Marsden added
that "no significance whatsoever should be attached to the value of the period
(5.5 years) of the elliptical solution, for this was merely adopted as a rea-
sonable lower bound." This comet's orbit bears some similarity to that of the
comet of 1585, which may also be periodic.
(Anderson)

1964 VI   Discovered: June 6   Discovery Magnitude: 6.0
    Tomita (Dodaira, Japan) discovered this comet in the morning sky and des-
cribed it as bright and diffuse, with a central condensation. Independent discoveries were made by Rev. Frederic William Gerber (Lucas Gonzales, Argentina)
1.5 days later, and by Honda (Kurashiki, Japan) 3 days later.
    The comet brightened after discovery and was widely observed between June
14 and 18 as it approached perihelion. The tail was visually described as
stubby and, when last detected before entering the sun's glare, the magnitude
was near 4.8. The comet reappeared in the evening sky on June 29, and passed
perihelion on June 30 (r= 0.50 AU). The greatest visual tail length was attained on July 1, when it appeared 2 degrees long. Photographs taken on that day
gave a length closer to 7 degrees and wide-angle photos taken between the 3rd
and 6th by McClure gave a length near 30 degrees. Curiously, McClure's photos
revealed a prominent kink in the tail at a distance of 24 degrees from the
head. The comet was brightest on July 2, when near magnitude 4.4, and began to
slowly fade thereafter, with magnitude estimates being near 5.5 on July 9. The
comet remained low in the evening sky during the first half of July as it moved
away from the sun and Earth and was lost in twilight after the 19th. The comet
arrived at conjunction with the sun in August, and was finally recovered on
October 11, by Roemer, who described the comet as 0.8 arcmin across, with a
rather sharp nucleus of magnitude 18.0 and a faint tail 5 arcmin long. Roemer
last detected the comet on January 26, 1965, when it appeared as a well-conden-

sed, but weak image of magnitude 19.2.   The orbit is elliptical with a period
near 1,364 years.
(Tomita-Gerber-Honda)

1964 VIII   Discovered: July 3   Discovery Magnitude: 8.0
    This comet was discovered near Gamma Tauri by Ikeya (Maisaka, Japan) as
a diffuse object, with a central condensation. It brightened afterwards, as
it neared both the sun and Earth, with magnitude estimates of 6 around July 21,
and 5 on the 30th. On the former date, the tail was already extending nearly
2 degrees on photographs.
    Perihelion came on August 1 (r= 0.82 AU), but the comet continued to
brighten rapidly as it neared Earth. On August 5, Roemer estimated the coma
as 3 arcmin across and the nuclear magnitude as near 12, and two days later,
observers were estimating the total magnitude as near 4. Perigee came on
August 12 (0.19 AU), and on the 14th, Dr. S. Archer (Wollongong University,
Australia) described the comet as near magnitude 3.2, with a coma 27 arcmin
in diameter and a visual tail 2 degrees long. Archer's extensive series of
observations continued into September. He indicated the longest tail length
as 4 degrees, which came on August 15, though, on a 3-minute exposure, it
actually extended 10 degrees and consisted of 4 branches. By August 26, Archer
described the comet as near magnitude 5.5, with a coma 7 arcmin across and a
visual tail of 50 arcmin. The comet continued to be observed until September
8, when Archer described it as near magnitude 7, with a coma 7 arcmin across.
Observations thereafter were impossible due to evening twilight. The orbit is
elliptical with a period near 391 years.
(Ikeya)

1964 IX   Discovered: August 5   Discovery Magnitude: 9.0
    Edgar Everhart (Professor of Physics, University of Connecticut) discovered
this comet on August 7, less than one degree southeast of Beta Librae, and described it as a diffuse, tailless glow. An independent discovery was made on
August 9, by John C. Bennett (Pretoria, South Africa) and a short time later,
Professor C. Hoffmeister (Sonneberg Observatory) reported finding two prediscovery photographs taken on August 5 and 6.
    The comet brightened slightly as it neared the sun and was near magnitude
8.5 during the latter half of August. Perihelion came on the 23rd (r= 1.26 AU),
and, thereafter, the comet slowly faded. On September 4, it was near magnitude
8.7, with a coma 3 arcmin across, and by the 12th, it was near 9.0. In early
October, the magnitude had dropped to 10 and one month later, it was near 12.
Dr. Roemer continued her observations of the comet thereafter, and estimated
the nuclear magnitude as 17.3 on November 26, and 18.0 on December 26. She
last observed the comet on February 25, 1965, as a small stellar object near
magnitude 19.5. The orbit is elliptical with a period near 6,860 years.
(Everhart)

1965 VIII   Discovered: September 18   Discovery Magnitude: 8.0
    Ikeya (Maisaka, Japan) and Seki (Kochi, Japan) independently discovered
this comet within 15 minutes of each other and described it as a diffuse, tailless glow just west of Alpha Hydrae. The earliest orbital calculations hinted
at a possible association with the sungrazers, and after this fact was firmly
established at the end of September, scientists around the world planned jet
flights and rocket launches to observe the comet to the fullest. Even the
National Aeronautics and Space Administration (NASA) decided to make observing

the comet part of the intended Gemini 6 mission, but problems caused the October 25th flight to be delayed for several months.

In true sungrazer fashion, the comet rapidly brightened as it approached perihelion with magnitude estimates reaching 6 on October 1, and 2 by October 15. On the latter date, the tail was 10 degrees long. The comet even became one of those rare comets to become visible in broad daylight. In fact, when the comet passed perihelion on October 21.18 (r= 0.008 AU), observers could see it just by blocking out the sun with their hands. G. de Vaucouleurs (McDonald Observatory, Texas) saw the comet with his naked eye on October 21.75 (noon local time) when it was only 2 degrees from the sun. He described it as possessing "a very bright nucleus with a silvery tail of 1 to 2 degree length." He also estimated the total magnitude as -10. On October 21.00, Roemer provided a similar description, but added that the tail showed marked curvature. Of major interest was an observation made by Japanese astronomers on October 21.16, when they used a coronograph at Mount Norikura to block out the sun. They then described a "disruption" of the comet into a possible three pieces just 30 minutes before perihelion passage.

After perihelion, the tail became the most spectacular feature as the comet faded. On October 28, estimates of the length were as high as 45 degrees, and on the 31st, they were as great as 60 degrees. As November began, the comet was quickly diminishing from its previous splendor. On the 4th, observers gave a magnitude near 5 and tail lengths near 20 degrees, but most interesting was the discovery of a secondary nucleus 14 arcsec from the main nucleus by Howard Pohn (U. S. Geological Survey, Arizona). A third condensation was suspected at a distance of 32 arcsec. Similar observations were made by Pohn on the following evening. The third nucleus was not observed thereafter, but the second was observed until January 14, 1966. Researchers calculated the time of splitting as October 21.20.

The comet continued to fade rapidly during November, and by the 26th, it was near magnitude 7.5, with a tail 15 degrees long. The comet was then only visible in the Southern Hemisphere, where Pereyra (Cordoba Observatory) continued to observe it until January 14, 1966. Possible images were obtained on plates exposed using Baker-Nunn cameras at Smithsonian observing stations as late as February 12, although larger telescopes failed to detect it. The orbit is elliptical with a period of 880 years. The secondary nucleus was moving in an orbit with a 1,056-year period.
(Ikeya-Seki)

1965 IX   Discovered: September 26   Discovery Magnitude: 10.0

Alcock (Peterborough, England) discovered this comet near Epsilon Herculis and described it as faint, with a fairly rapid southeastward motion. Although approaching both the Earth and sun, the comet faded and became more diffuse during the first week of observations. On September 29 and 30, Dennis Milon and van Biesbroeck (Steward Observatory, Arizona), as well as Milan Antal (Skalnate Pleso Observatory, Czechoslovakia), estimated a total magnitude of 11. On October 4, Milon and van Biesbroeck described the comet as extremely diffuse, with a magnitude near 12. According to Roemer, the nucleus did not fade so rapidly as she estimated it as near magnitude 16.5 on September 30, and near 16.7 on October 15. The comet was nearest Earth around October 21 (1.09 AU), and it passed perihelion on October 26 (r= 1.29 AU). Observations continued throughout November as the comet continued its southward motion, and

observations ceased on November 30, with Beyer (Hamburg Observatory) and B. Milet (Nice Observatory, France) being the final observers before the comet dropped below the horizon.
(Alcock)

1966 II   Discovered: August 15   Discovery Magnitude: 9.0
     Italian Astronomer Roberto Barbon (Mount Palomar Observatory, California) was conducting photographic studies of faint blue stars with the 122-cm Schmidt camera when he discovered this comet. On confirmatory photos exposed on August 17 and 18, Barbon described the comet as diffuse, with a central condensation and a tail 10 to 20 arcmin long. An independent discovery was made by Alan J. Thomas (Mount John University Observatory, New Zealand) on a plate taken for the Photographic Atlas of the Southern Sky on August 19. He confirmed Barbon's initial magnitude estimate.
     Perihelion had occurred on April 18, 1966 (r= 2.02 AU), and the comet faded after discovery, with magnitude estimates being near 11 at the end of August, and near 12 by mid-September. By October 7, the total magnitude was near 12.5 and the comet continued to move southward. It was situated low over the horizon when last detected on December 10, by Roemer. She then described it as possessing a well-condensed nucleus of magnitude 17. The orbit is elliptical with a period near 37,000 years.
(Barbon)

1966 IV   Discovered: September 8   Discovery Magnitude: 8.0
     Ikeya (Maisaka, Japan) discovered this comet during his regular program of comet hunting with his 20-cm reflector. The comet was in Coma Berenices and was described as diffuse and fairly bright. No confirmation was immediately available and the comet was independently discovered on September 12 by Edgar Everhart (Mansfield Center, Connecticut). He described the comet as circular, without a tail, and estimated the total magnitude as near 9.
     The comet had passed perihelion on August 5, 1966 (r= 0.88 AU), and faded fairly rapidly after discovery, with magnitude estimates being near 10 on September 17, and 11 by mid-October. Observations became very difficult as the comet moved into evening twilight and the last precise position was obtained on October 13 by N. S. Chernykh (Crimean Astrophysical Observatory, Russia). The comet was last recorded on Baker-Nunn films at the Smithsonian Florida station on October 16 and 17, and possibly at the Hawaii station as late as October 27.
(Ikeya-Everhart)

1966 V   Discovered: August 8   Discovery Magnitude: 10.6
     While engaged in photoelectric measurements of the variable star MM Herculis, with the 61-cm Lick Observatory reflector, Steven Kilston noticed a hazy object nearby. He described the new comet as diffuse, with a coma 30 arcsec in diameter and a central condensation. On August 13, James W. Young (Table Mountain Observatory, California) made a high-contrast print from a 60-minute exposure taken with a 250-mm telephoto lens and revealed a 10-arcmin-long tail. Throughout August and September, observers continually estimated the total magnitude as near 10 as the increasing distance from Earth countered the decreasing distance from the sun. After perihelion on October 28 (r= 2.38 AU), the comet slowly faded as it moved south-southeastward--allowing evening twilight to slowly catch up to it. Though observed throughout November, the comet was last detected near magnitude 11 on December 6, by Milet (Nice Observatory).

Although the comet was expected to be recovered after conjunction with the sun, searches during the summer and autumn of 1967 failed to reveal it. Van Biesbroeck's reported recovery on October 6, 1967, was later revealed to have been an object other than comet Kilston. The orbit is elliptical with a period near 236,300 years. It also bears a striking resemblance to the orbit of comet 1966 II.
(Kilston)

1967 II   Discovered: October 15, 1966   Discovery Magnitude: 13.5
    During a routine search for supernovae with Palomar Observatory's 122-cm Schmidt camera, Konrad Rudnicki discovered this comet near Gamma Ceti. It was then described as diffuse, with a central condensation and a tail 1 arcmin long.
    The comet brightened rapidly as it neared both the sun and Earth, with magnitude estimates of 13 around October 21, 11 in mid-November and 8.5 at the beginning of December. On December 9 and 20, Karl Simmons (Jacksonville, Florida) recorded a tail 20 arcmin long and, on the latter date, the total magnitude was 7.0. Perigee came on December 24 (0.40 AU), and the comet was near magnitude 6 when last detected in evening twilight on January 4, 1967. Perihelion came on January 21 (r= 0.42 AU), and the comet was recovered by several observers near magnitude 7 on January 28. One of the recoveries was made by J. C. Bennett (Pretoria, South Africa), who accidentally found it during a routine search for comets. The comet faded rapidly thereafter, with magnitude estimates of 9.5 on February 11, and 10.5 by the 20th. It was last detected at Tokyo on February 28. The orbit is hyperbolic with an eccentricity of 1.0004.
(Rudnicki)

1967 III   Discovered: February 11   Discovery Magnitude: 12.0
    Paul Wild (Zimmerwald, Switzerland) discovered this comet during a routine supernova search with the 40-cm Schmidt camera. Then located 8 degrees from the north celestial pole, the comet was described as diffuse, with a central condensation.
    The comet brightened fairly rapidly during the first days after discovery, as it neared both the sun and Earth, and when at perigee on February 16 (0.62 AU), it was near magnitude 10.5 or 11. Thereafter, the comet brightened more slowly as it continued to approach the sun and was near magnitude 10 on March 1, with a fan-shaped tail 4 arcmin long. Perihelion came on March 2 (r= 1.33 AU), and the comet faded rapidly thereafter, with Waterfield (Ascot, England) estimating a photographic magnitude of 12.5 on March 27. Roemer estimated a nuclear magnitude of 15.5 on April 10, and Milet (Nice Observatory) made the final observation on April 12.
(Wild)

1967 IV   Discovered: February 4   Discovery Magnitude: 11.0
    Seki (Kochi, Japan) discovered this comet using 20x120 binoculars less than one degree west of 106 Herculis. He described it as faint and diffuse, without a central condensation. Seki confirmed his discovery the following night and said the motion was to the northeast. Although the comet was then slowly moving away from Earth as February progressed, a rapidly decreasing distance from the sun caused it to brighten to magnitude 10 on the 13th, and 8.5 by the 18th. On February 12, Roemer described the comet as 0.6 arcmin in diameter, with a nuclear magnitude of 16.5 and a trace of a narrow tail to the

northwest. The comet was last detected on February 28, in Tokyo, as a difficult object in twilight. Perihelion came on March 13 (r= 0.46 AU), and no further observations were obtained thereafter.
(Seki)

1967 VII   Discovered: June 29   Discovery Magnitude: 5.0
     This comet went unobserved as it approached the sun from the side oppo-
site Earth, and passed perihelion on June 17 (r= 0.18 AU). Thereafter, it
moved slowly eastward and by the end of June it had emerged far enough from the
sun's glare to be briefly visible shortly after sunset. It was then independently discovered by Herbert E. Mitchell (Bowen, Australia) on June 29.39, M. V.
Jones (Maryborough, Australia) on July 1.35, and F. W. Gerber (Lucas Gonzales,
Argentina) on July 2.90. These men, as well as other observers during the first
days after discovery, estimated the total magnitude as near 5 and the tail
length as 3 to 7 degrees.
     The comet had been closest to Earth on July 1 (0.90 AU), and moved south-
ward after discovery. This made the comet a fine object in the Southern Hemisphere, but only 2 confirmed observations were made in the Northern Hemisphere:
the first was made by Michael McCants (Austin, Texas) on July 4, and the second
was made by Jose Olivarez (Mission, Texas) on July 8. With the distances from
both the sun and Earth increasing thereafter, the comet faded rapidly, with
magnitude estimates being near 6.5 on July 11, and 8.5 by the end of the month.
On August 24, Pereyra (Cordoba Observatory) found the total magnitude to be
10.8 and he last observed the comet on September 7, when near magnitude 12.0.
The comet was last detected on September 28, when the Smithsonian station at
Comodoro Rivadavia, Argentina,photographed it at a magnitude slightly fainter
than 12.
(Mitchell-Jones-Gerber)

1968 I   Discovered: December 28, 1967   Discovery Magnitude: 9.0
     This comet was independently discovered by Ikeya (Maisaka, Japan) and
Seki (Kochi, Japan) within 5 minutes of each other. The former observer had
used a 15-cm reflector, while the latter had used 20x120 binoculars. Both men
described the comet as diffuse, without a central condensation.
     As 1968 began, the comet slowly brightened as it approached both the sun
and Earth. At the end of January, observers described it as between magnitude
8 and 8.5, with a coma 3 arcmin across and a trace of a central condensation.
A yellow-light photograph taken by Milet (Nice Observatory) then revealed a
tail 7 arcmin long. Perihelion came on February 26 (r= 1.70 AU), and the comet
was then described as near magnitude 7, with a visual tail 30 arcmin long and
a coma nearly 12 arcmin across. Perigee came during the first half of March
(1.31 AU), and, thereafter, the comet faded slowly and moved northward. Curiously, an unexpected fading occurred after March 24. On that date, the comet
had been near magnitude 7, but 24 hours later, it was near magnitude 8, and by
the 30th, it was near 9.5. On the 31st, the comet had brightened back to its
expected brightness of 7. On the night of April 1 and 2, the comet passed 3
degrees from the north celestial pole. The slow fading continued during April,
as the comet moved southward, and it was near magnitude 8.5 by the end of the
month. Roemer photographed the nucleus on May 26, and described it as well
condensed, with a magnitude of 14.9 and a tail extending 10 arcmin to the
northeast. The final observation before conjunction with the sun was made by
Simmons (Jacksonville, Florida) on June 8, when the magnitude was near 9.5.

After conjunction, the comet was recovered on September 1 by C. Scovil and John E. Bortle (Stamford, Connecticut) on a photograph taken with the 56-cm Maksutov camera-telescope. On October 19, Roemer estimated the nuclear magnitude as 16.2 and three days later, Bortle estimated a total magnitude of 12.6. Roemer again photographed the comet on November 23, when its nearing opposition had increased the nuclear magnitude to 15. The next observations occurred at opposition in 1969: Tomita (Okayama, Japan) photographed it near magnitude 19 on October 9 and 10, and Roemer obtained the last observation on November 4, when the nucleus was near magnitude 21.5. The comet was then located 6.7 AU from the sun. The orbit is elliptical with a period near 89,500 years.
(Ikeya-Seki)

1968 III   Discovered: October 17   Discovery Magnitude: 15.0
Wild (Zimmerwald, Switzerland) discovered this comet on a photo taken with the 40-cm Schmidt camera and described it as faint and diffuse, without a central condensation. Other observers confirmed Wild's initial estimate in the following days, but at the end of October, both the discoverer and R. L. Waterfield (Ascot, England) reported the total magnitude as 14. There was then a distinct condensation and a faint tail 1 arcmin long.

The comet had passed perihelion on March 31 (r= 2.61 AU), and mathematicians indicated that the comet could have been found as early as November 1967, had photos been taken in the correct region of the sky. As November 1968 progressed, the comet continued its expected fading and by month's end, it was near magnitude 16. Interestingly, a pair of plates exposed by Roemer on November 23 showed two nuclei of magnitudes 18.0 and 18.4 and separated by 4 arcsec. The division seems to have occurred around August 3, 1968. The comet was last detected on December 23, when Roemer photographed only one nucleus of magnitude 18.4 and measured the tail as 0.4 arcmin long.
(Wild)

1968 IV   Discovered: April 25   Discovery Magnitude: 7.0
The uncanny ability of the Japanese to discover comets had been demonstrated several times during the early 1960's, but on the morning of April 30, 1968, this fact was reinforced greatly when this comet was independently discovered during a 75-minute period by 5 Japanese observers: Akihiko Tago (Tsuyama), Honda (Kurashiki), Hirobumi Yamamoto (Nangoku), Yasuo Sato (Nishinasuno) and Shigehisa Fujikawa (Onohara). The first three observers immediately notified the proper authorities and now have their names attached to this comet. There was only one other independent discoverer of this comet and he was also from Japan. Kimikazu Itaguki (Yamagata) had found the comet on April 25, but his report was delayed for several days.

At discovery the comet was near perigee (0.33 AU), and was described as fairly bright, without a condensation or tail. Widespread observations during the first mornings in May, gave coma diameters of 3 arcmin (visually) and 6 arcmin (photographically). As the month progressed, the comet moved rapidly northeastward and faded, with magnitude estimates being near 8 by mid-month. Perihelion came on the 16th (r= 0.68 AU), and the comet seemed to fade more rapidly thereafter. By the end of the month, the magnitude was near 9, with a coma 2 arcmin across. After reaching a maximum northern declination of 61 degrees on May 17, the comet moved to the southeast toward morning twilight. It was last detected on June 6, when the total magnitude was between 9.5 and 10.

After conjunction, Roemer attempted to photograph the comet on August 28, but her failure to find it on 60-minute exposures indicated a magnitude fainter than 20. The orbit is elliptical with a period near 1,911 years.
(Tago-Honda-Yamamoto)

1968 V   Discovered: June 15   Discovery Magnitude: 9.0
     After reading of the circumstances leading to the discovery of 1965 VIII, Mark A. Whitaker (Bishop, Texas) embarked on a regular program of comet hunting in June 1968. On his third night out, he spotted this comet with his 10-cm reflector near the globular cluster M5. The following night, he confirmed the cometary nature and, after establishing a rate of motion of nearly 3 degrees per day to the north, he reported his discovery. On June 17, Norman G. Thomas (Lowell Observatory) independently discovered the comet on a plate exposed for the asteroid Icarus with the 33-cm photographic telescope. The comet was described by Whitaker as diffuse, with a very small, but conspicuous central condensation.
     The comet had passed perihelion on June 4 (r= 1.23 AU), and perigee on June 7 (0.24 AU). Thereafter, it had faded slowly. On June 19, when near magnitude 8, McCants (Austin, Texas) estimated a coma diameter of 12 to 15 arcmin and Roemer (Catalina station of the Lunar and Planetary Laboratory, Arizona) gave a nuclear magnitude of 14.5. On the night of June 29-30, when near magnitude 9.5, R. L. Waterfield (Woolston Observatory, England) detected a secondary nucleus 30 arcsec from the primary on a photograph and his colleague, P. J. Keevil, suspected it on a photo exposed 1.4 hours later. No further observations were obtained. During the first week of July, the comet was near magnitude 10 and by the beginning of August, it had dropped to near 13. A photo by Waterfield on August 22, gave a total magnitude of 15 and a photo by Roemer on the 28th, gave a nuclear magnitude of 19.5 or 20. Roemer last detected the comet on September 22, when the nucleus was near magnitude 20.1.
(Whitaker-Thomas)

1968 VI   Discovered: July 6   Discovery Magnitude: 8.0
     Shortly before dawn on July 6.8, Honda (Kurashiki, Japan) discovered this fairly bright and diffuse comet midway between Eta and Lambda Aurigae, moving slowly northward. Twelve hours later, the comet was confirmed in the United States by four observers, the first of whom was Edgar Everhart (Mansfield Center, Connecticut). The other three observers were Milon (Massachusetts), Bortle (New York) and Simmons (Florida), who collectively described the comet as 3 to 3.5 arcmin in diameter, with a total magnitude between 8.0 and 8.3 and a central condensation of magnitude 11. A few days later, it was revealed that Fujikawa (Onohara, Japan) had independently discovered the comet about 20 minutes after Honda.
     As the comet approached both the sun and Earth during July, it steadily brightened, while the tail narrowed and lengthened. On the 31st, Antal (Skalnate Pleso Observatory) photographed the comet and estimated the tail length as 2 degrees, the coma diameter as 4 arcmin and the total magnitude as 6. Interestingly, on July 24, Waterfield (Woolston Observatory) photographed an antitail 2 arcmin long. Perihelion came on August 8 (r= 1.16 AU), but the comet changed little in appearance during the remainder of the month as its increasing distance from the sun was countered by a decreasing distance from Earth. Observers continually found the total magnitude near 6 and the visual tail length was near 15 arcmin. Photographically, the tail grew to a length of 3

degrees by month's end. After approaching to within 6 degrees of the north
celestial pole on August 28, the comet rapidly moved southward. Perigee came
on September 6 (0.65 AU), and several observers then reported the comet as
visible to the naked eye (magnitude 5.5). Fading was slow thereafter, with the
comet dropping to magnitude 7 by the end of September, and to magnitude 10 by
the end of October. The comet was last seen on November 10, when T. Urata
(Shimizu, Japan) detected it at a very low altitude. The orbit is hyperbolic
with an eccentricity of 1.0007.
(Honda)

1968 VII   Discovered: August 24   Discovery Magnitude: 11.5
    During the Southwestern Astronomical Conference at Las Cruces, New Mexico,
John Bally-Urban (Richmond, California) and Patrick L. Clayton (Springfield,
Missouri) discovered this comet while locating the planetary nebula M57 with
the latter observer's 25-cm reflector. The comet was then described as faint,
without a central condensation, but in better conditions on the following night,
a small condensation was detected in a coma 30 arcsec across.

    Perihelion had occurred on August 21 (r= 1.77 AU), and the comet slowly
faded thereafter as it moved away from both the sun and Earth. On August 27,
observers estimated the total magnitude as 11 and the coma diameter as slightly
less than 1 arcmin. Roemer then photographed the nucleus as stellar, with a
magnitude near 15.2. By mid-September, the total magnitude had dropped to near
11.5. On September 12, J. W. Young and J. Denman (Table Mountain Observatory,
California) visually observed a split nucleus with a 61-cm reflector. The
separation was estimated as 4 arcsec, with the magnitudes being 12 and 14. No
further observations of a second nucleus were made. On October 18, Beyer (Hamburg Observatory) estimated the total magnitude as 12.3 and on the next evening,
Roemer found the nuclear magnitude to be 17. The comet was last seen on November 24, when Roemer described it as 0.2 arcmin across, with a nucleus of magnitude 16.6. Twilight prevented observations thereafter, and the comet was not
recovered as expected during the spring months of 1969. It should have passed
within 5 degrees of the north celestial pole in mid-April 1969.
(Bally-Clayton)

1968 IX   Discovered: August 30   Discovery Magnitude: 10.0
    Honda (Kurashiki, Japan) discovered this comet in Monoceros as it moved
to the south-southeast at a rate of nearly 1 degree per day. He described it
as diffuse, without a central condensation. An independent discovery seems to
have been made by K. Ito (Hachinohe, Japan) on September 2, but the announcement was delayed for some time.

    The comet brightened and developed a tail as it approached both the sun
and Earth. The magnitude brightened to near 8 by the end of September, and
although observers began reporting a faint, short tail around mid-month, wide-
field photos taken at the Smithsonian station at Comodoro Rivadavia, Argentina,
revealed a split tail on September 17. Simmons (Jacksonville, Florida) seems
to have detected the split tail visually on September 23, and estimated the
lengths as 1 and 5 arcmin. The coma was then estimated as 4 arcmin in diameter.
Perigee came around October 5 (0.88 AU), and the comet slowly faded thereafter,
despite a steadily decreasing distance from the sun. A few days later, the
comet could no longer be observed from the Northern Hemisphere, due to its
southern declination, and on October 16, the comet passed 12 degrees from the
south celestial pole. Observers then estimated the total magnitude as between

8 and 8.5. Perihelion came on November 4 (r= 1.10 AU), and fading became more
rapid thereafter, with magnitude estimates of 9 on November 7, and 10.5 by the
30th. With the comet moving to the north, southern observers began experiencing problems in observing the comet due to low altitude and moonlight, and
no observations were reported after December 1, as the comet neared conjunction
with the sun. The comet was recovered by Roemer with a 154-cm reflector on
March 15, as a well-condensed object of magnitude 19.6. She last observed the
comet on April 19, 1969.
(Honda)

1969 I   Discovered: December 19, 1968   Discovery Magnitude: 13.0
     This comet was discovered on a large field proper motion plate by Thomas
(Lowell Observatory) and was described as diffuse, but strongly condensed. The
comet was then near the Cepheus-Camelopardalis border, less than 9 degrees from
the north celestial pole, and was moving slowly to the northwest. Observers
during the next few days indicated a total magnitude between 12 and 12.5 on
photographs and on January 11, Simmons (Jacksonville, Florida) detected the
comet visually with his 20-cm reflector. Simmons then described the comet as
1 arcmin in diameter with a total magnitude of 11.2. On the 8th, Waterfield
(Woolston Observatory) photographed two spikes extending from the central condensation and estimated their lengths as 1.5 and 2 arcmin.
     Perihelion came on January 12, 1969 (r= 3.32 AU), and the comet slowly
faded thereafter, as it moved away from both the sun and Earth. Waterfield's
magnitude estimates gave values of 13.0 for January 14, 13.3 for March 7 and
13.8 for April 8. Roemer gave a nuclear magnitude estimate of 17.5 for April
20. One month later, Roemer estimated the nucleus as near 18.5 and Waterfield
gave a total magnitude of 14.5. The coma was then 1 arcmin across and a perfectly straight spike extended 3 arcmin from the center of the coma. Roemer
reported the comet to be unexpectedly brighter on her photographs of June 23,
when she estimated a nuclear magnitude of 17.4, despite moonlight, and the comet
was last detected before entering twilight on July 5, when Giclas (Lowell Observatory) estimated the total magnitude as near 17.
     The comet was recovered by Roemer on January 5, as a well-condensed object
of magnitude 17.5 and when again observed on February 8, it had faded to 18.0.
The comet was last photographed on July 5, at a nuclear magnitude of 19.3, with
observations thereafter being impossible as the comet neared its second conjunction with the sun. It was again recovered near opposition on March 31, 1971,
when Roemer described it as fairly well-condensed images of magnitude 21.4.
Further observations were obtained on April 20, May 29 and 30, with the magnitude on the latter date being 21.0. The comet was then 8 AU from the sun. The
orbit is elliptical with a period near 18,750 years.
(Thomas)

1969 VII   Discovered: August 12   Discovery Magnitude: 11.0
     During a routine search for comets with a 16-cm reflector, Fujikawa (Ono-
hara, Japan) discovered this comet in the morning sky. Located in Taurus, the
comet was moving to the southeast and was described as diffuse, with neither a
condensation or tail.
     The comet brightened rapidly as it neared both the sun and Earth and by
August 24, Bortle (Mount Vernon, New York) estimated the magnitude as 9.0 while
using a 15-cm reflector and 16x50 binoculars. The coma was then 3.5 arcmin in
diameter. Bortle provided further estimates during September as the comet

brightened from magnitude 8.5 on the 10th, to 7.9 by the 21st. The coma was
between 3 and 3.5 arcmin across during this time and a tail extended 20 arcmin.
Perigee came around the 21st (1.35 AU), but the comet continued to brighten as
it neared the sun, and in the early days of October, it reached its maximum
brightness of 7. Perihelion came on October 12 (r= 0.77 AU), and the magnitude
had already declined to 8. Roemer then estimated the nucleus as near magnitude
13.5. With twilight affecting observations thereafter, the comet was followed
only until October 27, when Seki (Kochi, Japan) estimated the magnitude as near
9. Due to the lack of precise observations, only a parabolic orbit could be
calculated for this comet. Interestingly, I. Hasegawa (Nara, Japan) noted the
resemblance between the orbit of this comet and that of the comet of 1702,
though nothing further has developed on this possible identity.
(Fujikawa)

1969 IX   Discovered: October 10   Discovery Magnitude: 10.0
     Tago (Tsuyama, Japan) discovered this comet while searching for comets
with his 15-cm reflector. He described it as diffuse, without condensation, and
indicated a southeastern motion through southern Serpens Caput. Tago delayed
his announcement until he was able to confirm the comet two mornings later, at
which time it had apparently brightened to magnitude 9.5. Within the next 30
minutes, independent discoveries were made by Sato (Nishinasuno, Japan) and
Kozo Kosaka (Akasaka, Japan).
     The comet brightened as it neared both the sun and Earth, and was being
observed in both the Northern and Southern Hemispheres during the remainder of
October. By the end of the month, the total magnitude was near 8.5 and by November 8, it had increased to 8.0. After the latter date, northern observers
could no longer see the comet due to its southern declination and nearness to
the sun. In the Southern Hemisphere, observation of the comet was no easy
task, as its elongation from the sun decreased, but isolated observations still
came from F. Dossin (European Southern Observatory, Chile) on November 27, F. W.
Gerber (Lucas Gonzales, Argentina) on December 7, and A. Jones (Nelson, New
Zealand) on December 9. During this time the magnitude increased from 7 to 5.5.
     The comet was recovered on December 20, when observers estimated a total
magnitude of 3.5 and a tail length of 1 degree. After the comet passed perihelion on December 21 (r= 0.47 AU), the brightness remained virtually unchanged
from a value of 3 for the next month, as the comet approached Earth. At the
same time, the tail grew from 7 degrees on December 30, to 15 degrees after the
first week of January. On January 2, Jones observed a faint sunward jet 6 arcmin
long as the Earth passed through the comet's orbital plane. Of major interest
was the discovery on January 14 of a huge hydrogen cloud surrounding the comet.
The cloud was detected by the Orbiting Astronomical Observatory (OAO-2) and its
measured diameter of 1.5 degrees equalled a size 1.25 times the diameter of the
sun. This marked the first time a comet had ever been observed by a satellite.
     By mid-January, observations of the comet resumed in the Northern Hemis-
phere and on the 21st, the comet was at perigee (0.38 AU). Thereafter, the
comet began to fade and by the end of January, it was near magnitude 4.5, with
a tail about 2 degrees long. On February 2, Bortle (Stamford, Connecticut)
estimated the total magnitude as 5.2 in binoculars and the tail length as 30
arcmin in a 15-cm reflector. Other observers then reported a coma diameter of
6 arcmin. After magnitude estimates of 6 on February 5, the comet suddenly
brightened to nearly 5th magnitude by the 7th. Photographs on the latter date

revealed considerable tail activity and Roemer detected a jet 1 arcmin long
extending from the nuclear region. The brightness faded after the 7th, and was
back to normal by the 14th. By the end of February, the comet had faded to
near magnitude 7.5. During the first week of March, the comet was near 8th
magnitude and possessed no trace of a tail visually, although 3-minute exposures by Roemer with the 154-cm Catalina reflector soon afterward revealed a
weak trace of a tail extending nearly 12 arcsec to the east. A stellar nucleus
of magnitude 16.2 was also present. Curiously, on March 14, Roemer photographed a much fainter secondary nucleus 4 arcsec from the primary. Although no further observations of this nucleus were obtained, its position suggests a break-
up around February 9, 1970--very near the time of the comet's sudden outburst
in brightness. Further estimates of the comet's brightness thereafter gave
values of 9.5 on March 24, and 10.5 after the first week in April. On April 6,
Roemer estimated the nuclear magnitude as 17.2, and when she last observed the
comet on May 4, it had dropped to 18.1. The orbit is elliptical with a period
near 454,000 years.
(Tago-Sato-Kosaka)

1970 I   Discovered: January 26   Discovery Magnitude: 8.0
     The first comet discovered in the 1970's almost passed through perihelion
unobserved due to its being badly placed in the sky and observations were only
possible for 15 days before the comet became lost in the sun's glare. Takashi
Daido (Sendai, Japan) and S. Fujikawa (Onohara, Japan) independently discovered
the diffuse glow in Aquila within an hour of each other in the dawn hours of
January 27. Both observers reported their finds immediately to the Tokyo
Astronomical Observatory where the comet was confirmed the next morning by Dr.
H. Hirose. Soon after the announcement of the comet's discovery, Kiyotaka
Kanai, a high school student from Sakai, Japan, reported he had actually seen
the comet on the morning of January 26, and Honda (Kurashiki) found images on
a photo he had taken with a 5-cm patrol camera on January 27.
     The comet moved steadily southeastward at a rate of 1.5 degrees per day
and brightened rapidly as it approached perihelion. By February 1, its magnitude had increased to 7 and six days later, it was near 5. The tail increased
greatly from the time of discovery, despite the comet dropping deeper into the
brightening dawn as each day passed. Though no tail was noticed with the naked
eye by the discoverers, Seki (Kochi Observatory) and Urata (Nihondaira Observatory) both photographed a tail about 20 arcmin long on January 28. By February
5, observers were photographing the tail as 1 degree long, and on the 7th,
Roemer could see a tail at least 4 degrees long with the aid of binoculars.
The comet was last observed on February 9, when Seki reported a total magnitude
of 4. The comet was then 20 degrees from the sun.
     On February 15 (r= 0.066 AU), the comet passed perihelion and should then
have equalled Venus in brightness; however, even though the elongation was then
slightly less than 4 degrees, attempts to observe the comet visibly and in the
infra-red wavelengths were unsuccessful. By the time the comet had exited
from the sun's glare in March, it was too faint for further observations. The
orbit bears a striking resemblance to the orbit of the comet of 1577.
(Daido-Fujikawa)

1970 II   Discovered: December 28, 1969 Discovery Magnitude: 8.5
     Using a 12-cm Moonwatch apogee telescope, John Caister Bennett (Pretoria,
South Africa) was involved in searching for possible members of the Kreutz

sungrazing group of comets when he came upon a small, diffuse object only 24
degrees from the south celestial pole in Tucana. After Bennett reobserved the
comet on the following evening, it became obvious from the comet's direction of
motion that it was not a member of this group.

This comet quickly evolved into an impressive sight as perihelion approached. In mid-January, the magnitude was near 7, according to Pereyra (Cordoba
Observatory) and through photography he found a tail 25 arcmin long on the
20th. On February 9, A. F. Jones (Nelson, New Zealand) found the tail to extend 1 degree and he estimated the total magnitude as 5.5. By the 28th, Jones
described the comet as near magnitude 3.6, with a tail 2.2 degrees long. By
mid-March, the comet was becoming a spectacular object with a magnitude brighter than 2 and a tail near 10 degrees long. On the 19th, F. W. Gerber (Lucas
Gonzales, Argentina) estimated the total magnitude as 0.5, with a tail length
of 11 degrees.

Perihelion came on March 20 (r= 0.54 AU), but the comet changed little in
brightness during the next week as it neared Earth. On several occasions between the 20th and 27th, observers were reporting faint sunward jets and "pin-
wheel" streamers emanating from the nuclear region, and the tail was described
as 12 degrees long. By April 2, the magnitude had dropped to 1 and during the
first week of April, Bortle (Mount Vernon, New York) noted hoods similar to
those described in the coma of comet 1858 VI. Though the activity around the
comet's nucleus drew the attention of some observers, activity within the tail
attracted the attention of many others. Photographs continually showed the
dust tail as remaining fairly constant in length and shape; however, the gas
tail varied in intensity and structure from day to day. Observers at Woolston
Observatory obtained the most detailed observations of the gas tail during this
time. On April 4, they reported it as highly distorted, while on the 7th, it
was nearly straight with some irregular internal structure and was extending
13.5 degrees. On the 9th, it was again highly distorted and extended 11.5 degrees, but photos on the 10th, 11th and 14th showed very little trace of the
gas tail.

As with comet 1969 IX, this comet was observed by a satellite, OGO-5 (Orbiting Geophysical Observatory), and was shown to possess a hydrogen envelope
measuring 9 by 6 degrees on April 1 and 2. Between April 13 and May 13, the
OAO-2 satellite examined the comet with ultraviolet photometers and spectro-
meters and was able to help deduce the comet's production rates of key molecules
within the coma. This was the first time such information was gathered on a
comet.

Throughout April, the comet remained a conspicuous object as it slowly
faded. The tail remained near 20 degrees long during this time and the magnitude was near 4.5 by month's end. During May, the tail began to noticeably
shrink and by the end of the month it was near 9 degrees long. The total magnitude was then near 7. The comet faded slowly and remained visible in small
telescopes into August. On the 5th of that month, Bortle estimated the total
magnitude as 9.0 and by September 13, it had dropped to 10.6. Photographically,
the comet was much fainter, with Waterfield estimating the brightness as 10.5 on
July 14, and 12.8 by September 23. By late September, the comet was located
about 7 degrees from the north celestial pole and then began moving southward.
One month later it was no longer accessible throughout the night and then
possessed a visual magnitude near 12.5. By November 20, Beyer (Hamburg Obser-

vatory) estimated the total magnitude as 13.0, and 5 days later, Roemer (Lunar
and Planetary Laboratory) estimated the nearly stellar nucleus as of magnitude
18.1. The new year began with Roemer photographing the comet on January 21, as
a well-condensed object of magnitude 18.9. The coma primarily extended 6 arcsec to the southwest. The comet was last photographed on February 27, when it
was located 4.9 AU and 5.3 AU from the sun and Earth, respectively. Further
60-minute exposures by Roemer on June 28, while using the 229-cm reflector at
Kitt Peak Observatory, revealed no trace of the comet. The orbit is elliptical with a period near 1,680 years.
(Bennett)

1970 III   Discovered: July 23, 1969   Discovery Magnitude: 14.0
     On July 26, 1969, Lubos Kohoutek (Hamburg Observatory, West Germany) dis-
covered this comet on spectra plates exposed on the region of two novae discovered in 1968 in Vulpecula. Subsequently, he found two direct photos exposed
on July 23.97 and 24.00, which showed the comet as a 14th-magnitude, centrally
condensed object, with a tail 1 arcmin long.
     At discovery, the comet was still 8 months from perihelion and it slowly
brightened and developed as it approached the sun. Between early August and
mid-September, the comet brightened from magnitude 14 to 12.8 and the tail grew
from 2 arcmin to 4 arcmin. During the first half of September, Waterfield
(Woolston Observatory) continually photographed a tail consisting of 2 to 3
streamers; however, during November and December, only a single tail was visible. The comet was near magnitude 12 by the end of the latter month.
     By February 1970, this comet had reached magnitude 10 and remained vir-
tually unchanged during the next month, despite decreasing distances from both
the sun and Earth. Shortly before mid-March, Beyer (Hamburg Observatory) described the comet as near magnitude 9.9, with a tail 8 arcmin long. Perigee came
on March 13 (1.75 AU), and the comet passed perihelion on the 20th (r= 1.72 AU).
Thereafter, a slow fading set in with Bortle reporting the total magnitude as
10.8 on April 4, 11.0 on April 13, and 11.9 on May 8. On May 3, Roemer estimated
the nuclear magnitude as 16.7 and on May 24, Mrkos (Klet Observatory) observed
the comet for the final time before conjunction with the sun.
     The comet was recovered by Roemer on the morning of October 1, 1970, and
was described as possessing a sharply condensed nucleus of magnitude 17.5. As
the comet neared opposition, it brightened as expected, with nuclear magnitude
estimates of 17.2 on October 31, and 16.8 on November 29; however, unexpectedly,
Roemer photographed a secondary nucleus of magnitude 19.0 on October 31, which
was separated by 15 arcsec. By November 29, it had brightened to 18.8 and was
located 20 arcsec away. The comet remained observable during early 1971, as it
slowly faded, and when last observed on April 1, Roemer could see both nuclei
on the 60-minute exposure, which were separated by 33 arcsec. Researchers have
calculated the splitting date as April 29, 1970. The orbit is elliptical with
a period near 87,080 years.
(Kohoutek)

1970 VI   Discovered: May 18   Discovery Magnitude: 1.0
     This member of the Kreutz sungrazing group was hidden in sunlight as it
approached its perihelion date of May 14 (r= 0.009 AU); however, 4 evenings
later, it was discovered by Graeme Lindsey White (Barrack Point, New South Wales)
when only 12 degrees from the sun. It was then described as a star-like object
with a tail 1 degree long. Although White may have seen the comet again on the

19th, he had to wait until the 20th to confirm the sighting and then announce
his discovery. Independent discoveries were made by several observers after
May 19, although the first two announcements to arrive after White's were from
Emilio Ortiz (a pilot on the crew of Air Madagascar flight 281), who observed
the comet for 40 minutes while flying to Tananarive on May 21.63, and Carlos
Bolelli (Cerro Tololo Interamerican Observatory), who observed only the tail
on May 21.95, and the whole comet the next evening. The former observer estimated the tail length as 10 degrees.

The comet faded rapidly as it headed away from both the sun and Earth, with
observers reporting magnitudes of 4 on May 24 and 6 by May 31. On the latter
date, the tail was 3 degrees long. As June began, the comet began to fade more
rapidly than expected and observers described the coma as becoming more diffuse.
The comet was last observed on June 7, when both A. F. Jones (New Zealand) and
M. V. Jones (Australia) described it as near magnitude 9, with no distinct head.
The visual tail was then less than 6 arcmin long and the comet was located 22
degrees from the sun. The predicted magnitude for that date had been 7.2.
After conjunction with the sun, attempts to recover the comet were made by
Pereyra, in August, and Roemer, in October, but both were unsuccessful, despite
limiting magnitudes of 19.0 and 19.5, respectively.
(White-Ortiz-Bolelli)

1970 X   Discovered: October 19   Discovery Magnitude: 7.0
After having passed perihelion on October 2 (r= 0.41 AU), this comet was
independently discovered by Shigenori Suzuki (Kira, Japan) and Sato (Nishinasuno, Japan) within 5 minutes of each other on the evening of October 19. The
comet was confirmed the following evening by Tokyo Observatory, and shortly
thereafter, Seki (Kochi, Japan) announced his discovery of the comet. After
the first three announcements had been received, two further independent discoveries were announced. The first came from Toru Kobayashi (Imadate, Japan),
who spotted the comet on the 19th, and the second came from Tago (Tsuyama,
Japan), who found it on the 20th. All of these observers indicated a total
magnitude between 6 and 8, and described the comet as diffuse, without a central
condensation or tail.

The comet faded after discovery as it moved away from both the sun and
Earth, with Bortle describing it as 5 arcmin across and of magnitude 7.5 on
October 27. The next evening, Waterfield (Woolston Observatory) photographed
a faint, straight and narrow tail 10 arcmin long. As November began, observers
were remarking on how the comet was becoming more diffuse and estimates of the
total magnitude became very discordant. On the 2nd, Roemer estimated the center
of the poorly condensed coma as between magnitude 16 and 16.5, and on the 7th,
most observers found the total magnitude near 8.5. Beyer (Hamburg Observatory)
gave further total magnitude estimates of 9.8 on the 15th, and 11.4 on the
25th. The comet was last detected on November 30, when Seki estimated the
total magnitude as near 14.
(Suzuki-Sato-Seki)

1970 XV   Discovered: July 3   Discovery Magnitude: 10.0
Osamu Abe (Shinjo, Japan) discovered this comet with his 10-cm reflector,
while hunting for comets in the morning sky. He described it as diffuse and
uncondensed, but his magnitude estimate of 9 seems to have been an overestimate, since other observers in the following days found it to be closer to
magnitude 10.

The comet brightened after discovery as it approached both the sun and
Earth, with observers giving magnitude estimates of 8 at the end of July and
6 by the end of August. At the latter time, the comet had reached its maximum
northern declination of 74 degrees and, thereafter, moved to the southwest.
Perigee came during the first week of September (0.8 AU), and for the rest of
the month, the comet remained near magnitude 5.5--with some observers reporting it as faintly visible to the naked eye--as the decreasing distance from
the sun countered the increasing distance from Earth. In addition, two tails
were frequently noted throughout the month, with their maximum lengths being
reached on September 20, when the main tail was 2 degrees long and the fainter
tail was 45 arcmin long. On October 5, Roemer photographed a sharp, but definitely uncondensed nucleus of magnitude 14.5 and the comet passed perihelion
on October 20 (r= 1.11 AU). On October 28, Bortle estimated the total magnitude
as 7.5, and on the 31st, Roemer gave the nuclear magnitude as 13.4. The comet
seems to have undergone an outburst in brightness at the end of the first week
in November, when Bortle estimated the magnitude as 6.5 on the 7th, and Beyer
gave the magnitude as 6.6 on the 8th. This increased brightness persisted
through the 17th, when W. Singer (Keene, New Hampshire) gave an estimate of
6.5. The comet was lost in twilight after November 23, but was recovered at
magnitude 8.5 on December 6. Fading was slow thereafter, although a possible
outburst may have again occurred on January 7, when Bortle found the magnitude
to be 7.8. By the end of January 1971, the comet was near magnitude 8.5, which
it remained through the end of February. On February 28, Roemer estimated the
nuclear magnitude as 15.6, but by the end of March, it had apparently brightened to 15.2. Observers of the total magnitude continued to note a slower fading
than predicted up until the beginning of April, when on the 1st, Bortle gave
a value of 9.7, compared to the predicted value of 11.5. Curiously, as the
month advanced, a rapid fading set in, which dropped the total magnitude to
11.2 on April 23, and 13.4 by May 23. On the latter date, the predicted magnitude had been 13.5. The comet was last detected on May 27, 1971, when 30-minute
exposures by Roemer gave a nuclear magnitude of 16.9. The orbit varies little
from a parabola, with an eccentricity of 1.00005.
(Abe)

1971 I   Discovered: March 16, 1972   Discovery Magnitude: 16.0
     Having passed perihelion on January 6, 1971 (r= 3.28 AU), this comet was
not discovered until 14 months later, when Tom Gehrels (Palomar Observatory)
found it on photographs taken with the 122-cm Schmidt camera during a routine
search for new Apollo-type asteroids. The comet was then near opposition and
was described as diffuse and condensed, with a tail extending 10 arcmin to the
east-southeast.
     The comet steadily faded as it moved away from both the sun and Earth.
Roemer used a 154-cm reflector on April 13 to photograph well-condensed images
near magnitude 18.6, and on May 16, the comet had faded to 19.4. The comet was
last observed before the interference of twilight on July 4 and 8, when Pereyra
(Bosque Alegre) estimated the magnitude as 19.5. The comet remained lost in
the sun's glare until January 3, 1973, when R. E. McCrosky and C.-Y. Shao (Harvard Observatory's Agassiz station) estimated the nuclear magnitude as 20.0.
On the 8th, Roemer found the nuclear magnitude near 21.0 and detected a possible
tail extending 5 arcsec to the east. The comet was last observed on February
27, by Pereyra, when the heliocentric distance had increased to 7.4 AU. The

orbit is elliptical with a period near 1,072 years.
(Geherls)

1971 V   Discovered: March 7   Discovery Magnitude: 9.5
     After 2.5 years of searching for comets with an 11-cm altazimuth reflector,
Kenji Toba (Tsuchiura, Japan) discovered this comet in Pegasus on March 7.8,
1971. He described it as between magnitude 9.5 and 10, with a coma diameter of
3 arcmin and a slow movement to the southeast.
     The comet steadily brightened as it neared both the sun and Earth, with
total magnitude estimates increasing from 9 on March 22 to 8 by April 17. The
tail was first detected on March 27, with the use of photography, when Roemer
detected a trace extending to the west of a 14.5-magnitude nucleus. Waterfield
estimated the length as 1.5 arcmin on photos exposed on April 17, and Bennett
reported the first visual sighting on April 24.
     Perihelion came on April 17 (r= 1.23 AU), and the comet continued to
brighten and develop a tail afterwards, as it neared Earth. On May 7, several
observers gave the magnitude as 7.7, and the tail seems to have reached its
maximum length at that time as observers gave estimates near 1 degree. By
May 21, Bennett found the total magnitude to be 7.2 and the tail length was
near 15 arcmin. Due to the southern declination, Northern Hemisphere observers
could not observe the comet after May 26. Perigee came around June 8 (0.7 AU),
and the comet began to fade fairly rapidly. On June 14, it was near magnitude
9, and by the 26th, when only 4 degrees from the south celestial pole, it had
dropped to magnitude 10. Further estimates gave magnitudes of 10.5 in mid-July,
and 13.5 by August 20. The comet was last observed on September 9. The orbit
is hyperbolic with an eccentricity of 1.0008.
(Toba)

1972 III   Discovered: March 12   Discovery Magnitude: 10.0
     William A. Bradfield (Dernancourt, Australia) was searching for comets in
the morning sky on March 12.8, when he discovered this object in Piscis Austrinus. He described the comet as diffuse and uncondensed. Although Bradfield
continued to observe the comet on March 14, 16, 18 and 21, an unfortunate break-
down in communications prevented word of the discovery from reaching the proper
authorities until March 21. The comet was finally confirmed by another observer on March 22, when T. B. Tregaskis (Mount Eliza, Australia) estimated the
total magnitude to be near 9.
     The comet's southern declination prevented observations in the Northern
Hemisphere, but it was well observed by southern observers as it approached
perihelion. At the end of March, the magnitude was estimated as 8.6 by A. Jones
(Nelson, New Zealand), who also gave an estimate of 8.0 for April 9. On April
17, a photograph by Pereyra (Cordoba Observatory) showed a tail 1 degree long,
with indications of filamentary structure. Although perihelion had occurred
on March 28 (r= 0.93 AU), the comet had brightened until mid-April as it neared
Earth and after perigee at the end of April (0.83 AU), the comet began to fade
fairly rapidly. On May 3, the total magnitude was 8.6, but it dropped to 9.5
on May 11, and 10.5 by May 21. Northern Hemisphere observers could finally see
this comet shortly before mid-May, although the comet's faintness and low altitude allowed observations only by Roemer and Giclas, with the former observer
giving the nuclear magnitude as 15.7 on May 18. The comet was last observed on
June 2 and 7, by Bruwer (Johannesburg, South Africa). The orbit is elliptical
with a period near 11,000 years.
(Bradfield)

1972 VIII   Discovered: January 4, 1973   Discovery Magnitude: 12.0
     On January 11, 1973, Andre Heck (a Belgian astronomer from Liege Observa-
tory) was supervising three astronomy students from Liege University in the
operation of the 60-cm Schmidt camera at Haute-Provence Observatory in France.
While developing the photos, the French night assistant Gerard Sause noted an
unusual object on a plate exposed on a group of galaxies in Coma Berenices.
Heck immediately identified the object as a comet and two more plates were
taken to verify the object's movement. The comet was described as diffuse,
with a central condensation and a tail 30 arcmin long. Soon after the discovery announcement was made, prediscovery photos were found in Japan (January 4
and 11) and France (January 10).
     The comet had passed perihelion on October 5, 1972 (r= 2.51 AU), but it
slowly brightened after discovery as it neared Earth. Observers reported magnitudes of 11.5 on January 15, 11 on the 20th and 10.7 on the 26th, while
visual estimates of the tail length remained near 1 arcmin. Perigee came
around February 9 (1.98 AU), and Bortle reported the comet as at maximum brightness on the 10th, when at magnitude 10.6 with a visual tail length of 10 arcmin.
On February 6, Milet (Nice Observatory) photographed a tail 25 arcmin long.
Fading was slow thereafter, with observers estimating magnitudes of 12 at the
end of February, and 12.5 at the end of March. Curiously, on March 24, Bortle
found the visual tail length to be 24 arcmin. Mrkos provided the most extensive magnitude estimates thereafter, with values of 13.0 on April 5 and 14.2 on
May 3. The comet was last detected before being lost in the sun's glare on
May 25.
     After conjunction with the sun, the comet was recovered on November 21,
by Roemer. It was followed by observers in the United States, Japan and France
during December (magnitude 18) and January (magnitude 19). Roemer last detected the comet on February 26, 1974, when it possessed a nucleus of magnitude
19.6. The orbit is hyperbolic with an eccentricity of 1.0005.
(Heck-Sause)

1972 IX   Discovered: June 9   Discovery Magnitude: 13.0
     While photographing quasar fields with the 122-cm Schmidt camera, Dr.
Allan R. Sandage (Palomar Observatory) discovered this comet moving slowly in
Serpens Caput. The comet was described as uncondensed, with a coma 2 arcmin
across and a tail 30 arcmin long.
     The comet moved northwestward after discovery and changed little in bright-
ness in the next few months as its decreasing distance from the sun was countered by an increasing distance from Earth. On the other hand, Roemer indicated
a slow fading of the near stellar nucleus from a magnitude of 16.6 on July 10
to 17.1 in mid-August. Shortly after October began, the comet became a difficult object in evening twilight. It was last photographed on the 28th at
Woolston Observatory and was last detected visually on November 3 by Bortle,
who estimated the total magnitude as 12.9. Perihelion came on November 15 (r=
4.28 AU).
     The comet was recovered in the morning sky in January 1973, and on the 11th,
Urata (Nihondaira Observatory) reported a total magnitude of 13.5. The comet
again changed little in appearance during the next few months as the increasing
distance from the sun was countered by a decreasing distance from Earth. Observers then estimated a total magnitude near 13 and a nuclear magnitude near 16.5.
As the distance from Earth increased after April, the nucleus faded as expected

and was near 17.9 on July 2; however, observers of the total magnitude detected very little fading, and Bortle was still finding a value of 13 in early
September. During the latter half of July, the comet reached its maximum
northern declination of 72 degrees and moved southward thereafter. Observations continued into December, and on the 2nd, McCrosky and G. Schwartz (Agassiz station) noted a sharp, strong nucleus of magnitude 16.8. The comet remained lost in the sun's glare until June 17, 1974, when Roemer detected a nucleus
of magnitude 21.0 at the center of a coma 15 to 18 arcsec across. The comet
was last seen on November 10, when Roemer found the nucleus near magnitude 20.5
and the coma near 20 arcsec across. The orbit is hyperbolic with an eccentricity of 1.0063.
(Sandage)

1972 XII   Discovered: November 15   Discovery Magnitude: 12.5
    Gaston Araya (Cerro Tololo Interamerican Observatory) discovered this comet
on a plate taken with the Curtis-Schmidt telescope on December 9.2. The comet
was then located in Doradus and was described as diffuse, with a condensation
and a possible tail. Nearly a month afterwards, C. U. Cesco (El Leoncito,
Argentina) independently found the comet on a 2-hour exposure taken on November 15.2, 1972, when the magnitude had been 13.0.
    Perihelion came on December 19 (r= 4.86 AU), and the comet's large dis-
tances from both the sun and Earth caused it to fade slowly as 1973 progressed.
The comet's southern declination never allowed it to become visible in northern
skies and the comet was lost in the sun's glare at the end of April. After
conjunction with the sun, it was recovered on July 5, when Bruwer (Johannesburg,
South Africa) estimated the total magnitude as 13.3. Observations continued
through the end of the year, when the magnitude had dropped to near 14.
    Although the comet was detected by C. Torres (Cerro El Roble station of the
University of Chile) in January 1974, it was again nearing conjunction with the
sun and was soon lost in the sun's glare. When it was recovered in April, Torres
was again the first to see it. The fading continued as the year progressed and
in September 1974, observers at Carter Observatory (Wellington, New Zealand)
estimated the total magnitude as 15. On October 13, the nuclear magnitude was
estimated as 18.5 to 19.0. This seems to have been the final observation of
this comet, although Pereyra (Bosque Alegre) reported to have photographed it on
June 17, 1975, at a magnitude near 16. However, Torres found no trace of the
comet on 5 photos exposed in June, July and August, and the observation is
considered questionable. The orbit is elliptical with a period near 12.5 million
years.
(Araya)

1973 II   Discovered: October 31, 1972   Discovery Magnitude: 14.0
    Nobuhisa Kojima (Ishiki, Japan) discovered this comet on a plate he had
exposed for the periodic comet Giacobini-Zinner with a 31-cm reflector. The
new comet was then located nearly 1 degree north of the periodic comet and was
described as diffuse, with a condensation.
    The comet brightened as it approached both the sun and Earth, with magnitude
estimates of 12 around mid-November, and 11 in early December. With perigee
finally coming during the latter half of December (1.45 AU), the comet reached
a maximum brightness of 10.5 during the final days of December and the early
days of January. At the same time, the visual tail was extending up to 5 arcmin and Roemer was estimating the nuclear magnitude as 15.0. Although perihelion

arrived on February 12 (r= 2.15 AU), the comet slowly faded throughout January, and was at magnitude 11.0 on February 10. By the 27th, it had further dropped to magnitude 11.6. As March began, observations became very difficult as the comet approached conjunction with the sun. It was last detected by Mrkos (Klet Observatory) on March 5, at a magnitude of 12.8. The comet was recovered on August 1, at a magnitude near 14, and observers continued to estimate the brightness as 14 until the end of September. Roemer continued visual observations into November, but the comet's growing faintness prevented further observations until the next opposition in 1974, when it was recovered on May 25, at a magnitude of 19.3. Roemer and observers at the Agassiz station continued observations until August 21, when Shao (Agassiz) obtained faint, but measurable images of magnitude 20.0 to 20.5. The orbit is hyperbolic with an eccentricity of 1.0010.
(Kojima)

1973 III   Discovered: April 25   Discovery Magnitude: 13.0
     Having passed perihelion on March 11 (r= 2.38 AU,), this comet was discovered by John P. Huchra (Palomar Observatory) on a photo taken with the 122-cm Schmidt camera. It was described as diffuse and condensed, with no tail.
     The comet slowly faded as it moved away from both the sun and Earth. On May 5, K. Suzuki (Toyota, Japan) estimated the total magnitude as 13.5, and by the 23rd, Kojima (Ishiki, Japan) and Seki (Kochi, Japan) gave magnitudes of 14 and 14.5, respectively. Roemer obtained a 30-minute photo on May 31, which showed the nucleus to possess a magnitude of 18.4, and by June 5, it had faded to 18.7. On June 2, observers at Woolston Observatory gave the total magnitude as 14.5 to 15.0 and the coma diameter as 1.5 arcmin. Roemer detected the comet for the final time on July 1, when the nuclear magnitude was 19.1. The orbit is elliptical with a period near 821 years.
(Huchra)

1973 VII   Discovered: February 28   Discovery Magnitude: 14.0
     While searching for the minor planet 1971 UP$_1$, L. Kohoutek (Hamburg Observatory) discovered this comet near Delta Leonis. He described it as centrally condensed and tailless, with a nuclear magnitude of 14.5.
     Magnitude estimates varied greatly during March and April, due apparently to a two-staged coma. Observers at Woolston Observatory were the first to notice this unusual structure, when on March 5, they described an inner coma as highly condensed and 20 arcsec across, while an outer coma was faint and very diffuse, with a diameter of 3 arcmin. Short photographs to obtain positional measurements revealed only the inner coma which possessed a total magnitude between 14 and 15, while visual estimates with fairly large telescopes revealed a total magnitude of 13 or brighter. Using a 32-cm reflector, Bortle (Stormville, New York) estimated the total magnitude as 12.1 on March 24, and found the comet slightly brighter in early April. The brightness changed very little during April and May, as the increasing distance from Earth countered the decreasing distance from the sun: Bortle continued to estimate the total magnitude as 12 to 12.5, while Roemer obtained nuclear magnitudes near 17. The comet was last detected before conjunction with the sun on June 3.
     Perihelion came on June 7 (r= 1.38 AU), and the comet was recovered on August 1, when Kohoutek gave the total magnitude as 15. By September 22, the comet had faded to 17, and Roemer obtained the final observation on October 22. The orbit is elliptical with a period near 35,600 years.
(Kohoutek)

1973 IX   Discovered: November 24   Discovery Magnitude: 15.0

    During the first days of December 1973, J. Gibson (El Leoncito, Argentina)
announced his discovery of a moving object which he suspected was a comet; however, neither the discovery photo nor additional photos taken on November 29 and
30, gave definite proof of a cometary nature. Finally, on December 7, Gibson's
colleague C. U. Cesco photographed a short stubby tail, and Gibson himself
took simultaneous yellow- and blue-light photos on December 20, which showed
faint streamers nearly 15 arcsec long.

    At discovery, the comet was located in Horologium and never came north of
declination -50 degrees thereafter. Perihelion had occurred on August 10, 1973
(r= 3.84 AU), and after discovery, the comet slowly faded as it headed away from
both the sun and Earth. Between January 15 and 17, 1974, Pereyra (Bosque Alegre)
estimated the total magnitude as 16.0 and A. C. Gilmore (Carter Observatory,
New Zealand) photographed a tail 1 arcmin long on January 22. On January 29,
Gibson obtained the final observation prior to conjunction with the sun. The
comet was recovered by Gilmore on July 26, as a faint, diffuse spot 20 arcsec
across. On October 11, Torres (Cerro El Roble) obtained the last observation
when the comet was located only 1 degree from the south celestial pole. The
orbit is hyperbolic with an eccentricity of 1.0006.
(Gibson)

1973 X   Discovered: July 4   Discovery Magnitude: 15.0

    This comet was discovered by A. R. Sandage (Palomar Observatory) on a plate
taken of the same quasar field in which he had discovered comet 1972 IX 13
months earlier. It was described as diffuse, with a condensation and a tail
1.5 arcmin long. On July 7, McCrosky and Shao (Agassiz) confirmed both the
discovery and Sandage's initial description.

    The comet changed very little in brightness after discovery, as the de-
creasing distance from the sun was countered by the increasing distance from
Earth, and on September 26, McCrosky and Schwartz (Agassiz) estimated the total
magnitude as 15.7. The approaching conjunction with the sun temporarily ended
observations thereafter, but, after perihelion came on November 8 (r= 4.81 AU),
the comet was recovered on December 26, by Seki (Kochi Observatory). Seki and
other observers continually estimated the magnitude of the central condensation
as 17 between January and May 1974. The comet arrived at opposition in late
March, and was then described by Roemer as possessing a nuclear magnitude of
19.5. By June 16, the nucleus had faded to 20.0 and a tail was estimated as
extending 1.5 to 2 arcmin to the east. After a second conjunction with the
sun, Roemer recovered the comet on December 20, at a nuclear magnitude near
20. During January and February 1975, Seki estimated the brightness of the
central condensation as near magnitude 19, while on January 6, Roemer estimated
the nuclear magnitude as 20.6. At opposition in March, Roemer found the nucleus at magnitude 20.2, and she obtained the final observation on June 3,
when the comet was at a low altitude as it approached its third conjunction
with the sun. The orbit is hyperbolic with an eccentricity of 1.0001.
(Sandage)

1973 XII   Discovered: January 28   Discovery Magnitude: 16.0

    This comet was discovered on March 18 by Kohoutek (Hamburg Observatory)
while examining plates exposed on March 7 and 9 for the minor planet 1971 UG.
The comet was then described as diffuse and centrally condensed, and its slow
movement of 12 arcmin per day allowed Kohoutek to easily recover it on March

21. Kohoutek was later able to identify images of the comet on a plate exposed on January 28.9.

At discovery, the comet was located near the orbit of Jupiter and preliminary calculations by Marsden indicated it would be a bright object in late
December and early January. The comet brightened as predicted during the next
month as it neared conjunction with the sun, and was estimated as near magnitude 14.5 by Mrkos (Klet Observatory) at the end of April. Kohoutek was the
last observer to detect the comet before it entered the sun's glare, when he
photographed it on May 5. On September 23, Seki (Kochi Observatory) recovered
the comet in the morning sky at a magnitude near 10.5 and on the 29th, Roemer
photographed a tail at least 2 arcmin long and a nucleus near magnitude 14.5.
The comet steadily brightened during October, with Bortle (Stormville, New
York) giving estimates of 10.0 on the 1st, and 8.3 by the 27th. At the end of
October, observers were photographing a tail 15 arcmin long. During November,
the comet brightened to 7 at mid-month, and 5.5 by the 30th, and photographs
around the latter date were revealing a tail about 4 degrees long.

December was the month of the perihelion passage and the comet moved
steadily deeper into the dawn twilight as each day passed. During the first
week, observers reported a total magnitude near 5 and by the 13th, the comet
had brightened to 4th magnitude. The longest tail length seems to have been
observed on the 18th, when Walter Haas (Las Cruces, New Mexico) estimated it
as 18 degrees long. Thereafter, twilight steadily cut the tail shorter each
day and the comet itself could not be detected after an observation by J. C.
Bennett (Pretoria, South Africa) on the 22nd, when near magnitude 3. Although
observations were then impossible for ground-based observers, the comet was
followed by astronauts aboard the Skylab space station right through the perihelion passage on December 28 (r= 0.14 AU), and with the use of a coronograph,
they were able to photograph the comet 12 hours before perihelion passage when
it was located 30 arcmin from the edge of the sun at a magnitude near -3.

The Skylab astronauts again detected the comet on December 29.0, and
reported a spectacular sunward tail 22 hours later. The first visual observations began on Earth on December 31, when shortly after sunset it was visible
at a magnitude of -0.5. In the following days, as the comet moved out of twilight and into the western sky, it steadily faded from a magnitude of 3 on
January 4, to 5 on the 15th, 6 on the 22nd, and 7 by the 31st. Despite the
rapid fading, the comet managed to develop an impressive tail as it arrived at
perigee (0.8 AU) at mid-month, with estimates of the length being as great as
25 degrees. In addition, the anti-tail noted by the astronauts was observed
on Earth beginning on January 1.7, and remained visible throughout the month,
with a maximum length of 30 arcmin coming on the 17th.

At the end of the first week of February, observers indicated a total magnitude of 7.5 and a visual tail length of nearly 1 degree. On February 13, a
photograph by Waterfield (Woolston Observatory) gave a tail length of 4 degrees
and the anti-tail was still extending 15 arcmin. The anti-tail persisted during the remainder of the month and Bortle indicated the total magnitude faded
from 7.9 on the 19th, to 8.5 by the 25th. On March 11, Bortle estimated the
total magnitude as 9.0 and he still reported a trace of the anti-tail;
however, by the 18th, Waterfield could detect no trace of this feature on
photographs. After Bortle's observation on March 22, when the magnitude was
10.0, few additional visual observations were made. During April, as the comet

moved back toward evening twilight and became fainter, only one observation was made, and that came on the 26th, when Roemer obtained photographs which gave the nuclear magnitude as 18.0. After conjunction with the sun, several attempts were made to recover the comet in October and November, and, although these were thought to have been fruitless, a faint image was later found on Roemer's photos exposed on November 10. The image was estimated as near magnitude 22, and accurate measurement indicates it to be almost exactly in the predicted position. The orbit is essentially a parabola.
(Kohoutek)

1974 III   Discovered: February 12   Discovery Magnitude: 9.0
     Bradfield (Dernancourt, Australia) discovered this comet near Mu Sculptoris and described it as diffuse, without a condensation. On photographs taken by M. P. Candy (Perth Observatory) on February 14, a tail 5 arcmin long was detected and, at about the same time, Tregaskis (Mount Eliza, Australia) estimated the total magnitude as 8.5 and the coma diameter as 4 arcmin.

     The comet brightened after discovery as it approached both the sun and Earth, with observers giving estimates of 8 on February 20, and 7 on March 1. On the latter date, a tail was detected visually with a length of 3 arcmin, while on photographs it was 17 arcmin long. The comet was first detected in the Northern Hemisphere on March 9, when K. Simmons and M. Rogers (Switzerland, Florida) detected it at a low altitude with a coma diameter of 4 arcmin. By the 12th, P. Maley (Houston, Texas) estimated the total magnitude as 6.2, and on the 15th, Bortle (Stormville, New York) estimated the magnitude as 5.0.

     Perihelion came on March 18 (r= 0.50 AU), but the comet continued to brighten thereafter, as it neared Earth, and was widely reported as near magnitude 4.5 at the end of the month. Two tails were commonly detected on photographs taken during the latter half of March. On the 18th, Dr. R. G. Roosen (Joint Observatory for Cometary Research, New Mexico) reported that photographs taken with the 37-cm Schmidt camera showed a gas tail 5 degrees long and a dust tail extending 1 degree. Another observation, on March 22, gave values of 9 degrees and 3 degrees, respectively. Perigee came on April 1 (0.67 AU), and the comet began to fade fairly rapidly thereafter. On the 7th, it had faded to magnitude 5, while on the 14th, observers were referring to it as near magnitude 7. At the Joint Observatory for Cometary Research, photographs on April 9 gave 8 degrees as the length of the gas tail, while on the 14th, it was 1.5 degrees long, with a dust tail 30 arcmin long. At the end of April, the total magnitude was near 8.

     The comet passed 32 arcmin from the north celestial pole on May 15, when the magnitude had dropped to 8.5, and moved southeastward thereafter. On the 24th, Bortle reported a total magnitude of 9.0 and on the next night, Roemer estimated the nuclear magnitude as 16.5. On June 17, Bortle gave a total magnitude of 10.7 to 10.8 and by August 16, it was near 12.8. On the latter date, Roemer gave the nuclear magnitude as 19.2. On September 11, Roemer gave the nuclear magnitude as 18.8 and added that a trace of a fan-shaped tail extended to the northeast, and on the 15th, McCrosky (Agassiz) estimated the total magnitude as 17.5. The comet was last detected at Agassiz on October 12 and November 18. The orbit is elliptical with a period near 67,700 years.
(Bradfield)

1974 VIII   Discovered: July 26   Discovery Magnitude: 14.0
     Carlos U. Cesco and his son Mario R. Cesco (El Leoncito, Argentina) were

photographing the minor planet 1973 EE, when they discovered this comet. The
magnitude was estimated as 14 on the 51-cm double astrograph plates and the
comet possessed a central condensation, but no tail.

The comet became an elusive object soon after its discovery because of an
almost immediate change in the direction of motion. The discoverers were able
to obtain accurate positions on July 27 and 28, and, from the indicated southwestern motion, the Carter Observatory (New Zealand) exposed small-field plates
on the area of the extrapolated position on August 2. No images appeared, however, as the comet had reached its maximum southern declination of -35 degrees
on July 28, and, thereafter, moved to the northwest.

The discoverers obtained further precise positions on August 7 and 8,
which allowed Marsden to calculate a rough orbit which showed the comet to have
passed perihelion on May 13 (r= 1.37 AU). The orbit also indicated the comet-
Earth distance was increasing rapidly and that the elongation from the sun was
decreasing. The comet was next photographed by the discoverers on September
18, when the magnitude was very close to 17, and it was last detected by Carlos
Torres and S. Barros (Cerro El Roble) on October 8, when near magnitude 18. The
comet was then located 2.4 AU from the sun and 3.0 AU from Earth. The orbit is
elliptical with a period near 525 years.
(Cesco)

1974 XII   Discovered: November 12   Discovery Magnitude: 17.0
Canadian astronomer Sidney van den Bergh (Palomar Observatory) discovered
this comet near the galaxy M33 on photos taken with the 122-cm Schmidt camera.
He described it as faint and diffuse, with an uncondensed coma 8 arcsec in diameter and a tail 2 arcmin long.

Having passed perihelion on August 8, 1974 (r= 6.02 AU), the comet slowly
faded after discovery as it moved away from both the sun and Earth. Seki (Kochi
Observatory) estimated the total magnitude as 17 in mid-November, and 18 on
December 16, while Roemer estimated a nuclear magnitude of 19.2 on November 16,
and 19.4 on December 16. Curiously, Roemer found the nucleus near magnitude
18.0 on December 19, and she noted its fading to 18.5 on January 6, 1975, and
18.8 on February 5. During January, February and March, the total magnitude
was consistently found to be 18, and the comet was temporarily lost after March
7, as it neared conjunction with the sun.

Roemer recovered the comet on September 4 and 12, and estimated the nuclear
magnitude as 19.8 on the latter date. As the comet-Earth distance decreased
with the approaching opposition, the nucleus brightened to 19.2 on October 7,
and 19.0 on December 4. On November 5, van den Bergh photographed the comet
and found a tail 4 arcmin long. A few observations were made during the first
months of 1976, as the comet-Earth distance again began increasing, and the
comet was again temporarily lost in the sun's glare as it neared conjunction.
It was recovered late in the year and was seen for the final time on December
27, 1976, when the magnitude was near 20. The orbit is hyperbolic with an
eccentricity of 1.0039.
(van den Bergh)

1974 XV   Discovered: November 13   Discovery Magnitude: 9.0
J. C. Bennett (Pretoria, South Africa) discovered this comet shortly before
dawn with a 12-cm refractor. He described it as diffuse and uncondensed, and
on the following morning, he was able to establish its motion as 40 arcmin due
south.

As the comet approached both the sun and Earth, it brightened as expected, with Bennett estimating the total magnitude as 8.5 on November 14, and 8.3 on the 15th. Curiously, 18 hours after the latter observation, the comet was exhibiting a sudden decrease in brightness as Seki (Kochi Observatory) found the total magnitude to be near 9, and on the 17th, A. F. Jones (Nelson, New Zealand) estimated the total magnitude as 10.6. On the 18th, Jones found the magnitude near 10.7, while Bennett simply described it as fainter than 10. The comet seems to have stabilized in brightness from that date until the 25th, as observers at Carter Observatory continually estimated the total magnitude as 10 during this time. Their observation that the comet was becoming larger and more diffuse as each day passed probably accounts for the failure of many observers to detect it at that time.

Several observatories searched for the comet photographically during December, but reported their attempts as fruitless; however, in early February, Torres (Cerro El Roble) re-examined his plates exposed on December 8, 9 and 10-- when the comet was nearest Earth (0.3 AU)--and located faint nebulous images measuring 5 x 15 arcmin. Subsequently, Shao (Agassiz) was then able to find a faint, cloud-like structure measuring 10 arcmin in diameter on a photo taken on December 5--when the comet was 8 degrees from the south celestial pole. Despite further searches of plates exposed at Carter Observatory (December 12 and 15), Catalina (December 16) and Agassiz (December 17), no further images were found. The comet passed perihelion on December 1 (r= 0.86 AU).
(Bennett)

1975 II   Discovered: February 25, 1976   Discovery Magnitude: 15.0

Hans-Emil Schuster (European Southern Observatory, La Silla, Chile) discovered this comet on plates exposed with the 100-cm Schmidt camera on February 25, March 3, 4 and 5. He described it as possessing some tail structure to the north. On March 5, Roemer estimated the nuclear magnitude as near 17.5 and on the 20th, Schuster gave a tail length of 15 arcsec to the north-northeast.

Having passed perihelion on January 15, 1975 (r= 6.88 AU), this comet was discovered near opposition in Centaurus, and remained basically unchanged in brightness during the next couple of months, with observers at Carter Observatory estimating the nuclear magnitude as 17.5 on April 25. Observations continued during June, with Torres (Cerro El Roble) giving the tail length as 1 arcmin on the 2nd, and 20 arcsec on the 25th. On both days, the tail was described as narrow and faint. As the comet neared conjunction with the sun, it was last detected on July 28, when the nuclear magnitude must have been near 18. The comet was recovered on January 29, 1977, as it neared opposition, with Schuster then estimating a tail length of 8 arcsec to the north. Observations continued at La Silla and Carter observatories during February, and on May 13, Schuster estimated a nuclear magnitude of 19. As the comet neared its second conjunction with the sun, it was last detected on June 18, 1977, when Gilmore (Carter Observatory) estimated the nuclear magnitude as 18.9. The comet was again recovered on January 8 and 9, 1978, by Richard M. West (European Southern Observatory), who estimated the tail length as 15 arcsec. No further observations were obtained, although observers at Perth Observatory claimed to have again photographed this comet on April 7, 1978. Computations disproved the Perth object's identity with Comet Schuster. The orbit is hyperbolic with an eccentricity of 1.0022.
(Schuster)

1975 V   Discovered: March 12   Discovery Magnitude: 9.0

Bradfield (Dernancourt, Australia) discovered this comet in southern Cetus as a diffuse object without condensation. Observers confirmed his initial magnitude estimate of 9 in the following days, although Tregaskis (Mount Eliza, Australia) did detect some condensation on March 13. On the 16th, F. Dossin (European Southern Observatory) estimated the brightness of the central condensation as 12.5, and added that a fan-shaped coma extended 15 arcsec north of east. On the 19th, D. Herald (Woden, Australia) gave the coma diameter as 2 to 3 arcmin.

Although the comet moved towards the sun and Earth after discovery, the slow decrease in the distances caused it to change very little in brightness, with magnitude estimates of 9 being given throughout March and into early April. Perihelion came on April 4 (r= 1.22 AU), at which time Pamela M. Kilmartin (Carter Observatory) photographed a central condensation measuring 20 by 40 arcsec and possessing a magnitude of 11. Although the comet-Earth distance continued to decrease slightly during the following two weeks, the comet faded rapidly, with Herald giving the total magnitude as 9.5 to 10 on April 28. By May 19, Herald gave the magnitude as 11. As the comet neared conjunction with the sun, it was last detected at Carter Observatory on June 17 and 18, and the photographic images were then described as weak and diffuse.

The comet was recovered on January 2, 1976, by Shao (Agassiz) as it neared opposition. On the 8th, Seki (Geisei, Japan) estimated the total magnitude as 17, and on the 27th, Roemer and D. Daniels (Steward Observatory) estimated the nuclear magnitude as 18.7. Further observations were obtained at Agassiz on February 28 and March 7, and the comet was last detected on April 5, 1976. The orbit is hyperbolic with an eccentricity of 1.0014.
(Bradfield)

1975 VIII   Discovered: March 21, 1974   Discovery Magnitude: 13.0

While conducting a supernova-patrol program with the 60-cm Schmidt camera at the Konkoly Observatory (Budapest, Hungary), Miklos Lovas discovered this comet 2 degrees northeast of Gamma Virginis. The comet was described as diffuse with condensation, and was moving slowly to the south.

At discovery, the comet was located about 5.5 AU and 4.4 AU from the sun and Earth, respectively. Although still over a year from passing perihelion, the comet was due to pass opposition with Earth in late April, and, therefore, was expected to brighten slightly. Curiously, the comet faded, with Shao (Agassiz) giving a total magnitude of 14 for March 26, and observers at Klet Observatory and Japan giving it as 15 to 16 during mid-April. Similarly, the nuclear magnitude also declined, with Shao and Seki agreeing to estimates of 15 for March 26 and 27, and Roemer (Catalina) giving it as 17 on April 20. After opposition, the comet changed little during May and June as the decreasing distance from the sun countered the increasing distance from Earth. On May 15, observers at Agassiz estimated the nuclear magnitude as 17 and detected a faint, broad tail 10 to 15 arcsec long. As the comet neared conjunction with the sun, it was last detected on June 20, when Roemer gave a nuclear magnitude of 17.6.

The comet was recovered on January 8, 1975, by Gilmore (Carter Observatory) at a nuclear magnitude of 16.3. Due to the continued southern motion, Seki acquired the only observations from the Northern Hemisphere for the year on January 9 and 13, when the total magnitude was near 15.0. As the year progressed, it became very well placed for southern observers as it attained a declination

between -60 degrees and -70 degrees for most of the year. According to Gilmore
and Kilmartin, the nuclear magnitude steadily increased to 15.9 on March 18,
15.4 on April 16, and 15.0 on May 12. On the former date the coma was 30
arcsec across, while on the latter date it was 1 arcmin across. Although the
comet moved away from Earth after mid-April, the decreasing distance from the
sun caused it to continue to brighten and Gilmore and Kilmartin reported nuclear
magnitudes of 14.6 to 14.9 between June and August. During the same months,
Bruwer (Hartbeespoort, South Africa) gave the total magnitude as 13.0.

Perihelion came on August 22 (r= 3.01 AU), and the comet faded very slowly
thereafter, with Gilmore and Kilmartin giving nuclear magnitude estimates of
14.4 and 15.0 on October 31 and November 30, respectively. Further observations by Torres (Cerro El Roble) on December 4 and 6 gave a tail length of 5
arcmin to the southwest. With the comet moving northward after September 1975,
it was finally lost in the sun's glare at the end of the year and was not recovered until July 26, 1976, when Gilmore photographed it at a nuclear magnitude of 16.5. By September 19, Roemer photographed the nucleus and estimated
a magnitude of 18.3, as well as detecting a trace of tail. Gilmore continued
observations until November 21, shortly before the comet again entered the
sun's glare. It was recovered by Shao and Schwartz on August 15, 1977, and
they last recorded it on September 11. The orbit is elliptical with a period
near 658,200 years.
(Lovas)

1975 IX   Discovered: July 2   Discovery Magnitude: 8.0
While sweeping for comets with a 15-cm reflector, Toru Kobayashi (Takefu,
Japan) discovered this object about 2 degrees from Beta Aquarii. With the help
of three friends, Kobayashi determined the position of the comet, but bad weather prevented them from determining the direction of motion. The comet was
described as 12 arcmin across, without any condensation. Kobayashi's announcement reached the Central Bureau for Astronomical Telegrams on July 4, but without a clue as to the comet's direction of motion, all searches on the 4th and
5th failed. Subsequently, beginning on the 5th, numerous independent discoveries were made around the world, and the first two observers to report their
independent discoveries were D. Douglas Berger (Union City, California) and
Dennis Milon (Mount Washburn, Wyoming).

The comet brightened rapidly as it neared both the sun and Earth, with
Bortle estimating a total magnitude of 5.6 on July 14. The coma was then described as 25 to 30 arcmin in diameter, with a bright inner coma 18 arcmin across.
A few hours later, Roemer (Steward Observatory) photographed a central condensation of magnitude 14.0 to 14.5. Photographs by some observers shortly before
and after that date revealed a tail about 1 degree long. The comet moved rapidly northward after discovery and was located in Draco on July 21, when near
perigee (0.25 AU). Bortle, then visiting in Japan, described it as 18 arcmin
in diameter on that date--despite a full moon--and estimated a total magnitude
of 4.9. Although the distance from Earth increased thereafter, a decreasing
distance from the sun caused a slight brightening to magnitude 4.5 during August. The tail developed rapidly beginning at the end of July, and by mid-
August, photographs revealed a length of 13 degrees. Several observers then
reported several straight and narrow streamers flanking the sides, as well as
screw-like twists in the main tail. At the end of August, when the comet was
involved in strong twilight, observers were still estimating a visual tail

length of nearly 8 degrees.

Perihelion came on September 5 (r= 0.43 AU), and the comet rapidly diminished in both size and brightness thereafter. On September 29, Bortle estimated a total magnitude of 6.4 and by October 16, A. F. Jones (Nelson, New Zealand) found it near 8.0. On the latter date, the tail was less than 1 degree
long on photographs. By mid-November, the comet had faded to magnitude 10.5
and during the first days of December, observers at Carter Observatory estimated the nuclear magnitude as 15.2. Observers at Cerro El Roble were still
photographing a tail 25 arcmin long and a coma 10 arcsec across on December 30.
The comet was last seen on January 28, 1976. The orbit is hyperbolic with an
eccentricity of 1.0001.
(Kobayashi-Berger-Milon)

1975 X   Discovered: October 5   Discovery Magnitude: 8.5
Within 30 minutes on October 5, this comet was independently discovered by
5 Japanese observers: Shigenori Suzuki (Kira), Yoshikazu Saigusa (Kofu),
Hiroaki Mori (Mukegawa), Kiyomi Okazaki (Kahoku) and Shigeru Huruyama (Tone).
Each described the comet as diffuse, without condensation or tail. It should
be noted that only 70 minutes prior to this discovery on October 5, Mori had
discovered 1975 XII. Ten hours later, Giovanni Casari (Novi di Modena, Italy)
independently found this comet with his 25-cm reflector.

The comet moved at a rate of only 8 arcmin per day after discovery due to
a nearly direct approach toward Earth and it brightened rapidly as October
progressed. Bortle estimated the total magnitude as 8.2 on October 8, 7.0 on
the 21st, and 5.5 on the 28th. During the same period of time, he described
the coma as increasing in diameter from 3 arcmin to 12 arcmin, although on the
latter date, a very faint, diffuse cloud 1 degree in diameter was suspected as
surrounding the coma. Although perihelion had occurred on October 15 (r= 0.84
AU), the comet had continued to brighten as it neared Earth and was at perigee
(0.104 AU) on October 31. Between October 30 and November 1, the comet was at
inferior conjunction with the sun, and when recovered on November 2, by observers in the Southern Hemisphere, it was near magnitude 4.5. On the 4th, B.
Nikolau (Palmerstown, New Zealand) estimated the total magnitude as 4.1. Fading was rapid thereafter, with A. F. Jones (Nelson, New Zealand) estimating the
total magnitude as 5.6 on November 7, 7.5 on the 11th, 9.2 on the 23rd, and 9.5
on December 2. Nuclear magnitude estimates by Gilmore (Carter Observatory) on
December 5 and 6, were near 13.6. By the end of December, Jones remarked that
averted vision was needed to see the comet in his 32-cm reflector when the
total magnitude was 12.8. The comet was last seen on January 4, 1976, when
Gilmore gave the nuclear magnitude as 16.1. The orbit is elliptical with a
period near 446 years.
(Suzuki-Saigusa-Mori)

1975 XI   Discovered: November 11   Discovery Magnitude: 9.5
Bradfield (Dernancourt, Australia) discovered this diffuse and uncondensed
comet in Antlia after having spent 106 hours comet hunting since finding 1975 V.
The position and motion at discovery led many to believe the comet was a member
of the Kreutz sungrazing group, but the comet soon deviated from the expected
sungrazer path.

The comet brightened fairly rapidly as it neared both the sun and Earth,
with T. B. Tregaskis (Mount Eliza, Australia) estimating the total magnitude as
9 on November 13, 8 on the 28th, and 7.5 on December 1. He also detected a tail

5 arcmin long on the latter date. On the 5th, D. Goodman (Nelson, New Zealand) estimated the visual tail length as 45 arcmin and photographic estimates by Torres (Cerro El Roble) on December 6 and 7 gave values of 50 and 80 arcmin, respectively. After December 8, the comet was lost in twilight.

Perihelion came on December 21 (r= 0.22 AU), and the comet was recovered shortly after sunset on December 24, 25 and 26 by four Japanese observers, who estimated the total magnitude as between 1 and 3. Curiously, when next observed on December 31, by S. Furia (Varese, Italy), the comet had dropped to a magnitude of 6.7--nearly 3 magnitudes fainter than had been predicted. Furia's estimate was confirmed by other observers during the next few days, but the comet then began to fade slower than had been predicted, with estimates of 8 on January 9, 1976, 9 on the 18th, and 10 by the 31st. On the latter date, the observed brightness was less than 0.5 magnitude fainter than the predicted brightness. The comet was last observed on February 5, when Bortle estimated the total magnitude as 10.9. The orbit is parabolic.
(Bradfield)

1975 XII   Discovered: October 5   Discovery Magnitude: 10.5
This comet was independently discovered in northwestern Hydra by H. Mori (Mukegawa, Japan), Yasuo Sato (Nishinasuno, Japan) and Shigehisa Fujikawa (Onohara, Japan) within 70 minutes of each other. It was described as diffuse, with neither a condensation nor a tail and was moving fairly rapidly toward the south.

The comet brightened steadily in the following weeks as it neared both the sun and Earth, with Bortle estimating the total magnitude as 10.0 on October 8, 9.5 on the 21st, 8.9 on November 1, and 8.4 on the 16th. On November 3, Roemer obtained a magnitude estimate of 14.3 for the well-condensed nucleus. Northern Hemisphere observations ended shortly after mid-November, as the comet continued to head almost directly south, and southern observers were reporting a total magnitude near 8 at the end of November. Perigee came during the first days of December (1.3 AU), and the comet changed little in brightness during the remainder of the month as it neared perihelion. Although a visual tail was not detected until the end of December, when A. F. Jones (Nelson, New Zealand) gave an estimated length of 10 arcmin, a photograph at the European Southern Observatory on December 8, revealed a tail 1 degree long. On December 30, S. Barros (Cerro El Roble) photographed a tail 2 degrees long.

Perihelion came on December 26 (r= 1.60 AU), and on the 27th, observers were estimating the total magnitude as 8.5. At the end of the month, the comet was located only 9 degrees from the south celestial pole and began moving almost directly northward thereafter. During January 1976, the comet faded by only 0.5 magnitude and on the 31st, R. R. D. Austin (Mount John Observatory) observed a tail 5 arcmin long in his 15-cm reflector. By the end of February, the comet had reached a total magnitude near 10 and one month later, it had dropped an additional one-half magnitude. At the beginning of April, the comet slowed its movement toward the north and began drifting toward the southwest as it passed its superior conjunction with the sun. With the comet-Earth distance decreasing thereafter, the comet was expected to fade very slowly over the next two months; however, although Jones estimated a total magnitude of 10.8 on April 17, his estimate for May 7 was 12.7--1.7 magnitudes fainter than predicted. Curiously, Jones gave magnitude estimates of 12.4 for May 9 and 12.2 for the 10th, and this may indicate that the comet had undergone a temporary drop in brightness shortly before May 7.

The comet was at opposition during the first half of June, and by the 25th, Kilmartin (Carter Observatory) estimated the nuclear magnitude as 14.6. Gilmore (Carter Observatory) gave further nuclear magnitude estimates of 16 on August 22 and 16.9 on September 18. The comet was last observed on October 24, 1976, when Gilmore estimated the nuclear magnitude as 17.1. The orbit is elliptical with a period near 15,775 years.
(Mori-Sato-Fujikawa)

1976 I   Discovered: November 30, 1975   Discovery Magnitude: 9.5
Sato (Nishinasuno, Japan) discovered this comet in Coma Berenices on December 5.7, and described it as diffuse, without a central condensation. E. A. Harlan (Lick Observatory) obtained a confirmatory photo on December 7, on which he noted some condensation, but still no trace of a tail. Shortly after mid-January 1976, a Japanese publication revealed that S. Utsunomiya (Oguni, Japan) had actually discovered the comet on November 30.8, at a magnitude of 9.5 and with a coma 4 to 5 arcmin across; however, the comet could not then be confirmed.

After discovery, the comet moved southward at a rate of nearly 2 degrees per day, but this increased as the comet neared Earth, and was slightly greater than 7 degrees per day when it passed perigee (0.28 AU) on December 19. At that time, the total magnitude peaked at 8, and photographs taken a few days earlier gave a tail length of 15 arcmin and a nuclear magnitude of 13.8. Although the comet continued to near the sun thereafter, it faded to a magnitude of 8.5 by December 27, as the distance from Earth rapidly increased, and three days later, it passed slightly less than 5 degrees from the south celestial pole. On January 1, 1976, T. B. Tregaskis (Mount Eliza, Australia) estimated the total magnitude as 9, and the comet passed perihelion on the 4th (r= 0.86 AU). Also, on the 4th, Gilmore determined the nuclear magnitude to be 12.8. The comet faded fairly rapidly as January progressed and on the 31st, Austin (Mount John Observatory) estimated the total magnitude as 10.6. The comet was last observed on February 4, by Candy (Perth Observatory). The orbit is hyperbolic with an eccentricity of 1.0012.
(Sato)

1976 IV   Discovered: February 19   Discovery Magnitude: 10.0
Bradfield (Dernancourt, Australia) discovered this comet in Fornax and described it as diffuse, but without condensation. After discovery, the comet moved to the northeast and also brightened slightly as it approached both the sun and Earth.

Perihelion came on February 25 (r= 0.85 AU), and although the comet was predicted to brighten an additional one-half magnitude as it neared Earth, the comet's continued diffuseness caused observers' estimates to vary greatly. For instance, on February 26, estimates of the total magnitude ranged from 9.5 to 11. Bortle (Brooks Observatory, New York) began his observations on March 8, and while using his 32-cm reflector, he determined the total magnitude as 8.6 and the coma diameter as 4 arcmin. Although the comet's brightness was predicted to change slowly as it passed through perigee (0.4 AU) in mid-March, Bortle's estimates displayed the difficulty in determining the comet's total magnitude as a result of the diffuseness. On March 18, he found the total magnitude to be 10.1, and this changed to 9.9 on the 24th, 10.0 on the 26th, and 9.8 on the 31st. The comet faded rapidly during April, although estimates

of the total magnitude indicated a slightly faster fading than predicted. On
the 5th, A. Hale (Alamogordo, New Mexico) determined the total magnitude as
10.5, while on the 8th, it had dropped to 11.2. By April 20, M. R. Dykes
(Woolston Observatory) estimated the total magnitude as 12.5, and on the 24th,
K. A. Haddow (Woolston) determined it as near 13.0. On May 1, Seki (Geisei)
described the comet as faint and very diffuse and the comet was last seen on
May 28, when Roemer (Steward Observatory) estimated the nuclear magnitude as
near 20.3. The orbit is elliptical with a period near 1,600 years.
(Bradfield)

1976 V   Discovered: March 3   Discovery Magnitude: 9.0
     Bradfield's second comet discovery in 12 days was made while sweeping for
comets near the Grus-Indus border. It was described as fairly bright and diffuse, without condensation, and was moving slowly to the southeast.
     The comet had passed perihelion on February 25 (r= 0.68 AU), and slowly
brightened as it neared Earth. After attaining a maximum southern declination
of -50 degrees on March 8, the comet slowly faded as it turned northeastward.
When perigee came on March 12 (0.53 AU), the total magnitude had dropped to
magnitude 9. Thereafter, the comet rapidly became a difficult object to observe due to its nearness to evening twilight. With an elongation of 52 degrees
from the sun on March 19, the total magnitude was judged to be 9.5 and the coma
diameter was near 4 arcmin. With the elongation increasing by only 6 degrees
by the 25th, the comet's decline in magnitude to 10.5*made it very difficult
to observe. The comet seems to have been last detected on March 28, when J.
Armstrong (Powder Springs, Georgia) estimated the total magnitude as slightly
fainter than 11. Borderline images of the comet were reported on photographs
taken by Candy (Perth Observatory) between March 23 and April 4; however, the
images obtained after March 27 are presently considered very doubtful.
(Bradfield)

1976 VI   Discovered: August 10, 1975   Discovery Magnitude: 16.5
     On November 5, 1975, Richard M. West (Geneva, Switzerland) reported his
discovery of a comet while examining Sky Atlas plates exposed with the 100-cm
Schmidt camera at the European Southern Observatory in La Silla, Chile. The
comet was described as between magnitude 14 and 15, with a coma 2 to 3 arcsec
across and a possible tail to the north. West subsequently found two further
trails on photos exposed on August 10 and 13, when the comet had been somewhat
fainter. Word of the discovery reached the Central Bureau (Cambridge, Massachusetts) the next day, and Marsden was unsure whether the August photos and
the September photos were of the same object. Nevertheless, a parabolic orbit
was calculated and confirmatory photos were obtained at La Silla on November 8,
9, 10 and 11. The comet then appeared as a well-condensed object 20 arcsec
across and of total magnitude 13. The accurate positions not only confirmed
Marsden's preliminary orbit, but they also confirmed Marsden's supposition
that the comet would become a naked-eye object during February and March 1976.
     After discovery, the comet continued to brighten as it approached both the
sun and Earth. The first visual observation was made on November 25, when the
Reverend Leo Boethin (Bangued, The Philippines) estimated the total magnitude
as 12.7, while the tailless coma measured 4 arcmin across. On December 6, C.
Torres (Cerro El Roble) photographed a faint, broad tail 1 arcmin long. With
the comet moving northeasterly, it finally became accessible to Northern Hemisphere observers on December 1, when Seki (Geisei, Japan) estimated the total

magnitude as 12.5. Giclas (Lowell Observatory) photographed the comet on December 6, but northern observations ceased thereafter as the comet became badly
situated with respect to the sun. Southern observers, however, continued their
observations, and by the end of December, the total magnitude had reached 9--
one magnitude brighter than predicted. As observations continued during January 1976, the comet attained a magnitude of 7 by the 20th, and a magnitude of
6 by the 30th. On the former date, the tail was visually measured as 2 arcmin
long. By mid-February, the total magnitude was slightly brighter than 4 and
estimates of the tail length were between 30 arcmin and 1 degree. On February
22, the magnitude reached -1 and by the 24th, it was near -1.5. Perihelion
came on the 25th (r= 0.20 AU), and several observers detected the comet in
broad daylight with optical aid. Descriptions then indicated a total magnitude
of -3, with a coma 15 arcsec across and a tail 15 arcmin long.

Although optical observations of the comet began immediately after perihelion, naked-eye observations did not begin until March 1. The comet then
rose a few minutes before the sun and possessed a very prominent nucleus which
brightly glowed through the twilight. The total magnitude was then estimated
as between -1.0 and -0.5, while the tail extended at least 2 degrees. The
comet was nearest Earth on March 4 (0.8 AU), and, although the distances from
the sun and Earth steadily increased, the comet remained a naked-eye object
throughout March, with magnitude estimates being near 4.7 by the 31st. Visual
observers were also treated to a spectacular tail during the first week or so
of March, as it grew from 10 degrees on the 2nd, to 30 degrees by the 8th. As
many as three branches were noted to the fan-shaped dust tail and photographs
revealed it to consist of numerous parallel streamers. Photos also revealed a
narrow, straight and bluish gas tail.

Of major significance, was the discovery of a secondary nucleus (B) 3 arcsec
from the main nucleus (A) on March 5. This discovery was followed on March 11
by the discovery of two more nuclei (C and D). Although nucleus C rapidly faded
and was last detected on March 24, the remaining nuclei were well observed during the next several months, with B and D being last observed on September 25 by
Shao (Agassiz). A study published in 1981, and written by L. Kresak (Astronomical Institute, Czechoslovakia) gave splitting dates of February 27, 1976, for
nucleus B, March 6 for nucleus C and February 19 for nucleus D.

Visual observations at the beginning of April gave the total magnitude as
4.8 and a tail length of 4 degrees. By the end of the month, the total magnitude had declined to 7 and the tail length had decreased to 1 degree. The
comet continued to be widely observed by amateurs throughout May and June, as it
continued its slow fading and grew more diffuse. Both of these characteristics
caused discrepancies in magnitude estimates, with observers on June 30 giving
values of 9 in moderate-sized reflectors and 7.5 in binoculars. The number of
observers dwindled rapidly during July and August, with the last visual observation coming from Bortle on August 25, when he estimated the total magnitude
as 11.0. The comet continued to be photographed thereafter, with the last definite observation coming on September 25. Six days earlier, Roemer had given
the nuclear magnitudes as 19.3 for A, 20.1 for B and 20.3 for D. There is a
strong possibility that nucleus A was detected on a photo exposed on October
23, though some uncertainty still exists. The orbit is elliptical with a period near 558,300 years.
(West)

1976 IX   Discovered: October 27   Discovery Magnitude: 17.0

Having passed perihelion on July 6, 1976 (r= 5.86 AU), this comet was discovered in November 1976, by Lovas (Konkoly Observatory) on a plate he had exposed on October 27. The comet was then described as near magnitude 17, with
a slightly condensed coma. On November 19, he managed to recover the comet
about 4 degrees to the northwest and described its appearance as the same as
in October.

Although the comet was moving away from the sun when discovered, it changed
very slowly in brightness during the next few weeks as it neared opposition.
On November 30, Shao (Agassiz) gave the nuclear magnitude as 19 and by December 27, Roemer and C. A. Heller (Steward Observatory) gave it as near 19.2.
The latter observers also reported a possible tail to the south-southwest. After
opposition came in early January 1977, the comet was predicted to begin fading
more rapidly than before; however, between January and March, Seki (Geisei)
and T. Furuta (Tokai, Japan) continually estimated the total magnitude as between 17 and 17.5. With the comet nearing conjunction with the sun, Schwartz
and Shao obtained one last photo of the comet on April 20. The comet was recovered on November 19, by observers at Agassiz, who also detected it on February 3 and 10, 1978. No further observations were made. The orbit is hyperbolic with an eccentricity of 1.0039.
(Lovas)

1976 XII   Discovered: February 17, 1977   Discovery Magnitude: 16.0

Lovas discovered this comet as it moved slowly southwestward in Leo Minor.
He described it as diffuse, with a condensation and a short tail; however, his
total magnitude estimate of 15 seems to have been an overestimate since other
observers in the days following discovery noted the total magnitude as 16.

The comet had passed perihelion on November 1, 1976 (r= 5.72 AU), and at
discovery it had just passed opposition and was slowly fading. An observation
on March 12 by H. Kosai (Kiso, Japan) indicated a total magnitude of 16, while
a nuclear magnitude estimate by McCrosky (Agassiz) on March 20 gave a value of
17. On March 22 and 24, Mrkos (Klet Observatory) estimated the total magnitude
as 16.5 and 16.7, respectively. As the comet headed for conjunction with the
sun, it was last detected on April 12, 1977, by Shao (Agassiz).

The comet was recovered on December 10, 1977, by observers at Agassiz, who
again detected the comet on January 11 and February 3, 1978. On February 4, J.
D. Mulholland (McDonald Observatory) photographed it as a slightly diffuse object with a possible fanned tail. The comet was last detected on April 7. The
orbit is hyperbolic with an eccentricity of 1.0032.
(Lovas)

1976 XIII   Discovered: April 27   Discovery Magnitude: 16.0

Eugene A. Harlan (Lick Observatory) discovered this comet in Canes Venatici on a plate exposed on May 3 with the 51-cm double astrograph. He described
it as small and diffuse, with a strong condensation, but no tail. Based on this
one photograph, Harlan was unsure whether the comet was moving southwestward or
northeastward, but an official announcement was made. After word of the discovery reached Palomar Observatory, Charles T. Kowal found a prediscovery image
on a plate exposed on April 27. Although Kowal's object was believed to be the
same as Harlan's comet, confirmation had to await a photo exposed by Harlan on
May 6. This photo established the comet's motion as southwestward and proved
its identity with Kowal's object.

Although Harlan's original announcement gave the comet's discovery magnitude as 15, he revised it to 16 shortly thereafter. Other observers also gave discordant total magnitude estimates during the next few months which ranged from 14 to 16. Although the comet was approaching both the sun and Earth during this time, no apparent brightening could be detected among the accumulated observations; however, estimates of the nuclear magnitude were more consistent, with Mrkos (Klet Observatory) giving estimates of 17.0 on May 24, 16.8 on May 29, 16.3 on June 21 and 16.2 on July 2. At the beginning of July, the comet passed perigee (2.144 AU), but the brightening continued as the distance from the sun continued to decrease. With the comet continuing its southward motion, observations in the Southern Hemisphere began in September, and on the 18th, Gilmore (Carter Observatory) gave the nuclear magnitude as 14.7. Observations continued until October 25, and the comet was lost in twilight thereafter. Perihelion came on November 3 (r= 1.57 AU), and although the comet was expected to be recovered when it exited the sun's glare in April 1977, no further observations were made. The orbit is elliptical with a period near 368,920 years.
(Harlan)

1977 VIII   Discovered: April 16   Discovery Magnitude: 15.0
Eleanor Helin (Palomar Observatory) discovered this comet about one degree from the galaxy NGC 4697, in Virgo, on exposures made with the 46-cm Schmidt camera. She described the photographic image as a rather wide, diffuse trail, with some condensation and a faint tail to the east or northeast.

The comet moved fairly rapidly to the southwest after discovery as it approached both the sun and Earth, and also brightened slowly. On April 25, Gilmore (Carter Observatory) said the comet possessed a fine condensation and a tail 40 arcsec long; however, his nuclear magnitude estimate of 14.1 seems to have been more of a total magnitude estimate since other observers around this time gave the total magnitude as 14 to 15. During the first week of May, the comet passed perigee (0.56 AU) and on the 8th, Furuta (Tokai, Japan) gave the total magnitude as 14. The comet was followed until at a very low altitude on May 15, and became lost in twilight thereafter. Perihelion came on June 30 (r= 1.12 AU), and attempts to locate the comet in Southern Hemisphere skies about one month later failed.
(Helin)

1977 IX   Discovered: March 30, 1976   Discovery Magnitude: 17.5
West (Geneva, Switzerland) discovered this comet on 100-cm Schmidt plates taken at La Silla on January 12 and 13, 1978. The comet was described as near total magnitude 17 and possessed a fan-shaped tail extending 6 arcmin to the south. Shorter exposures taken on the 15th and 16th revealed a tail 10 arcmin long. Two months later, prediscovery images were found by Russell D. Eberst (Royal Observatory, Scotland) on photos exposed at Siding Spring between March 30 and April 2, 1976. The comet had then been moving southwestward and was within 13 degrees of the south celestial pole. The total magnitude was then between 17 and 18.

The comet had passed perihelion on July 21, 1977 (r= 5.61 AU), and was discovered as it headed for opposition with Earth. Although West had given the comet's discovery magnitude as 17, observers during the next couple of months indicated a brightness near 16. Interestingly, Mrkos (Klet Observatory) consistently gave the total magnitude as 16.0 between March 28 and April

7, but on April 8, he estimated it as 16.6 and on the 9th, he found it to be
near 15.8. Unfortunately, no other observers photographed the comet at this
time, and it is currently undecided on whether this was an outburst in brightness or an observational error.

Opposition came in mid-April, and by the end of the month, the total magnitude had dropped to 17. On June 2, observers at Carter Observatory estimated
the tail length as 1 arcmin. With the comet nearing conjunction with the sun,
observations became more difficult and it was last detected on August 5, 1978.
The comet was recovered on January 26, 1979, by Kosai (Kiso, Japan) and was
described as near magnitude 17, with a tail 10 arcmin long. Observers at
Agassiz photographed the comet on February 4, 23, and March 23, and it was
last observed on June 26, when observers estimated the total magnitude as near
19.5 and gave a tail length of 1 arcmin. The orbit is hyperbolic with an eccentricity of 1.0026.
(West)

1977 X   Discovered: November 3   Discovery Magnitude: 13.5

Observers at Purple Mountain Observatory (Nanking, China) discovered this
comet near f Piscium and described it as diffuse, with a fairly rapid motion to
the southwest. Although nothing was reported about a tail, other observers
during the next two weeks reported one measuring 2 to 5 arcmin in length. The
comet was named Tsuchinshan after the observatory at which it was found.

Perihelion had occurred on July 24, 1977 (r= 3.60 AU), and when discovered
the comet had already passed opposition and was steadily fading as it approached
conjunction with the sun. By mid-November, observers were estimating the total
magnitude as 14 and during the first week of December, it had dropped to 15.
On January 3, 1978, Furuta (Tokai, Japan) estimated the total magnitude as near
16 and the comet was last detected on January 16 by observers at Agassiz, before becoming lost in the sun's glare. The comet was recovered on August 11,
1978, by Torres (Cerro El Roble). Then located near opposition, the comet was
described as possessing a nucleus 3 arcsec across, a coma 20 arcsec across and
a faint, narrow tail 10 arcmin long. It was last detected on November 1.
(Tsuchinshan)

1977 XIV   Discovered: September 2   Discovery Magnitude: 10.0

Merlin Kohler (Quincy, California) discovered this comet near Eta Coronae
Borealis on the evening of September 3 (local time), while observing deep-sky
objects with his 20-cm catadioptric telescope. He described it as diffuse, with
a daily motion near 30 arcmin to the southeast. After confirming the comet's
existence on the following evening, Kohler sent a telegram to Lick Observatory,
where Harlan was able to photograph it on the evening of September 5 (local
time). It was later learned that Michel Verdenet (France) had independently
found the comet on September 2, but did not make a formal announcement.

The comet brightened as it approached both the sun and Earth, with the
magnitude reaching 9.5 around September 15, and 8.0 by the 30th. On the latter
date, Bortle (Stormville, New York) estimated the coma diameter as 7 arcmin and
the tail length as 25 arcmin with his 32-cm reflector. During October the
comet was widely observed as it slowly brightened from magnitude 8 to 7. The
coma remained very near a diameter of 7 arcmin and during the last days of the
month, the tail divided into a narrow straight tail 40 arcmin long and a faint,
broad tail extending 30 arcmin.

The comet was brightest during the first week of November, with Bortle giving a total magnitude of 6.7. Stephen O'Meara (Cambridge, Massachusetts) estimated the length of the narrow tail as 1 degree long at that time, and added that no less than 6 jets were visible within the coma. Other observers generally gave a coma diameter of 7 to 8 arcmin. Perihelion came on November 11 (r= 0.99 AU), and perigee came a few days later (0.95 AU). Subsequently, the comet began fading during the latter half of November, and Bortle gave magnitude estimates of 6.8 for November 12, 7.1 for the 19th, 7.4 for December 8, and 8.2 for the 28th. During this same time, observers indicated the coma shrank from 8 to 3 arcmin. As the comet headed southeast, it became difficult to observe in the Northern Hemisphere and observations temporarily ceased on January 8, 1978, when located low over the horizon at a magnitude slightly fainter than 9. After mid-January, the comet turned to the northeast and this allowed a further observation on January 29, by Furuta (Japan). Thereafter, the comet remained nearly south of the sun and was very nearly in the same right ascension, thus making northern observations impossible.

Observations in the Southern Hemisphere had begun in October 1977, and continued well into 1978. On January 30, A. F. Jones (Nelson, New Zealand) gave the total magnitude as 11.9, and by February 8, he reported it as 12.4. After February, photographic observations were made only by observers at Perth and Carter observatories, with the latter reporting nuclear magnitudes of 15.2 on March 12, and 15.5 on April 8. After further observations at Perth on April 10 and 27, Gilmore (Carter) estimated the nuclear magnitude as 16.4 on May 7. The comet was possibly seen again on May 9, by observers at Perth and a possible image may have been found on photos taken at Carter on June 4, although this is currently considered doubtful. The orbit is elliptical with a period near 102,000 years.
(Kohler)

1978 VII   Discovered: February 4   Discovery Magnitude: 7.5
Bradfield (Dernancourt, Australia) discovered this comet near Zeta Tele-scopii and described it as diffuse, with condensation. Observers during the next few days consistently gave the total magnitude as between 7 and 8, and described a tail extending nearly 3 arcmin.

The comet brightened throughout February as its distance from the sun and Earth decreased. On the 20th, the total magnitude was near 7, while on the 28th, it had brightened to near 6. During this same time, the tail grew from a length of 5 arcmin to near 6 arcmin. As March began the comet was being reported as about magnitude 5.5, while the visual tail length was still near 6 arcmin. Photographs by Candy (Perth Observatory) on the 3rd revealed a main tail 70 arcmin long, with narrow spines extending 10 and 15 arcmin. Shortly thereafter, the comet was nearest Earth (1.18 AU), but it continued to slowly brighten as it neared the sun. On March 9, Brazilian amateur Vincente Assis Neto estimated the total magnitude as 5.2 and on the 10th, he estimated the tail length as 17 arc-min in 10x70 binoculars.

Although the comet's northeastern motion had allowed observations in the Northern Hemisphere as early as March 5, it became more widely observed during the latter half of March, as the distance above the horizon increased; however, a new obstacle began hampering observations as the comet moved into morning twilight. After passing perihelion on March 17 (r= 0.44 AU), the comet was last observed on the 25th near magnitude 6.
(Bradfield)

1978 XIII   Discovered: September 12   Discovery Magnitude: 10.7

On the morning of September 12, Don E. Machholz (Los Gatos, California) was conducting his 691st comet-hunting session from Loma Prieta Mountain when he came across a "faint patch of diffuse light" two degrees south-southwest of Sirius. Although he was unable to detect any motion before sunrise, Machholz located it the next morning about 1 degree south of the discovery position and estimated the total magnitude as 10.7.

The comet had passed perihelion on August 13 (r= 1.77 AU), and slowly brightened after discovery as its distance from Earth decreased. Machholz remained the most prolific observer of the comet until it dropped below the southern horizon in October. On September 19, he found it at magnitude 10.5 and by October 3, it was at magnitude 10.7. On the 4th, Machholz observed the comet when it was 15 degrees above the horizon and he last detected it on the 11th, when the altitude was only 6 degrees. The comet was nearest Earth (1.51 AU) about mid-October, and began to fade more rapidly thereafter. As the comet continued to move southward, observers at Perth and Carter observatories began acquiring photographs during October and November. After reaching a declination of -66 degrees on November 9, the comet began moving northwestward. On November 20, Herald (Kambah, Australia) estimated the total magnitude as near 12, and on December 24, when the comet had dropped to near magnitude 13, Gilmore (Carter Observatory) estimated the nuclear magnitude as 14.9. The comet was again detected on December 28, by observers at Perth, with further attempts being impossible due to evening twilight. After conjunction with the sun, the comet was recovered and observed until September 25, 1979.
(Machholz)

1978 XV   Discovered: October 1   Discovery Magnitude: 5.0

David A. J. Seargent (The Entrance, New South Wales) discovered this comet in Centaurus as a bright telescopic object involved in morning twilight. He described it as diffuse, with a condensation, and detected a tail less than 1 degree long.

The comet was located at a declination of -37 degrees when discovered and moved southward in the following days. It also faded, despite a decreasing distance from Earth, since it had passed perihelion on September 15 (r= 0.37 AU). There is a strong indication that Seargent's magnitude estimate may have been slightly overestimated, since observers on October 2 and 3 gave estimates between 5.5 and 6. On October 3.8, B. Sumner (Perth, Australia) estimated the total magnitude as 6.4, and by the 14th, K. Harrison (Durban, South Africa) found it at 7.0. On the former date, the tail extended 2 degrees. The comet was nearest Earth on October 27 (0.73 AU), and passed less than 5 degrees from the south celestial pole two days later. During this time the total magnitude was near 8, and some observers found a coma diameter near 10 arcmin. After the first week of November, the comet had faded to magnitude 8.5 and was moving almost directly northward. As the month progressed, the comet continued to fade and also seemed to remain somewhat large. By the 29th, Seki (Geisei, Japan) obtained the only Northern Hemisphere observation as he made short-exposure photos when very low over the horizon. His estimate of the total magnitude indicated a more rapid fading than predicted, as it was near 13.5, instead of 11. The comet was last detected on December 26, when observers at Perth Observatory photographed it for the final time. The orbit is elliptical with a period near 3,100 years.
(Seargent)

1978 XVIII   Discovered: October 10   Discovery Magnitude: 9.0
     Bradfield (Dernancourt, Australia) discovered this comet near Alpha
Crateris as it moved southeastward. He described it as diffuse, with a condensation, but he did not detect a tail. The comet's initial magnitude estimate
was 9 and, although a few other observers gave estimates nearly 1 magnitude
brighter, Bradfield's later proved to have been most accurate.
     After having passed perihelion on September 29 (r= 0.43 AU), this comet
was first detected in the morning sky. It faded rapidly thereafter, despite a
decreasing distance from Earth, and by October 24, Herald (Kambah, Australia)
estimated the photographic magnitude as 10.5. Perigee came on the 28th (0.40
AU), and the comet reached its maximum southern declination of -67 degrees on
October 30. Thereafter it moved northeastward and observations continued during November as the comet faded. Seki (Geisei, Japan) obtained the only Northern Hemisphere observation on November 20, when he described the comet as diffuse, with a total magnitude near 16. The comet was not detected after this
date.
(Bradfield)

1978 XXI   Discovered: April 27   Discovery Magnitude: 10.0
     This comet was discovered in Lynx by Rolf Meier (Ottawa, Canada), who was
sweeping for comets at the observatory of the Ottawa Center of the Royal Astronomical Society of Canada. Meier was using a 40-cm reflector and found the
comet in the evening sky. He described it as diffuse, with a condensation, and
estimated the daily motion as 0.4 degrees southward. Although the total magnitude was then given as 10, observers in the following days indicated a value
near 10.5.
     The comet slowly brightened during the next several months as it neared
both the sun and Earth. During May, observers remarked on the coma being exceptionally condensed and by the end of the month, the coma was being commonly
described as 1 arcmin across, with a total magnitude of 10 and a faint trace of
a tail. By mid-June, the magnitude had brightened to 9.5 and the coma was 2 to
3 arcmin across. During the next month, the comet's elongation from the sun
began decreasing to the point of making it a difficult object to observe. Subsequently, the comet was last seen on July 13, when Bortle (Stormville, New
York) estimated the total magnitude as 7.8. The comet was then only 6 degrees
above the horizon.
     The comet remained lost in the sun's glare for the next few months, passing
less than 4 degrees from the sun during September, and being located at a declination of -24 degrees by the end of October. It was finally recovered on
November 4, when Herald (Kambah, Australia) estimated the total magnitude as
near 6. Several other Southern Hemisphere observers detected the comet during
the next 5 days and reported a similar magnitude. Perihelion came on the 11th
(r= 1.14 AU), and the comet passed perigee (2.00 AU) a few days later. Seargent
(The Entrance, Australia) reported the comet as brightest during the latter half
of November, when he gave a total magnitude of 5.5. Thereafter, the comet slowly faded, with observers giving magnitude estimates of 6 in mid-December, and 7
by the end of the month.
     During the first days of 1979, the comet began moving to the northeast.
The most prolific observer during the next few months was Seargent, who reported that the comet continued its slow fading and finally reached magnitude
9.5 by the end of March. One month later, Seargent reported a magnitude of 10

and by June 4, it had dropped to 10.8. As June progressed, the comet moved into Northern Hemisphere skies and was first detected on the 21st, when Bortle found it 14 degrees above the southeastern horizon. The total magnitude was then estimated as 11.0, but as the altitude increased during the next two weeks, the comet seemed to brighten, and by July 5, Bortle gave the magnitude as 10.8. On the latter date, the coma was estimated as 1.8 arcmin across and a 13th-magnitude, nearly stellar nucleus was visible in Bortle's 32-cm reflector. Bortle continued his extensive series of observations during the next two months and obtained magnitude estimates of 11.2 on August 17, 11.4 on September 2, and 11.7 on September 24. On the latter date, the coma was 1.2 arcmin across with "very indistinct" outer boundaries. Bortle's failure to detect the comet on October 15 indicated a total magnitude fainter than 12. Photographic observations continued until December 9, 1979.
(Meier)

1979 VI   Discovered: August 23, 1978   Discovery Magnitude: 19.5

Carlos Torres (National Observatory, University of Chile) found this comet on photographs exposed between June 26 and July 2, 1979. Then near a total magnitude of 18, the comet was located near its closest approach to Earth (3.7 AU) and arrived at perihelion on July 15 (4.69 AU). Observations made by Torres and C.-Y. Shao (Harvard College Observatory) during the remainder of July showed the comet to be small and round, with no trace of a tail. Due to the comet's great distance from the sun, fading was slow and when last seen on October 23, shortly before entering the sun's glare, it had faded to 19[*]. Although the comet was not detected during the summer of 1980, a prediscovery image was found on a plate exposed on August 23, 1978 which allowed a precise orbital calculation that gave an eccentricity of 1.001.
(Torres)

1979 VII   Discovered: June 24   Discovery Magnitude: 10.0

Using a 15-cm refractor, William A. Bradfield (Dernancourt, Australia) was sweeping for comets in the evening sky on June 24, when he found an unexpected diffuse object in the head of Hydra. The object was estimated as moving northward at a rate of 70 arcmin per day. The next evening T. B. Tregaskis (Mount Eliza, South Australia) confirmed the comet and estimated the coma diameter as 3 arcmin.

The comet rapidly became a difficult object to observe as it entered the sun's glare and most observations ceased after July 3, when the magnitude had brightened to 9. A final precise position was obtained on July 9, when P. Jekabsons (Perth Observatory) found the comet when it was only 23 degrees southeast of the sun. Perihelion came on July 23 (r= 0.41 AU) and the comet was finally recovered on August 3, when Don Machholz (Los Gatos, California) estimated the total magnitude as 8.8 or 8.9. After the August full moon, numerous other observations were made. On August 17 John Bortle (Brooks Observatory, New York) estimated the total magnitude as 10.1 and the coma diameter as 2.2 arcmin. Using his 32-cm reflector, he spotted a very faint, nonstellar nucleus. By the end of August the magnitude had dropped to 11.

The comet was closest to Earth on September 13 (0.43 AU) and Bortle described it as "a nebulous smudge in the rich Milky Way field." He estimated the total magnitude as 11.2 and roughly measured the coma diameter to be 2.5 arcmin. The last visual sighting of the comet occurred on the 17th when A. Hale (Annapolis, Maryland) found it near the limit of his 41-cm reflector. The comet was

last detected photographically on September 24, 1979. Calculations revealed
the comet to be traveling in an almost perfect parabolic orbit with an eccentricity of 1.00006.
(Bradfield)

1979 IX   Discovered: September 20   Discovery Magnitude: 11.5
     Rolf Meier (Ottawa Center of the Royal Astronomical Society of Canada)
discovered this comet while using a 40-cm reflector. He described the comet
as diffuse with a condensation. The comet was then a circumpolar object with
a declination of +69 degrees and confirmation was made by H. L. Giclas (Lowell
Observatory, Arizona) less than four hours later.
     On September 21 Charles S. Morris (Prospect Hill Observatory, Massachusetts)
visually observed the comet using his 25-cm reflector and determined the total
magnitude to be 12.0 and the coma diameter to be 1 arcmin. Slightly over one
hour later, J. H. Bulgar and C.-Y. Shao (Harvard College Observatory's Agassiz
Station) obtained photographs of the comet which showed a definite condensation
and a possible tail. Other visual observers continued to estimate the comet's
brightness as near 12 during the next few days, but at the beginning of October
it was evident that the comet was slowly brightening as it approached its October 17 perihelion (r= 1.43 AU). On October 15 the total magnitude was estimated as 11.5, while the coma diameter was 2 arcmin. The comet was lost in the
sun's glare thereafter.
     The comet was recovered in the morning sky on October 25 by J. Morgan
(Beloit, Wisconsin). Morgan was using a 32-cm reflector and described the
comet as 0.8 arcmin in diameter with a total magnitude of 12.1. On the 26th
G. Keitch (Wrington, England) used a 30-cm reflector to determine the coma
diameter as 1 arcmin, the total magnitude as 11.9 and the tail length as 3 arcmin. Although the comet steadily moved away from the sun thereafter, the distance from Earth steadily decreased throughout November and December. The
result was only minor changes in the total magnitude, so that by January 1,
1980, estimates were still near 12. The comet was then near its closest point
to Earth (0.9 AU). Thereafter, fading was rapid as the distances from the sun
and Earth increased. On January 30 A. Hale (Annapolis, Maryland) described
the comet as a 13th-magnitude glow. The comet was last seen on March 7, when
the nuclear magnitude was then near 20. The orbit is elliptical with a period
of 391 years.
(Meier)

1979 X   Discovered: December 24   Discovery Magnitude: 6.0
     During a routine comet-hunting session with his 15-cm refractor, William
A. Bradfield (Dernancourt, Australia) discovered his 10th comet about 10 degrees
south of Antares. The comet was described as possessing a tail over 1 degree
long and the magnitude was estimated as 5; however, more precise estimates of
the brightness during the following days (5.6 on December 29 and 5.3 on December 31) indicated the comet was actually closer to 6th magnitude when discovered.
The comet moved rapidly southward.
     . The comet had passed perihelion on December 21 (r= 0.55 AU), but instead
of fading, it slowly brightened in the following weeks as its distance from
Earth decreased. On January 9 Dennis Goodman (Wellington, New Zealand) estimated the total magnitude as 5.2, the coma diameter as 10 arcmin and the tail
length as 1.5 degrees. On January 21, Alan C. Gilmore and P. M. Kilmartin
(Wellington, New Zealand) estimated the total magnitude as 4.7, the coma diameter

as 20 arcmin and the tail length as 2 degrees. The comet was farthest south on
January 23, when at a declination of -80 degrees. Thereafter, it moved northward at a rate which quickly increased to 10 degrees per day as the comet passed
closest to Earth on January 26 (0.198 AU). Also on that date, observers estimated the coma diameter as 30 arcmin and the total magnitude as 4.5.

The comet was first detected in the Northern Hemisphere on January 27,
when P. Maley (Houston, Texas) roughly estimated the total magnitude as 4 and
found the coma diameter to be 15 arcmin. Nearly 24 hours later, John E. Bortle
(Brooks Observatory, New York) obtained a precise magnitude estimate of 4.8 and
found the coma to be 12 arcmin across. By January 31 the total magnitude had
dimmed to 5.5. On February 6 C. S. Morris (Harvard, Massachusetts) reported
the magnitude as 6.5, the coma diameter as 6 arcmin and the tail length as nearly 1 degree. A slightly shorter secondary tail was also noted. On February 20
the comet had faded to magnitude 8.8, according to Morris, with a coma diameter
of 4 arcmin, and by March 4 J. De Young (Palmyra, Virginia) estimated the magnitude as 11.0 and the coma diameter as 3.5 arcmin. The comet was last detected
on March 17 at a magnitude between 12 and 13[*], with further observations being
impossible due to the comet's entrance into evening twilight. Further attempts
to recover the comet in the summer of 1980 proved fruitless.

Orbital calculations indicate the comet travels an elliptical path with a
period of 293 years. There is a similarity between the orbit of this comet and
the orbit of comet 1770 II, though they are not identical since only 209 years
elapsed between their respective appearances.
(Bradfield)

1979 XI   Discovered: August 30   Discovery Magnitude: -3.5

In September 1981, during routine examinations of solar corona photos made
by the U. S. Defense Department satellite P78-1 in 1979, Dr. Russ Howard (Naval
Research Laboratory, Washington, D. C.) found images of a comet on photos made
on August 30, 1979. Detection of the comet was made possible due to instrumen-
tation developed and operated by Dr. Martin Koomen and Dr. Don Michels.

The comet first appeared on a photo exposed at 18:56 Universal Time and was
located nearly 6 solar radii from the sun's center. Interestingly, Venus was
then near superior conjunction and was also on the photograph, which allowed a
very precise estimate of the comet's brightness. Seven more photos had been
taken by 21:15 UT at which time the comet had moved to within 3 solar radii and
was about to pass behind the occulting disk used to block the sun. After per-
forming other experiments aboard the satellite, coronagraph photos were resumed
at 23:44 UT and showed the tail to be very prominent as it extended to the
photo's edge toward the west-southwest. Eleven more photos were taken by 3:36
UT on August 31 which showed the tail to steadily brighten. Thereafter, a
break of four hours occurred.

When the camera resumed operation at 7:46 UT it was met with an unusual
sight--material was streaming out over an entire hemisphere of the sun centered
to the northwest. This material brightened until about 11 hours UT and then
faded until last detected at 20:36 UT. Curiously, the comet's head never reappeared.

Orbital calculations have firmly established this comet as a member of the
Kreutz sungrazing family, though it possesses the smallest perihelion distance
on record. This distance of 0.00164 coincides with a solar radius of 0.35, thus
indicating that the comet collided with the sun and was obliterated.
(Howard-Koomen-Michels)

1980 II   Discovered: June 13   Discovery Magnitude: 15.0

Carlos Torres (Department of Astronomy, University of Chile) discovered
this comet on exposures made of southeastern Sagittarius on June 13 and 14
at the Cerro El Roble station. It was described as diffuse, with condensation and a tail less than 1 degree long.

The comet moved slowly northwestward in the following days, with no
noticeable change in brightness, and precise positions were obtained at
Cerro El Roble, Harvard Observatory's Agassiz station and Perth Observatory.
From the latter location, M. P. Candy calculated parabolic elements on June
25 from Perth observations spanning June 16 to 24. He found the comet already
two months past perihelion and noted that the distance from Earth steadily
decreased from about 2.3 AU on that day, to 1.67 AU on June 23. Thereafter,
the comet began to slowly fade.

Torres was the comet's most prolific observer and by mid-July he described it as magnitude 16, with a nuclear magnitude of 19. The coma was
then 15 arcsec across and the condensation was 5 arcsec in diameter. No
tail was visible. The comet showed some unusual activity on plates exposed
in early August. On the 2nd, Torres estimated the nuclear magnitude as
19.5. Three days later he found it at 19 and said the 5 arcsec condensation was located in a "broad, faint 20 arcsec coma." On the 13th, he estimated the nuclear magnitude as 18 and said the coma was broad, eccentric
and 15 arcsec across, with most of the material located between position
angle 130 and 170 degrees.

On September 4, Torres estimated the nuclear magnitude as 18.5 and by
the 16th it was at 20.5. The comet was last seen on October 9, with further
observations being impossible as it entered the sun's glare.

A more precise orbit was computed by Brian Marsden using observations
obtained between June 13 and October 6. This yielded a perihelion date of
April 20 (q=2.58 AU). Due to the large perihelion distance and the relatively short observed arc, only a parabolic orbit could be computed.
(Torres)

1980 IV   Discovered: July 31   Discovery Magnitude: 9.0

From Maidanak Mountain in Uzbekistan, U.S.S.R., came a cable on August
4th announcing the discovery of a 9th-magnitude comet by Kazimeras Cernis and
Jovaras Petrauskas (Vilnius University Observatory). The notice indicated
the comet was in southernmost Ursa Major with a daily motion of 20 arcmin to
the east. The Central Bureau for Astronomical Telegrams requested several
observers to confirm the comet's existence, but after a few days no observations were reported and the comet did not receive a designation. No
further reports were received until shortly after mid-August, when Paul Wild
(Zimmerwald station of the Astronomical Institute of Berne, Switzerland)--
unaware that the comet could not be detected elsewhere--submitted his observations of August 2, 4, 6, 7 and 14. These indicated the comet's daily
motion to actually be 1 degree, thus explaining observers' difficulty in
finding the comet. Later, it was reported that E. P. Belserene (Maria
Mitchell Observatory, Nantucket, Massachusetts) had seen the comet on August
10 and had estimated the coma diameter as 0.8 arcmin.

Besides the comet's actual discovery, Vilnius University was also the
first to note that the comet was developing an unusual feature. Photos on
August 15 and 19 by V. Straizys showed a narrow, straight tail about 12
arcmin long pointing toward the sun. An independent discovery of this fea-

ture was made on August 29 by Edgar Everhart (University of Denver, Colorado) when he described it as conspicuous and 14 arcmin long. Several additional observations around the world were noted up through September 7--all photographic. Zdenek Sekanina (Harvard-Smithsonian Center for Astrophysics) said the Earth had been less than 2 degrees from the comet's orbital plane on the 29th of August and that the emission times for the particles forming the anti-tail ranged from 120 to 10 days prior to perihelion.

The comet faded rapidly following discovery as it moved away from both the sun and earth, and, at the same time, it became more diffuse, thus, making magnitude estimates difficult. During the first week of September, observers reported total magnitudes ranging from 11.5 to 14.5, while estimates of the coma diameter varied from 1.5 to 5 arcmin.

The final observation of the comet came on September 12, when John E. Bortle (Stormville New York) caught a glimpse of the comet in his 32-cm reflector. He described it as about 2 arcmin across and added that it "was extremely pale and uncondensed." The magnitude was estimated as 11.6. A first-quarter moon ended observations of this early evening object and attempts to locate the comet in late September were unsuccessful.

With perihelion having come on June 22 (q=0.523 AU), many astronomers were puzzled as to why the comet had not been seen earlier. As it turned out, the comet stayed within 30 degrees of the sun from mid-February until shortly after the first week of July. From early July until shortly after mid-month, the moon was brightening up the evening sky.
(Cernis-Petrauskas)

1980 XII   Discovered: November 6   Discovery Magnitude: 10.5

On November 6, 1980, Rolf Meier (Ottawa, Ontario) discovered his third comet in only two years. There are three things which make this feat particularly noteworthy. First, Meier used a 40-cm reflector for his comet-sweeping, which is unusually large for what is generally accepted. Second, the comets were found in a total of only 105 hours of searching and, finally, each comet was found in the northwestern evening sky.

This particular comet was located about 6 degrees northwest of Vega at discovery and was described as diffuse, with a condensation. Meier's magnitude estimate of 10.5 was found to be about 0.5 magnitude fainter than the estimates of other observers in the next few days, but this is probably due to the use of a larger aperture, since other observers were typically using binoculars and telescopes smaller than 25 cm.

An orbit published by Daniel W. E. Green (Harvard-Smithsonian Center for Astrophysics) on November 10 showed the comet to then be about 1.5 AU from both the sun and Earth. For the next month the distance from the sun decreased as the distance from Earth increased, which resulted in little change of the total magnitude, which, for the most part, remained between 9.5 and 9.8. The coma was consistently estimated as between 2 and 3 arcmin across and no tail was detected.

A more precise orbit computed by Brian Marsden (Harvard-Smithsonian Center for Astrophysics) indicated the perihelion date to be December 9 (q=1.52 AU). Shortly thereafter, observations of this comet temporarily ended as it moved into evening twilight on its way to conjunction with the sun. Observers began sweeping above the horizon in the early morning sky during the last days of December, but the comet was not found until January 2, 1981, when Don Machholz (San Jose, California) spotted it in his 25-cm

reflector. At that time the distances from the sun and Earth were increas-
ing--as they had been since perihelion--and the predicted brightness was
expected to be 0.5 magnitude fainter than when seen in early December. However, both Machholz and other observers reported magnitudes ranging from
8.7 to 9.2 throughout the month of January.

By the end of January the distance from Earth had begun to decrease
and from then until it reached its minimum value of less than 1.4 AU in mid-
April the comet slowly faded. During this time, John Bortle (Stormville,
New York) estimated total magnitudes of 8.6 on February 6, 9.2 on March 2
and 9.8 on March 30--all with the use of a 20-cm reflector. Thereafter,
Charles S. Morris (Harvard, Massachusetts) used a 25-cm reflector to determine total magnitudes of 10.0 on April 7, 10.4 on April 23 and 10.7 on May
8. The coma slowly shrank from 6 arcmin in January to 4 arcmin in May and
no tail was ever detected.

The comet was last detected on June 3, 1981, by observers at Harvard
Observatory's Agassiz station.

(Meier)

1980 XV   Discovered: July 18   Discovery Magnitude: 16.0

William A. Bradfield (Dernancourt, Australia) discovered his eleventh
comet on December 17, 1980, during a routine comet-sweeping session. Then
located about 7 degrees east of Epsilon Scorpii, the comet was described as
diffuse, with condensation and a tail less than 1 degree long. The magnitude
was given as 6. The comet was confirmed the next morning by D. Herald (Kambah,
Australia), who estimated the total magnitude as between 5 and 6 and reported
a tail extending nearly one-half degree.

The comet moved quickly across the sky after discovery as it headed
towards conjunction with the sun, and was temporarily lost after December
20. On the 22nd, Brian Marsden was able to compute parabolic elements from
Herald's 5 precise positions and found the perihelion date to be December
29 (q=0.26 AU). The orbit was accurate enough to allow the comet to be
observed by E. P. Ney (University of Minnesota) on January 1, 2 and 3, while
still in evening twilight. This observer was using equipment to measure
the infra-red wavelengths of the comet and said they showed "the enhanced
visual brightness due to forward scattering previously seen in the case of
comet 1976 VI." He also noted a strong similarity to comet 1973 XII at
similar distances from the sun and Earth.

As January progressed, the comet faded from magnitude 4 on the 5th to
5.5 on the 12th, according to several observers, while the tail was constantly
estimated as between 3 and 4 degrees. On the 13th, however, the comet
underwent a sudden increase in brightness, which amounted to about one
magnitude, and on the 19th, Stephen O'Meara (Cambridge, Massachusetts) detected a secondary nucleus about 2.5 arcsec from the primary nucleus.
O'Meara continued to detect the second nucleus with the 23-cm Clark refractor
until the 21st, at which time John E. Bortle (Stormville, New York) independently discovered the same feature. No further sightings of this new
nucleus were reported and its reality is currently in doubt. By month's
end the brightness was back to normal with observers' estimates being
near 7.5.

The comet was temporarily lost after the first week of February as it
entered the sun's glare, but it was recovered at the end of June by observers at Harvard Observatory's Agassiz station. It was then described as

"a small, condensed object as faint as magnitude 18.5." These observers
last detected the comet on August 27, when fainter than magnitude 20.

It should be noted that while observations were still continuing, K. S.
Russell found prediscovery images on an exposure made with the 120-cm U. K.
Schmidt telescope at Coonabarabran on July 18, 1980. The magnitude was then
estimated as 16.

(Bradfield)

1981 I   Discovered: January 26   Discovery Magnitude: 0.0

This comet was discovered in July 1982 on 15 photos taken of the sun's
corona by the SOLWIND coronagraph aboard the United States Department of
Defense satellite P78-1. The discovery team, led by Donald J. Michels,
operates out of the Naval Research Laboratory and included Neil Sheeley, D.
Roberts and F. Harlow. Due to the number of people involved in the discovery,
the comet was named after the satellite--hence, SOLWIND 2 (SOLWIND 1 was the
unofficial name given to 1979 XI).

The comet was under observation for less than 4 hours (January 26.860
to January 27.021) and moved from a distance of 8 solar radii to 3 solar
radii before disappearing behind the occulting disk used to study the sun's
corona. When first seen, the magnitude of the head was about 0.0 and when
last seen it was near magnitude -2.5. Interestingly, the comet experienced
a sudden decrease in brightness when at 5.5 solar radii as it reached magnitude 1.0. The tail was straight and narrow and remained visible some 7
hours after the final observation of the comet's head. No trace of the
comet's head or even debris (as in the case of 1979 XI) ever appeared after
perihelion.

Brian Marsden said the comet was a member of the Kreutz sungrazing group
and based "on the assumption that the comet's perihelion direction agrees with
the Kreutz group" he computed the perihelion date to be January 27.076 and
the perihelion distance to be 0.00488 AU (1.05 solar radii). As it turns out,
this comet was well placed in the evening sky prior to January 27, but must
have then been too faint for discovery.

(SOLWIND 2)

1981 II   Discovered: October 9, 1980   Discovery Magnitude: 14.0

Roy W. Panther (Walgrave, England) discovered this 10th-magnitude
comet on December 25, less than 4 degrees west of Vega. He described it
as diffuse, with condensation. It was confirmed later that evening by fellow
countrymen M. J. Hendrie and G. E. D. Alcock and was noted to be moving to
the north-northeast. On the 27th, C. S. Morris (Harvard, Massachusetts) more
precisely described the comet as magnitude 9.7 with a coma 2.3 arcmin across.

The comet was routinely estimated as between magnitudes 9.3 and 9.5 by
the end of December, while the coma was described as 3 arcmin across. Although the comet was slowly approaching the sun and Earth throughout January, very little brightening occurred, with estimates still being between
9.2 and 9.4 by the 31st. On January 8, B. Milet became the first to detect
a tail when his photos from Nice Observatory revealed it to extend about
15 arcmin.

Perihelion came on January 27 (q=1.66 AU), and the comet brightened
slightly thereafter as it continued to approach Earth. Observers commonly
estimated the total magnitude as 8.5 by mid-February and the comet held this
brightness throughout March as it passed 1.39 AU from our planet. On the
12th, it passed within one-half degree of the north celestial pole.

During the first half of April observers estimated the total magnitude
as about 9.0 and by May 1 it had dropped to 9.5. Observations continued
throughout May, but as June dawned the comet was moving into evening twilight.
On the 3rd, Morris estimated the total magnitude as 10.4 and the comet was
involved in the sun's glare after the 4th.

After a couple of months, the comet was finally recovered near a magnitude of 14[*] and continued to be observed until December 9, when it again
dropped into the sun's glare. No further observations were obtained; however, prediscovery images were identified on photos exposed on October 9,
1980.

(Panther)

1981 V   Discovered: September 6, 1980   Discovery Magnitude: 17.0

This faint comet was discovered by Kenneth S. Russell (U. K. Schmidt
Telescope Unit at Siding Spring, Australia) on photographic plates exposed
on September 6 and 7, 1980. He described it as a fuzzy trail on the photos,
with signs of a tail.

The comet was closest to Earth at the beginning of October at a distance
of 2.2 AU. Thereafter, as this distance increased, the comet changed little
in brightness as it continued its trek towards perihelion. In November it
reached its farthest southern declination of -63 degrees and, thereafter,
began moving slowly north-northwestward. By December 6, the total magnitude had increased to near 16.5, but observations ceased thereafter as the
comet approached conjunction with the sun.

Perihelion came on March 6 (q=2.11 AU) and the comet was apparently
recovered by A. C. Gilmore (Mt. John University Observatory, Australia) on
April 5, though the comet was described as a "very faint image."

(Russell)

1981 VII   Discovered: June 29   Discovery Magnitude: 15.0

Shortly after mid-July, Luis E. Gonzalez (University of Chile) discovered a diffuse cometary trail on two plates obtained at Cerro El Roble on
June 29. The comet was then at a declination of -51 degrees, the motion was
to the southeast and there was a condensation, but no tail. Shortly thereafter, C. Torres (University of Chile) confirmed the comet's existence on a
plate exposed on July 22 and estimated the magnitude as 16.

The comet had been closest to Earth in late July and was found to have
passed perihelion on March 25 (q=2.33 AU). During August, observers continued to estimate the total magnitude as 16, as the comet passed less than
4 degrees from the south celestial pole, and on the 30th, A. C. Gilmore
(Mt. John Observatory) estimated the nuclear magnitude as 16.6. Gilmore
obtained the final observation of the comet on November 27, when at a
nuclear magnitude of 17.3, with observations thereafter ending as the comet
entered the sun's glare.

(Gonzalez)

1981 XIII   Discovered: July 19   Discovery Magnitude: -0.8

This comet was discovered on 34 photos taken of the sun's corona by the
SOLWIND coronagraph aboard the United States Department of Defense satellite
P78-1. The discovery team at the Naval Research Laboratory included Donald
J. Michels, R. Seal, R. Chaimson and W. Funk. Due to the rule of attaching
no more than 3 names to a comet, the official name became SOLWIND 3.

The photos were taken at 5-minute intervals during the course of nearly
4 hours on July 19 and 20 UT. When first detected the comet was located

9.56 solar radii from the sun and when last seen at 3:54 UT on July 20 it was 6.30 solar radii from the sun's disk. The total magnitude was estimated as -0.8 when about 8 solar radii away and the tail always appeared short, with a length not exceeding 30 arcmin.

Photos resumed at 8:03 UT, at which time the comet's head had gone behind the occulting disk; however, the tail was still easily visible. This was about the time of the comet's perihelion, which Brian Marsden computed to be July 20.32. Based on the apparent direction of the perihelion, the comet was identified as a Kreutz sungrazer with the perihelion distance being about 0.00427 AU (about 0.92 solar radii), or beneath the surface of the sun.

No trace of the comet was detected after perihelion and a close scrutiny of photos taken as late as 22:25 UT on July 20 revealed no debris as in the case of 1979 XI.

(SOLWIND 3)

1981 XIV   Discovered: April 26   Discovery Magnitude: 16.5

This comet was discovered by Schelte J. Bus (California Institute of Technology) on a plate he had taken with the 46-cm Schmidt at Mt. Palomar. It was described as diffuse, with condensation and no tail and was confirmed by Charles Kowal using the 122-cm Palomar Schmidt on April 28.

On April 30, A. C. Gilmore (Mt. John Observatory) and T. Seki (Geisei, Japan) estimated total magnitudes of 16 and 15.5, respectively. Estimates for the first half of May continued to be within this range as the comet passed closest to Earth (1.6 AU). Thereafter, a very slow fading set in as the increasing geocentric distance countered the decreasing heliocentric distance.

Occasional observations continued during late May and throughout June, but a nearing conjunction with the sun temporarily halted observations after July 4. Perihelion came on July 31 (q=2.46 AU) and the comet was finally recovered after leaving the sun's glare. It was followed until last photographed on January 16, 1982, by J. Gibson (Palomar Observatory), who described the image as "very weak."

(Bus)

1981 XV   Discovered: April 3   Discovery Magnitude: 15.0

Jonathan H. Elias (Cerro Tololo Interamerican Observatory) discovered this comet on plates exposed with the 60-cm Curtis Schmidt telescope on April 3 and 4, 1981. At that time it was located near its closest distance to Earth (4.4 AU) and was described as diffuse with neither a condensation nor tail.

From a declination of -77 degrees at discovery, the comet moved slowly north-northeastward and slowly faded as the increasing distance from the Earth countered the decreasing distance from the sun. Perihelion came on August 18 (q=4.74 AU) and observations continued into September before the comet was lost in the sun's glare. The magnitude was then near 15.5.

The comet was recovered after conjunction with the sun on January 28, 1982, by Edgar Everhart (Chamberlin Observatory field station, Colorado) and was again observed by him on March 24. The magnitude on the latter date was then near 16[*]. Observers at Oak Ridge Observatory observed the comet on May 26, with further observations being impossible due to another conjunction with the sun.

R. E. McCrosky and G. Schwartz (Oak Ridge) obtained the next observation on January 16, 1983, and, at Palomar, J. Gibson described the comet as a

"well-condensed, faint coma" on April 2. A short tail was also noted. On
the 14th, observers at Oak Ridge again detected the comet before it entered
the sun's glare on its way to a third conjunction. The brightest magnitude
at this opposition must have been between 17 and 18[*].

   The orbit is hyperbolic with an eccentricity of 1.0007.

(Elias)

NOTES

NOTES

Arend

Discovered: October 1, 1951   Discovery Magnitude: 14.0

Silvio Arend (Royal Observatory, Uccle, Belgium) discovered this comet on
routine minor planet plates exposed on October 4, 1951, with the 40-cm Zeiss
double astrograph. Confirmatory plates were exposed the following evening.
The comet was described as a small, but conspicuous nucleus surrounded by a
round coma. The coma diameter was 14 arcsec and no tail was then visible.
Shortly thereafter, a prediscovery image was found on a plate exposed at Yerkes
Observatory on October 1.

Each succeeding apparition of this comet has steadily worsened in terms
of observability. The discovery apparition was the best observed since the
comet was found shortly before passing 0.86 AU from Earth--very nearly the
closest possible approach. Subsequently, the discovery apparition still remains
the brightest, with a maximum magnitude of 14 and a maximum tail length of 2
arcmin. During the 1959 return, the comet's maximum magnitude nearly reached
15 during an unusual outburst shortly before November 25, and had this event
not occurred the comet would not have become brighter than 17th magnitude. The
returns of 1967 and 1975 were even more unfavorable, with the maximum magnitude failing to exceed 18.

Comet Arend lies at the 2/3 commensurability and, thus, encounters Jupiter
after every third perihelion passage. The latest of these encounters occurred
in 1969, when the comet passed only 0.64 AU from Jupiter, but, as in past instances, the orbit changed little. Computer studies by Brian Marsden indicates
that the comet's latitude of perihelion is so large (12 degrees) that the comet
has remained relatively undisturbed since before 1725.

	T	q	P	Max. Mag.	Max. Tail
1951 X	Nov. 23	1.82	7.76	14.0	2 arcmin
1959 V	Sep. 1	1.83	7.79	15.5	--
1967 VI	Jun. 14	1.82	7.76	18.0	--
1975 VI	May 25	1.85	7.98	18.0	--

Arend-Rigaux

Discovered: January 8, 1951   Discovery Magnitude: 10.5

This comet was discovered by Arend and F. Rigaux (Uccle, Belgium) during
an examination of routine minor planet plates exposed on February 5 and 6,
1951. The comet was described as diffuse, with a central condensation, and had
a total magnitude of 11. Prediscovery images were later found on survey plates
exposed at McDonald Observatory on January 8 and February 4, and on plates exposed at Tokyo Observatory on January 28.

The discovery apparition was very favorable for observations, since the
comet was located at nearly its closest distance from Earth (0.4 AU) in early
January. Although a coma and tail were noted on several occasions, the comet's
true character was not revealed until its recovery in 1958. In that year, it
took astronomers nearly three weeks to firmly establish that the object being
observed was the expected comet. The problem rested on the similarity of the
comet's orbit with that of an asteroid and the comet's unexpected stellar appearance. Since that time, the comet has become noted for its near-asteroidal
appearance, since neither a coma nor a tail was noted in 1958, 1963 and 1970.
During these returns the comet was never observed when less than 1.5 AU from
Earth or 2.0 AU from the sun. At the 1978 return, however, another close ap-

proach to Earth (0.83 AU) allowed the observation of both a weak coma and a
faint tail.

Computer studies have revealed that this comet's orbit is primarily in-
fluenced by minor perturbations of the planets. In fact, the last time the
comet ventured within the critical distance (0.9 AU) of Jupiter was nearly 900
years ago; thus, the comet has shed its material at a relatively constant rate
for the last 9 centuries. One major finding by astronomers during the 1970s
was this comet's lack of nongravitational forces--a decelerating or accelera-
ting effect caused by the outgassing of the nucleus. This discovery, plus the
comet's observed lack of a strong coma or tail, has led several astronomers to
believe that the comet is on the verge of entering an asteroidal phase.

	T	q	P	Max. Mag.	Max. Tail
1950 VII	Dec. 19	1.39	6.71	10.5	42 arcsec
1957 VII	Sep. 7	1.39	6.71	19.0	none
1964 V	Jun. 6	1.44	6.82	19.4	none
1971 IV	Apr. 6	1.44	6.84	18.8	none
1978 III	Feb. 2	1.44	6.83	13.0	4 arcmin

Ashbrook-Jackson

Discovered: August 26, 1948   Discovery Magnitude: 11.5

Joseph Ashbrook, a Yale astronomer, discovered this comet on August 26.3,
1948, while visiting Lowell Observatory (Arizona) to observe minor planets. The
comet was found on a photo exposed for 1327 Namaqua, and was described as diffuse, with a nucleus and a short tail. Twelve hours later, an independent discovery was made by Cyril Jackson (Yale-Columbia Station, Johannesburg, South
Africa) while experimenting with a 50-cm focus camera to determine its ability
to photograph fast-moving minor planets.

Prior to its discovery apparition in 1948, this comet had been ejected
into a smaller orbit by a close approach to Jupiter (0.178 AU) in 1945. The
comet's orbit changed from one of small eccentricity, with a perihelion distance
of 3.89 AU, to one of moderate eccentricity, with a perihelion distance of 2.31
AU. The comet is one of the intrinsically brightest short-period comets known
and, although astronomers have thus far failed to follow it thoughout its orbit,
some researchers still believe the comet is a good candidate for the "annual"
class of comets. Comet Ashbrook-Jackson has been detected at each return since
its discovery, although its 7.5-year period makes every second return an unfavorable one. In 1948, perihelion occurred one month after opposition with Earth
and the comet was as bright as 10th magnitude. In 1956, perihelion occurred
when the comet was nearly in conjunction with the sun, and the total magnitude
barely reached 12. The 1963 return was the most favorable return possible for
this comet, with perihelion and opposition occurring nearly at the same time;
unfortunately there were few observers of faint comets in this year and only
one total magnitude estimate is available for the optimum observing time. This
estimate indicated a magnitude between 11 and 12. The comet again arrived at
perihelion when near conjunction with the sun in 1971, and failed to become
brighter than magnitude 15. The 1978 return was again a very favorable one,
with perihelion occurring 5 weeks before opposition. The comet was well observed at that time and reached a maximum magnitude of 11.5. In addition, the
comet became one of the first comets beyond 2 AU to have its spectrum examined.

	T	q	P	Max. Mag.	Max. Tail
1948 IX	Oct. 4	2.31	7.48	10.0	10 arcmin
1956 II	Apr. 6	2.32	7.51	12.0	"short"
1963 VI	Oct. 2	2.31	7.49	11.5	4 arcmin
1971 III	Mar. 13	2.28	7.43	15.0	15 arcsec
1978 XIV	Aug. 20	2.28	7.43	11.5	10 arcmin

Barnard 1

Discovered: July 17, 1884   Discovery Magnitude: 9.5

Edward Emerson Barnard (Nashville, Tennessee) discovered this comet on
July 16, 1884, as it moved slowly eastward through Lupus. He described it as
diffuse and estimated the coma diameter as 2 arcmin. The comet was then situated only 0.42 AU from Earth and moved slightly closer within the next two
weeks, thus making this apparition very favorable for observations.

The comet brightened to between magnitude 8 and 8.5[*]by the time of the
mid-August perihelion with the only unusual feature being the apparent lack of
a nucleus. During the first couple of weeks after discovery, observers noted
a complete lack of both a nucleus and a condensation; however, by the end of
July, some condensation began to appear and some observers began to notice a
very faint, but star-like, point within the coma. Curiously, on August 10,
William Tempel (Arcetri) described a double structure of the condensation and
for the next couple of weeks observers again reported a complete lack of the
central condensation. Around mid-September, Finlay (Cape Observatory, South
Africa) reported an apparent outburst in brightness which caused the comet to
continue to brighten through the end of the month, when the total magnitude
may have been brighter than 8[*]. The comet was observed to possess a sharp nucleus at the end of September, which Perrotin (Nice, France) described as
containing a narrow luminous jet. The comet was followed until November 20,
when the magnitude had apparently dropped below 11[*].

This comet's short-period nature was first established by Finlay and
Morrison while the comet was still under observation; however, it was not until
1889, that a fairly accurate orbit was calculated. At that time Berberich indicated a period of 5.40 years and said that the comet's chances for recovery in
1890 were not good, due to an unfavorable return. Berberich's calculations for
the 1895 return indicated a good chance for recovery since the magnitude was
expected to reach 9 or 10; however, searches proved to be unsuccessful in that
year and later returns also went by unobserved. During 1919, the comet should
have approached within 1 AU of Jupiter, which probably caused an increase in
the period and maybe even an increase in the perihelion distance. This factor
will no doubt make the comet a difficult object to recovery since it is an intrinsically faint comet. In addition, the unusual activity displayed in 1884
has on several occasions been observed in comets which later broke up into two
or more pieces; thus, there is a possibility that the comet may no longer exist.

	T	q	P	Max. Mag.	Max. Tail
1884 II	Aug. 17	1.28	5.38	8.0[*]	none

Barnard 2

Discovered: June 24, 1889   Discovery Magnitude: 9.5

Barnard (Lick Observatory, California) discovered this comet while comet-
hunting with a 16-cm refractor on June 24, 1889. He found it in Andromeda as

it moved northeastward and described it as very faint, with no condensation or
tail. The coma was 2 arcmin across.

The comet was found 3 days after perihelion passage and steadily faded
thereafter as it moved away from both the sun and Earth. It was last detected
on August 7, when Barnard indicated a total magnitude slightly fainter than 12.
Despite a duration of visibility of only 45 days, mathematicians found the orbit
to be decidedly elliptical--though the exact period has been difficult to establish. In 1889, Berberich calculated it to be 128 years and in 1971, Marsden
and Sekanina calculated a value of 145 years, with an uncertainty of 10 years.
The most recent calculation was by Muraveva in 1975, who arrived at a value of
130 years.

	T	q	P	Max. Mag.	Max. Tail
1889 III	Jun. 21	1.10	145.4	9.5*	none

Barnard 3

Discovered: October 13, 1892   Discovery Magnitude: 11.5

This comet holds the distinction of being the first comet to be discovered
on a photographic plate exposed during a stellar survey. Barnard (Lick Observatory) took the photo on October 13, 1892, and upon examination he located a
suspicious-looking trail which gave evidence of a slight tail. The next evening, he searched the region with a 30-cm refractor and found a faint comet to
the southeast of the previous night's position. He described it as 1 arcmin
across, with a condensation between magnitude 12 and 13.

The comet was discovered two months prior to perihelion passage, but during its 57 days of visibility it slowly faded as its increasing distance from
Earth countered the effects of the decreasing distance from the sun. When the
comet was last observed on December 8, the total magnitude was near 12.5. Orbital calculations revealed that the comet had a very short period of roughly
6.3 years and several mathematicians remarked on the close similarity between
the orbits of this comet and Comet Wolf, which had been first seen in 1884. L.
Schulhof was the main instigator of this theory and he stated that the comets
may have separated from one another just prior to Comet Wolf's close approach
to Jupiter in June 1875 (0.117 AU). However, in 1975, D. K. Yeomans (Computer
Sciences Corporation, Silver Spring, Maryland) investigated the problem. He
first calculated the period of Comet Barnard 3 to be 6.52 years, with a probable
error of only two weeks. He then calculated the motions of the two comets back
to 1867, and found the orbit of Comet Wolf to change drastically from that of
Comet Barnard 3 so that a common origin seems to be impossible. Interestingly,
Comet Barnard 3 was never detected after the discovery apparition, due primarily to both the comet's faintness and the uncertainties in its orbit. Yeomans'
orbit indicated a close approach to Jupiter in September 1922 (0.09 AU), which
when "coupled with the uncertainties in the initial conditions, would prevent
realistic computation of Comet Barnard 3's motion beyond 1922," according to
Yeomans.

	T	q	P	Max. Mag.	Max. Tail
1892 V	Dec. 11	1.43	6.52	11.0	trace

Biela

Discovered: March 8, 1772   Discovery Magnitude: 6.5

This comet was discovered at three different apparitions before its periodic

nature was established. The first discovery came on March 8, 1772, when Montaigne (Limoges, France) found the comet at a total magnitude near 6.5*. The
comet was then past perihelion, but as it faded a tail 4 to 5 arcmin long
was observed for a few days after discovery. Observations ceased after 29
days. On November 10, 1805, Pons rediscovered the comet in Andromeda and indicated a magnitude near 4.5*. During the next 12 days independent discoveries
were made by Bouvard and Huth as the comet brightened to magnitude 4*and by
December 10, as the comet arrived at perigee (0.04 AU), the comet reached 3rd
magnitude. On this occasion, the coma was observed with a diameter greater
than 6 arcmin during early December; however, no tail was ever reported. The
comet's duration of visibility of only 36 days prevented the establishment of
a short-period orbit. The comet was discovered for a third time on February
27, 1826, when Wilhelm von Biela (Josephstadt) found it in Aries at a magnitude
near 8.5*. Ten days later, Gambart (Marseilles, France) independently discovered
the comet and a few days later, its visibility in moonlight indicated a total
magnitude near 5.5*. While the comet was still under observation, Biela and
Gambart independently expressed their beliefs that this comet was identical to
those of 1772 and 1806, and after the comet's 72 days of visibility, orbital
calculations proved the identity beyond a doubt.

Comet Biela became the third comet to have a return successfully predicted
when it was recovered by John Herschel on September 24, 1832. The discovery
magnitude was then near 8*, and with the distance from Earth decreasing to 0.56
AU during late October and early November, it brightened to magnitude 7*. On
this occasion, no tail was ever observed and the coma was never much larger
than 3 arcmin. The comet was missed at the unfavorable 1839 apparition, but
it was recovered on November 26, 1845, by de Vico (Rome, Italy). The initial
magnitude was near 10.5* and, despite the fainter than normal discovery magnitude,
nothing unusual was noted as the comet rapidly brightened during December. In
mid-January 1846, the astronomical community was caught off guard by the sudden
reports of the comet having split into two parts. Matthew Fontaine Maury (U. S.
Naval Observatory, Washington) was the first to note the double nucleus on the
13th, but observations by other observers quickly followed. As the comet continued to near both the sun and Earth, the comet brightened and the nuclei moved
farther apart. In early March, the total magnitude was near 5.5*and by the end
of the month, the nuclei were 14 arcmin apart. The actual distance, however,
remained nearly constant at 1.6 million miles. Later computations by researchers have established the actual date of splitting as falling between 1840 and
1844. The comet was recovered by Secchi (Rome) on August 26, 1852, at a magnitude near 7.5*and the fainter satellite was finally recovered on September 15.
This apparition was not especially favorable since the perihelion occurred when
the comet was in conjunction with the sun and observations ceased after September 29. The 1859 return was very unfavorable for observations and the comet
completely passed through the sun's vicinity unnoticed. Predictions indicated
that the 1865-66 return would be very favorable; however, elaborate searches
failed to locate the comet and after several months of fruitless attempts, it
was believed that the comet had broken up completely and was no longer observable. This theory seems to have been proven correct during the comet's next
approach in 1872. Predictions again indicated a close approach to Earth, but
the few searches that were made proved fruitless; however, on November 27,
an unexpected shower of meteors suddenly appeared with an estimated rate of

3,000 per hour. This shower has been proven over and over again to have been caused by Comet Biela and, on one occasion, the shower was used to try and locate a remnant of the famous double comet. This occurred on November 30, 1872, when the German astronomer Klinkerfues calculated a possible position of the comet based on the apparent radiant of the meteor shower. Coincidentally, a comet was found and observed on two days by the English astronomer Pogson, before bad weather prevented further observations from his observatory in Madras, India; however, over the years, it has been proven that this object was not Comet Biela. Further strong meteor showers were observed at the next three expected returns of this comet in 1885 (15,000 per hour), 1892 (6,000 per hour) and 1899 (150 per hour). Thereafter, no further trace of Comet Biela ever appeared and the most recent attempt to locate a possible small asteroidal remnant of the comet occurred in 1971, and again ended in failure.

	T	q	P	Max. Mag.	Max. Tail
1772 I	Feb. 18	0.99	6.87	6.0[*]	5 arcmin
1806 I	Jan. 2	0.91	6.74	3.0[*]	none
1826 I	Mar. 19	0.90	6.72	5.5[*]	small
1832 III	Nov. 27	0.88	6.65	7.0[*]	none
1846 II	Feb. 11	0.86	6.60	5.5[*]	45 arcmin
1852 III	Sep. 24	0.86	6.62	7.0[*]	small

Blanpain

Discovered: November 28, 1819   Discovery Magnitude: 6.5

     Blanpain (Marseilles, France) discovered this comet in Virgo on November 28, 1819--about 8 days after it had passed perihelion. Then located at perigee (0.26 AU), the comet was near magnitude 6.5[*], with neither a tail nor a nucleus, and possessed a coma diameter near 6 arcmin. The brightness steadily faded during December and January, as the comet moved away from both the sun and Earth, and, when last detected on January 25, 1820, it had faded to near 9th magnitude.

     Comet Blanpain's duration of visibility of 59 days allowed the calculation of a periodic orbit; however, the probable error in the period seems to amount to several months. Currently, orbits by Encke (1821) and Lagarde (1907) are usually cited as representing the observations the best, with the latter being most preferred. This gives a period of 5.10 years (as opposed to Encke's estimate of 4.81 years). The fact that such a large error exists in this period probably best accounts for the failure to locate Comet Blanpain at later apparitions. It is interesting to note that if Lagarde's period is accurate, the comet would have experienced no major perturbations from Jupiter before the 20th Century began, with the smallest separation being larger than 1.5 AU. The result of this would be a return of the comet in 1870, under conditions very similar to those of the discovery apparition. The fact that no comet matching the description and orbit of Comet Blanpain was found during a time when the successful comet-hunters Tempel, Coggia and Swift were present, leads one to believe that another orbital period is likely. Curiously, by decreasing the period by two months (4.93 years) the comet should have passed about 0.4 AU from Jupiter in 1842, thus preventing the comet from experiencing another close approach to Earth similar to that of 1819.

	T	q	P	Max. Mag.	Max. Tail
1819 IV	Nov. 21	0.89	5.10	6.5[*]	none

Boethin

Discovered: January 4, 1975   Discovery Magnitude: 12.3

During a routine comet-hunting session, the Reverend Leo Boethin (Bangued, The Philippines) discovered this comet with a 20-cm reflector on January 4.5, 1975, near Phi Aquarii. The comet was described as very faint, with a slightly condensed coma measuring 3 arcmin across. As the comet moved slowly to the east-northeast, Boethin reobserved it during the next three days before mailing a letter to the Central Bureau (Cambridge, Massachusetts) announcing his discovery. The letter arrived on January 17, but bright moonlight made confirmation impossible. In addition, the Central Bureau refrained from making an announcement, since Boethin had belatedly announced the discovery of two other comets in previous years, which could not be found after the receipt of the letter giving discovery details. Subsequently, no further news of the new comet was received during the remainder of January 1975; however, on February 3, a cablegram was received from Boethin giving a rough observation obtained on February 1. The comet was finally confirmed on February 4, when C. Scovil (Stamford, Connecticut) spotted it with the aid of a 56-cm Maksutov telescope.

The comet was nearest Earth around mid-January (1.1 AU), but continued to brighten until it reached a maximum of 11.0 shortly before mid-February. It never exhibited a tail and, as February progressed, the coma diameter shrank to 2 arcmin and the total magnitude fell below 12. On March 7, Roemer (Steward Observatory) estimated the nuclear magnitude as 17.3 and, 3 days later, Bortle gave the total magnitude as 12.6. Fading continued to be fairly rapid as the comet moved away from both the sun and Earth, and it was last detected on June 3, when Roemer (Catalina) gave the nuclear magnitude as near 19.8.

Orbital calculations are considered to be fairly accurate, despite the loss of possible accurate positions in January since the discovery was not immediately announced. Marsden gives a period of 11.05 years and considers a probable error of only one week. This period indicates that the comet has not experienced an appreciable perturbation from Jupiter in this century; however, it appears likely that an approach to within 0.5 AU from that planet is due in mid-1984. It should also be noted that, based on the given period, the comet should have been very favorably placed for discovery during the autumn of 1930, with a total magnitude of between 6 and 8--depending on how much the period is in error.

	T	q	P	Max. Mag.	Max. Tail
1975 I	Jan. 6	1.09	11.05	11.0	none

Borrelly

Discovered: December 28, 1904   Discovery Magnitude: 10.0

Borrelly (Marseilles, France) discovered this comet on December 28, 1904, in Cetus during a routine search for comets. He described it as 1 to 2 arcmin in diameter with a small, faint nucleus. The comet was moving almost due north.

By January 1, 1905, the comet was being widely observed; however, the ill-defined boundaries of the nearly 2-arcmin-diameter coma caused problems in the estimates of the total magnitude. These values ranged from 9 to 11, with the brighter values usually coming from observers using low-power instruments and the fainter values coming from high-power instruments. A tail nearly 10 arcmin long was also observed at this time. The comet reached a maximum brightness of 9 after the first week of January, and thereafter it began to fade fairly rapidly. On March 8, the total magnitude had dropped to 12.5 and by May

9, it was near 13.5. The comet was last seen on May 25, when near magnitude
14.5.

An extensive study of this comet's orbit was carried out by E. I.
Kazimirchak-Polonskaya in the mid-1960's, and it was revealed that between
1660 and 2060, the comet would approach moderately close to Jupiter nine times.
These passages have never been through Jupiter's sphere of influence and so it
was concluded that the orbit "evolves smoothly and experiences insignificant
changes over the 400 years." The comet was placed in its discovery orbit in
February 1889. The period of 6.9 years caused the comet to arrive at perihelion one month earlier each time it approached the sun. Consequently, the
comet passed perihelion during December of 1911, and was then more favorably
placed than at its discovery apparition, since it came within 0.53 AU from
Earth. The total magnitude reached a value of 8.4 at that time, while the
coma and tail attained dimensions of 4 and 30 arcmin, respectively. The next
perihelion came in November 1918, and the comet again came within 0.53 AU of
Earth and reached a magnitude of 9. Thereafter, the comet's 6.9-year period
caused observing conditions to steadily worsen at future perihelion passages
with maximum magnitudes of 10 being reached in 1925, and 11 being reached in
1932. During April of 1936, another moderate approach to Jupiter caused the
period to increase to almost precisely 7.0 years and the 1939 and 1946 perihelion passages were so badly situated that the comet was not recovered. In
1953, Roemer managed to recover the comet nearly 7 months after perihelion at
a magnitude of 18.5. With the orbit again known with more precision, the comet
was recovered much earlier during the 1960 and 1967 apparitions, despite the
conditions being almost identical to those of 1953. At both apparitions, observers recovered the comet at magnitudes between 15 and 16, and were able to
follow the comet during the next 5 months. In February 1972, the comet again
passed near Jupiter and had its period reduced to 6.8 years. The 1974 return
was not a particularly favorable one with perihelion again coming when the
comet was in conjunction with the sun, but future returns will become more
favorable as the perihelion date comes two to two and a half months earlier at
each return until the next close approach to Jupiter in 2019.

	T	q	P	Max. Mag.	Max. Tail
1905 II	Jan. 17	1.40	6.91	9.0	10 arcmin
1911 VIII	Dec. 18	1.40	6.93	8.4	30 arcmin
1918 IV	Nov. 17	1.40	6.91	9.0	4 arcmin
1925 VIII	Oct. 8	1.39	6.89	10.0	2 arcmin
1932 IV	Aug. 27	1.39	6.87	11.0	5 arcmin
1953 IV	Jun. 9	1.45	7.01	18.5	"short"
1960 V	Jun. 13	1.45	7.02	15.0	4 arcmin
1967 VIII	Jun. 18	1.45	6.99	16.0	none
1974 VII	May 13	1.32	6.76	18.0	2 arcmin
1981 IV	Feb. 20	1.32	6.77	8.7	1 degree

Brooks 1

Discovered: May 23, 1886   Discovery Magnitude: 8.5

William R. Brooks (Phelps, New York) discovered this comet (his third in
25 days) on May 23, 1886, near Omicron Virginis, during a routine search for
comets. Brooks described the comet as moving slowly to the southeast, with a
circular coma and a slight condensation also visible. No tail was reported.

The comet became widely observed during the next few days, with observers confirming Brooks' magnitude estimate and consistently measuring the coma diameter to be 2 arcmin. Again, the coma possessed a circular appearance, with no sign of a tail. The comet was nearest Earth (0.54 AU) at the end of May, and began to fade thereafter. By mid-June, the total magnitude had declined to 9.5[*] and by the end of the month, it was near 10[*]. The comet was last seen on July 3.

The orbit of this comet was found to be elliptical, but the short duration of visibility (42 days) made it difficult for mathematicians to pinpoint the period. Although one mathematician had obtained a period of 9 years in 1886, Dr. J. Hind found a period of 6.30 years during the same year, which was immediately recognized as being closer to the truth. This period indicated a return during the autumn of 1892, and early in that year Dr. S. Oppenheim recalculated the orbit to more accurately decide when and where the comet could be recovered. His result gave a period of 5.60 years, thus indicating a perihelion at the very beginning of 1892, for which the comet had not been well placed for recovery. Later returns of the comet also went unobserved and it is now believed that the comet will only be recovered by accident. Nevertheless, in 1976, Buckley recalculated the orbit and obtained a period of 5.44 years, with a probable error of 3 months. Despite this new orbit, the comet is still considered lost.

	T	q	P	Max. Mag.	Max. Tail
1886 IV	Jun. 7	1.33	5.44	8.5[*]	none

Brooks 2

Discovered: July 7, 1889   Discovery Magnitude: 11.0

Brooks (Geneva, New York) discovered this comet in Aquarius while sweeping in the southwestern sky on July 7, 1889. He described it as faint, with a diameter of 1 arcmin and a tail length of 10 arcmin. No motion could be detected before sunrise, but on the following morning, Brooks reobserved the comet and found it to have moved slightly northward.

The comet steadily brightened as it approached both the sun and Earth and by August 1, it had reached magnitude 10[*], with a coma 3 arcmin across and a tail 15 arcmin long. Also, on August 1, Barnard discovered 2 small and nebulous companions (designated B and C) located 1 and 4.5 arcmin from the main nucleus (A). Each companion was described as a perfect miniature of the main comet and possessed a small coma, with a small nucleus and a small tail. On August 2, Barnard again detected the three nuclei of the previous morning as well as four or five additional objects which were described as nebulous and were not detected on the 3rd. On the 4th, two more objects appeared (D and E), in addition to the original three nuclei. Nucleus E was not seen again, and nucleus D vanished after about one week of observation. In mid-August, B began to grow larger and more diffuse and was finally lost after September 5, while C continued to be observed until November 26. Studies of the motions of these nuclei have led several researchers to believe the comet disrupted during an extremely close approach to Jupiter in 1886. Details of this close encounter will be discussed shortly as the orbit itself is examined.

Comet Brooks 2 reached its maximum brightness of 8 during September 1889, as it passed both perigee (0.95 AU) and perihelion (1.95 AU). Fading was somewhat slow thereafter, as the comet dropped to magnitude 10[*] during October, and

11[*]by the end of November. Although near magnitude 11.5 during December, the comet seemed to brighten to near 11 during January 1890. By February, it had declined to magnitude 12.5 and when last observed on April 7, shortly before becoming lost in the sun's glare, it was near magnitude 13[*]. After conjunction with the sun, the comet was recovered on November 21, 1890, by Barnard, while using the 91-cm Lick telescope, and he continued to observe the comet until January 13, 1891.

Interestingly, the luminosity of the comet at the discovery apparition has never been repeated, nor have the minor nuclei ever been reobserved. At the comet's next return in 1896, it was again very favorably placed for observation with a minimum distance from Earth of 1.01 AU; however, although this distance was only slightly greater than that achieved in 1889, the comet never became brighter than magnitude 10.5. In addition, no tail was detected. Each apparition thereafter became steadily worse for observing as the 7.10-year period moved the perihelion date back one month at each appearance. In 1903, the maximum magnitude was only 12.5, and in 1911, it was 15.5. The 1918 apparition was missed, but the comet was again seen in 1925, when it brightened to magnitude 12 and possessed a photographic tail 1 arcmin long. In 1932, it reached magnitude 10.5 and possessed a tail 4 arcmin long. After the latter apparition, the comet again started a series of badly placed appearances, with maximum magnitudes of only 12.5 in 1939 and 1946. From 1953 on, observations of the comet's total magnitude nearly ceased as the preferred nuclear magnitude was achieved on photographs made to determine the comet's precise position. In 1953, the nucleus was never brighter than 17, while in 1960, it was never brighter than 17.8. The comet was missed in 1967, due to unfavorable conditions, but when again detected at the somewhat unfavorable 1974 apparition, it did not become brighter than 18.7.

Prior to its first appearance in 1889, this comet experienced the closest known approach (0.001 AU) of a comet to Jupiter. For two days in 1886, this comet was within the satellite system of Jupiter--just inside the orbit of Io. The effects of this approach not only disrupted the comet--as noted by the observations mentioned earlier--but it also rearranged the orbit, with the period decreasing from 29.2 years to 7.1 years and the perihelion distance decreasing from 5.48 AU to 1.95 AU. The comet was then traveling in a 3:5 resonance with Jupiter, the result of which was another close approach 35 years later in 1921. Although the distance from Jupiter was much larger than in 1886, the orbit still experienced much change with the reversal of the nodes and a rotation of the line of apsides. The period, however, declined by only 0.2 years, so that the comet's arrivals at perihelion were not considerably changed.

	T	q	P	Max. Mag.	Max. Tail
1889 V	Sep. 30	1.95	7.07	8.0[*]	15 arcmin
1896 VI	Nov. 5	1.96	7.10	10.5	none
1903 V	Dec. 7	1.96	7.10	12.5	"short"
1911 I	Jan. 9	1.96	7.10	16.0	none
1925 IX	Nov. 2	1.86	6.92	12.0	1 arcmin
1932 VIII	Oct. 10	1.87	6.94	10.5	4 arcmin
1939 VII	Sep. 15	1.87	6.95	12.5	2 arcmin
1946 IV	Aug. 26	1.88	6.96	12.5	3 arcmin
1953 V	Aug. 7	1.87	6.93	17.0	1 arcmin
1960 VI	Jun. 17	1.76	6.72	17.8	none
1974 I	Jan. 4	1.84	6.88	18.7	3 arcsec
1980 IX	Nov. 25	1.85	6.90	13.5	trace

Brorsen

Discovered: February 26, 1846   Discovery Magnitude: 7.5

T. Brorsen (Kiel, Germany) discovered this comet on February 26, 1846, during a routine comet-hunting session. It was then moving rapidly north-ward through Pisces and was between magnitude 7 and 8[*]. The comet was discovered the day after perihelion passage, but instead of growing fainter, it brightened as its distance from Earth decreased. The comet reached a maximum brightness of 6[*] shortly after mid-March, and was described as 8 to 10 arcmin across on March 25. It was nearest Earth (0.48 AU) at the beginning of April, and, thereafter, faded. The comet was last detected on April 22, when near magnitude 8.5[*].

This comet was placed into its discovery orbit by a very close approach (0.06 AU) to Jupiter in 1842. Although it was discovered at its first perihelion passage in 1846, the comet's period of nearly 5.5 years caused alternate returns to be unfavorable and it was missed in 1851. After a minor approach (0.84 AU) to Jupiter in 1854, the comet was recovered by Bruhns (Berlin, Germany) in 1857. This recovery was purely accidental, since predic-tions were three months off due to a small inaccuracy in the 1846 orbit. The comet brightened to near magnitude 5[*] and was frequently observed to possess a thin tail. The unfavorable 1862 return was missed and the comet made another minor approach (1.3 AU) to Jupiter in 1866. In 1868, the comet was recovered very near the predicted position at a magnitude near 7[*]. In mid-May, the comet reached its maximum brightness of 6.5[*] and possessed a tail 40 arcmin long. Although the 1873 return was an unfavorable one, observers did manage to recover the comet and observe it for 57 days as it slowly brightened from magnitude 9[*] to 7[*] during its approach to perihelion. The comet entered the sun's glare shortly after reaching its maximum brightness and no tail was ever reported. In January 1879, the comet was recovered at a magnitude near 11 and brightened during the next three months as it neared both perihelion and perigee. By mid-April, it was slightly brighter than 7[*] and possessed a tail 30 arcmin long. Curiously, an unexpected fading occurred near the beginning of May, which caused the comet to drop to magnitude 11[*]--three magnitudes fainter than predicted by mid-May. The comet was lost after May 23.

Astronomers are currently undecided as to the fate of Comet Brorsen, for after the 1879 appearance it was never seen again. After the unfavorable 1884 apparition, exhaustive searches were made at the very favorable 1890 apparition --but no comet was detected. After the unfavorable 1895 apparition, intensive searches were made at the favorable 1901 return, but, again, no comet was found. Based on these searches, Marsden concluded in 1963 that the possibility of finding Comet Brorsen was remote and he felt it safe to assume that it had "faded out of existence." Nevertheless, he carried out a computer study of the orbit which showed that close approaches to Jupiter had occurred in 1913 (0.39 AU), 1925 (0.35 AU), 1937 (0.91 AU) and 1949 (1.23 AU). He said that the comet would be well placed for a recovery in 1973. Japanese observers carried out several searches for the comet in 1973, but failed to locate any trace, thus supporting the theory that the comet no longer exists.

	T	q	P	Max. Mag.	Max. Tail
1846 III	Feb. 26	0.65	5.57	6.0[*]	none
1857 II	Mar. 30	0.62	5.54	5.0[*]	30 arcmin
1868 I	Apr. 18	0.60	5.48	6.5[*]	40 arcmin

	T	q	P	Max. Mag.	Max. Tail
1873 VI	Oct. 11	0.59	5.47	7.0*	none
1879 I	Mar. 31	0.59	5.46	6.9*	30 arcmin

Brorsen-Metcalf

Discovered: July 20, 1847   Discovery Magnitude: 9.5

Brorsen (Altona Observatory, Germany) discovered this comet on July 20, 1847, in Aries. He described it as very faint and diffuse, without any trace of a nucleus. The comet brightened as it approached both the sun and Earth, and reached a total magnitude of 6.5* in mid-August. At this time, the tail extended at least 15 arcmin. With the distance from Earth increasing thereafter, the comet faded slowly as it continued to near the sun. When last seen on September 13, the total magnitude was near 9.5*.

Orbital calculations revealed that the comet was periodic, though the duration of visibility of only 56 days made it difficult to determine the exact period. Values generally ranged from 71 years to 81 years, with d'Arrest's calculation of 75 years being most accepted. Thus the comet was to be next expected around 1922. Quite unexpectedly, the first preliminary orbit for a comet found by the Reverend Joel H. Metcalf (Camp Idlewild, Vermont) on August 21, 1919, indicated it was identical to Comet Brorsen, thus establishing the comet's period at 72 years. During this return the comet was independently discovered by 5 observers in 12 days as it rapidly brightened. On September 5, the comet passed within 0.19 AU of Earth and was described as near magnitude 5.3, with a coma at least 15 arcmin across. Although the distance from Earth increased thereafter, the comet's decreasing distance from the sun caused it to continue to brighten, although the coma rapidly shrank. During the first week in October, the comet reached magnitude 4.5, and the coma was estimated as less than 8 arcmin across. The tail was longest in late September, when it extended 1.5 degrees. Interestingly, Barnard (Lick Observatory) reported that his photographs on October 22 revealed that the tail separated from the coma. Within the next few days, a new tail formed at a 12 degree angle to the old tail.

	T	q	P	Max. Mag.	Max. Tail
1847 V	Sep. 10	0.49	73.12	6.5*	15 arcmin
1919 III	Oct. 17	0.48	71.97	4.5	1.5 degrees

Bus

Discovered: February 9, 1981   Discovery Magnitude: 20.0

During March 1981, Schelte J. Bus (California Institute of Technology) was observing at the U. K. Schmidt Telescope Unit at Siding Spring, Australia. While examining a photographic plate exposed by Kenneth Russell on March 2 he detected a 17.5-magnitude comet. A confirmatory plate was exposed on March 3 and the new comet was found to possess a daily motion of 9 arcmin northwestward. Bus described the comet as having a central condensation and a faint tail extending about 20 arcsec to the northwest.

By March 9, precise positions for March 4 and 5 had been received from Harvard Observatory's Agassiz Station and Tokyo Observatory's Geisei Station, respectively. At the latter observatory, T. Seki estimated the total magnitude as 18. These positions, as well as precise positions for March 2 and

3, allowed Brian Marsden to compute a parabolic orbit, as well as a preliminary short-period orbit. The periodic orbit revealed a perihelion date of
June 29.9, 1981 and a period of 6.3 years. An ephemeris from this orbit
allowed Bus to find and measure two prediscovery plates of the comet which
had been taken in February. The first plate had been taken by Russell on
the 9th, and displayed a weak image estimated as between magnitude 19.5 and
20. The second plate had been taken by Malcolm Hartley on the 13th, and
showed a weak image of magnitude 20. These two positions confirmed Marsden's
suspicion that a new periodic comet had been found and eventually the orbit
was shown to have a period of 6.52 years and a perihelion date of June 11
(q=2.18 AU).

The comet's total magnitude brightened to 16.5 by March 7, but by the
end of that month the comet's distance from earth began to increase. A very
slow fading set in during the next two months, with the total magnitude reaching about 17[*]by early June. On May 24, observers at Agassiz estimated the
nuclear magnitude as 17.5 and they were the last to see the comet on June 27--
with further observations being impossible as the comet entered the sun's
glare.

	T	q	P	Max. Mag.	Max. Tail
1981 XI	Jun. 11	2.18	6.52	16.5	20 arcsec

Chernykh

Discovered: August 19, 1977   Discovery Magnitude: 14.0

During the examination of routine minor planet survey plates exposed on
August 19 and 22, 1977, Nikolaj Stepanovich Chernykh (Crimean Astrophysical
Observatory, Russia) discovered this comet as a fairly bright, diffuse object,
with a slight condensation. Then located near the Pisces-Cetus border, the
comet was moving slowly to the southwest.

The comet was discovered as it approached both the sun and Earth, and
during the months of September and October it was intensively observed. The
comet was brightest at the end of September, when it was at perigee (1.74 AU),
with visual observers giving the total magnitude as about 12.5. The coma was
frequently described as 1 arcmin across and, throughout September, the tail
appeared forked on photographs and extended up to 2 arcmin. After early
October, the comet faded slowly as the increasing distance from Earth was
nearly countered by the decreasing distance from the sun. By the time perihelion arrived in February 1978, the distance from Earth had grown to 3 AU and
the total magnitude was near 15. The comet was lost in the sun's glare after
March 3, and was too faint for recovery after conjunction with the sun.

Although the comet has made just one appearance, a preliminary investigation of the evolution of its orbit between 1660 and 2060 was carried out by
Kazimirchak-Polonskaya and Chernykh in 1978. The study showed that the comet
had experienced 11 encounters with Jupiter and 9 with Saturn and the authors
commented that "it happens more than once that it (the comet) is not able to
end the encounter with one planet before the encounter with the other begins."
Interestingly, the comet's closest approach to Saturn (0.36 AU in 1749) is only
the second known case of a comet entering that planet's sphere of influence.
A close approach (0.35 AU) to Jupiter in January 1980, reduced the period to
14.0 years.

	T	q	P	Max. Mag.	Max. Tail
1978 IV	Feb. 15	2.57	15.93	11.8	2 arcmin

Churyumov-Gerasimenko

Discovered: September 9, 1969   Discovery Magnitude: 13.0

On October 22, 1969, Klim Ivanovic Churyumov was measuring the positions of P/Comas Sola from photos taken in September by Svetlana Ivanovna Gerasimenko at the Alma-Ata Observatory in Russia. Suddenly, his measurement of the prominent 13th-magnitude image on the September 11.9 plate proved to be two degrees off the expected course and he realized that he had found a new comet. Soon afterward, Churyumov located a prediscovery image on a September 9 plate, and further images were later found on a September 13 photo exposed at Nice Observatory and on a September 14 photo exposed at Stamford, Connecticut.

With perihelion having occurred on September 11, the comet slowly faded after discovery, despite the steadily decreasing distance from Earth; however, the tail grew rapidly and by early November, it was nearly 10 arcmin long. The comet reached perigee (1.15 AU) in January 1970, and was then described as near magnitude 14. Thereafter, the brightness dropped by one magnitude per month and the comet was lost in the sun's glare after May 8. The comet's next return in 1975 proved to be less favorable than the previous apparition with perihelion coming while the comet was in conjunction with the sun. Nevertheless, the comet was recovered by Elizabeth Roemer and R. A. McCallister (Steward Observatory) on August 8, 1975 at a nuclear magnitude of 19.5. Roemer proved to be the only observer of this comet as she followed it until December 7, when the nuclear magnitude reached 18.8.

Computer studies of the comet's orbit indicate a close approach to Jupiter in 1959, which decreased the perihelion distance from 2.77 AU to 1.28 AU. Prior to this, an approach to Jupiter in 1840 had reduced the perihelion distance from 4.0 AU to 3.0 AU.

	T	q	P	Max. Mag.	Max. Tail
1969 IV	Sep. 11	1.28	6.55	13.0	10 arcmin
1976 VII	Apr. 7	1.30	6.59	18.8	6 arcsec

Clark

Discovered: June 1, 1973   Discovery Magnitude: 13.0

Michael Clark (Mount John University Observatory, New Zealand) discovered this comet on June 9.7, 1973, near the edge of a photo exposed for the southern variable-star patrol. The next morning, Clark confirmed that the 13th magnitude object had moved to the southeast and described it as diffuse, with a tail 1 arcmin long. He then managed to find a prediscovery image on another patrol plate exposed on June 1.7.

The comet remained a very small object during this apparition, despite its reaching a minimum distance of 0.65 AU from Earth in the first days of July. On June 11 and 20, A. C. Gilmore (Carter Observatory, New Zealand) estimated the coma as 10 arcsec across and by July 3, it was measuring 8 by 4 arcsec. During this same time, the total magnitude remained near 13. The comet faded slowly thereafter, with the total magnitude finally reaching 15 in late September. By this time the comet was moving northward, thus allowing Roemer to determine the nuclear magnitude as 18.8 on September 22. This observer managed to follow the comet until November 21. Orbital calculations gave a period of 5.5 years, which not only indicated a possible close approach to Jupiter in 1954, but would also cause every second return to be unfavorable for Earth-based observations. Subsequently, the favorable 1973 apparition

was followed by an unfavorable 1978 return with the comet only brightening
to a total magnitude of 18 before becoming lost in the sun's glare.

	T	q	P	Max. Mag.	Max. Tail
1973 V	May 25	1.56	5.52	13.0	1 arcmin
1978 XXIII	Nov. 26	1.56	5.51	18.0	none

Comas Sola

Discovered: November 6, 1926   Discovery Magnitude: 12.0

Josep Comas Sola (Fabra Observatory, Spain) found this comet on a plate
exposed on November 4, 1926, in the search for minor planets. Then located
near Alpha Ceti, the comet was described as of the 12th magnitude, with a coma
2 arcmin across and a motion to the northwest. Neujmin later found the comet
on a photo exposed only two hours prior to Comas Sola's plate.

The comet was discovered shortly before one of its closest possible
approaches to Earth (1.15 AU on November 26) and remained near magnitude 12
until January 1927. During the months in between, a tail was frequently
observed, which at one time measured nearly 4 arcmin in length. After January,
the comet slowly faded, despite a decreasing distance from the sun, and was last
photographed on May 31, when van Biesbroeck (Yerkes Observatory) estimated the
total magnitude as 14. Attempts to photograph the comet after conjunction with
the sun failed.

This comet has been recovered at every return since its discovery, due to
its large perihelion distance (1.8 AU), and the period of 8.5 years between
1927 and 1969 caused perihelion to fall in either the spring or autumn months.
This situation always kept the comet's closest distance from Earth at between
1.1 and 1.4 AU and, therefore, the maximum magnitude has always been between 12
and 13. A moderately close approach (0.73 AU) to Jupiter in 1971 increased the
period by 0.4 years, but the 1978 return was still favorable with a maximum
magnitude of 13.

In 1967, N. A. Belyaev (Institute of Theoretical Astronomy, Leningrad,
Russia) published a study of this comet's orbit for the years 1660 to 2060. He
found that prior to 1710, the comet was a member of Saturn's family of comets
with a period of 13.1 years and a perihelion distance of 3.2 AU. An approach to
within 0.14 AU of Jupiter in December 1710 decreased the period to 11.0 years
and the perihelion distance to 2.6 AU. In December 1781, the comet passed
0.50 AU from Jupiter, the effects of which dropped the period to 9.4 years and
the perihelion distance to 2.2 AU. The comet's present orbit was formed when
passing 0.18 AU from Jupiter in May 1912 and it will again change in April 2007,
when it passes 0.32 AU from Jupiter. The period will then increase to 9.5 years
and the perihelion distance will increase to 1.9 AU.

	T	q	P	Max. Mag.	Max. Tail
1927 III	Mar. 22	1.77	8.52	12.0	4 arcmin
1935 IV	Oct. 7	1.78	8.53	13.0	10 arcmin
1944 II	Apr. 11	1.77	8.50	12.5	3 arcmin
1952 VII	Sep. 11	1.77	8.54	13.0	20 arcmin
1961 III	Apr. 4	1.78	8.58	12.5	2 arcmin
1969 VIII	Oct. 29	1.77	8.55	13.0	20 arcmin
1978 XVII	Sep. 24	1.87	8.94	13.0	3 arcmin

Crommelin

Discovered: February 23, 1818   Discovery Magnitude: 7.5

   The history of Comet Crommelin has been a sketchy one, with the comet
being discovered at three separate apparitions before its periodic nature was
finally established. Jean Louis Pons (Marseilles Observatory, France) was the
first discoverer of this comet when he spotted it on February 23, 1818. He
described it as a condensed, elongated nebulosity, but its westward movement in
the evening sky soon took it into the sun's glare after only 5 days of visibility. When last seen, the comet was 0.49 AU from Earth. The comet was not
seen again until November 10, 1873, when Jerome Eugene Coggia (Marseilles,
France) discovered it moving southwestward through Hercules in the evening sky.
The next evening, an independent discovery was made by F. A. T. Winnecke
(Strasbourg, Germany), who described it as a pale disk, with a bright inner
coma 3 arcmin across. Both observers indicated a magnitude near 8.0[*]. By the
12th, Winnecke was estimating the coma diameter as 6 arcmin. The comet was
followed only until the 16th before it dropped into the evening twilight. Its
closest distance from Earth during those days had been 0.25 AU.

   Unfortunately, the 1873 apparition afforded little advantage over the
1818 return in that only a rough parabolic orbit could be calculated. In the
following years, several mathematicians noted a resemblance between the parabolic orbits of 1818 I and 1873 VII. Schulhof concluded, with some reserve,
that both comets were distinct short-period comets having a common origin, which
he believed was the comet 1457 I. However, he later reversed this belief and
concluded that the comets were the same and possessed a 55-year period. Weiss
also believed the two comets were the same, but he was unsure whether the period
was 55.8, 18.6 or 6.2 years--though the latter appeared most probable.

   The comet remained a mystery until it was finally rediscovered as a 6th-
magnitude object on November 19, 1928, by A. F. I. Forbes (Rosebank, South
Africa) in the morning sky. When the first preliminary orbits were computed
in early December, the controversy of 1873 was restarted, but this time the
problem would finally be solved. A. C. D. Crommelin began computing a periodic
orbit and concluded that the period was probably 27 years. Shortly thereafter,
it was learned that M. Yamasaki (Kyoto, Japan) had first observed the comet on
October 27, but had lost the object due to bad weather. He and Mr. Kinoshita
(Tokyo Observatory) were independently engaged in trying to recover the comet
when it was learned that Forbes had found it. The comet faded rapidly and was
last detected on December 24, when near 13th magnitude. Shortly thereafter,
F. Quenisset (Flammarion Observatory, France) found prediscovery images on
photos he had taken of the Zodiacal light on October 25. Crommelin then took
the available observations and tackled the problem of seeking a definitive
answer. His conclusion was that the comets 1818 I, 1873 VII and 1928 III were
the same object with a period of revolution near 28 years. He then predicted
that the comet's next return would be in 1956. The comet became known as
Pons-Coggia-Winnecke-Forbes, but due to Crommelin's efforts it was decided in
1948 to name the comet after him.

   The comet was recovered by Ludmilla Mrkosova-Pajdusakova (Skalnate Pleso
Observatory, Czechoslovakia) on September 29, 1956. Then near magnitude 10,
the comet brightened to nearly magnitude 7 before it was lost in the sun's
glare after November 29. The only controversy that still remains about Comet
Crommelin is the supposed identity to comet 1457 I and 1625. The latter

object was so badly observed that no independent orbit has ever been calculated
and although the former object has a parabolic orbit very similar to Comet
Crommelin, the inclination varies by 16 degrees—a difference difficult to
explain. The current consensus is that these two comets are unrelated to
Comet Crommelin.

	T		q	P	Max. Mag.	Max. Tail
1818 I	Feb.	6	0.75	27.70	7.5*	none
1873 VII	Dec.	2	0.75	28.07	8.0*	none
1928 III	Nov.	5	0.75	27.92	6.0	none
1956 VI	Oct.	25	0.74	27.89	7.0	3 arcmin

Daniel

Discovered: December 7, 1909   Discovery Magnitude: 9.0

Zaccheus Daniel (Princeton University) discovered this comet while sweeping for comets on December 7, 1909. He described it as near 9th magnitude,
with a 13th-magnitude nucleus, and gave the direction of motion as almost due
north.

The comet was discovered about one week after perihelion passage and was
then near its maximum brightness of 9. During the following weeks it slowly
faded and by March 1 the brightness was near 13th magnitude. During the same
period of time, the coma decreased from 3 arcmin to 2 arcmin in diameter. The
comet was last detected on April 11. The comet was unfavorably placed for
observation at its next three returns; however, the orbit had been calculated
accurately enough in 1910 to allow an easy recovery at the favorable 1927
return—despite a moderately close approach to Jupiter in 1911-12, which
increased the perihelion distance by 0.16 AU and the period by 0.3 years. The
comet reached a maximum magnitude of 12.5 in 1937, and 13.0 in 1943, and, after
a moderate approach to Jupiter in 1946—which decreased the period by 0.1 year—
the comet was recovered in 1950 and brightened to maximum magnitude of 15.

The comet was again missed in 1957, due to unfavorable conditions, and in
1959 it passed 0.53 AU from Jupiter. This latter passage increased the period
to almost exactly 7 years and placed the comet's arrival at perihelion at a
time when it was exactly opposite the sun from Earth. Despite short glimpses
in 1964 and 1979—the latter being at a record distance—the comet remains a
very difficult object for observation.

	T		q	P	Max. Mag.	Max. Tail
1909 IV	Nov.	29	1.38	6.48	9.0	short
1937 I	Jan.	28	1.54	6.83	12.5	none
1943 IV	Nov.	23	1.53	6.80	13.0	none
1950 V	Aug.	24	1.46	6.66	15.0	1 arcmin
1964 II	Apr.	21	1.66	7.09	20.0	none
1978 XII	Jul.	8	1.66	7.10	19.0	none

d'Arrest

Discovered: June 28, 1851   Discovery Magnitude: 10.0

Heinrich Louis d'Arrest (Leipzig, Germany) discovered this comet as it
moved eastward through Pisces on June 28, 1851. It was then described as very
faint, and, during the next two months, observers consistently described it as
a diffuse, circular object with neither a nucleus nor a tail. With the perihelion distance being fairly large, the comet faded slowly after discovery and
when last detected on October 7, the total magnitude was still near 10.5.*

This comet was placed into its discovery orbit by a close approach (0.13 AU) to Jupiter in 1695. The orbit was then slightly modified by further approaches in 1706 (0.90 AU), 1742 (0.53 AU), 1778 (0.20 AU), 1790 (0.37 AU), 1801 (0.68 AU) and 1849 (1.13 AU). Following the discovery apparition, the comet was reobserved in 1857-58, when the maximum magnitude reached nearly 9.5. Another close approach (0.34 AU) to Jupiter occurred in 1861, which increased the perihelion distance to 1.28 AU, and the comet was not observed in 1864, since perihelion occurred when the comet was in conjunction with the sun.

The comet was reobserved in 1870 and brightened to magnitude 8.5[*] near the day of perihelion passage. The 1877 return, however, was not as favorable with the minimum distance from Earth being almost twice that of the 1870 apparition and, thus, the magnitude only brightened to 10[*]. After a missed return in 1884, the comet was next detected in 1890, when the total magnitude reached 9.5[*] and this was followed by the slightly less favorable 1897 appearance when the maximum magnitude reached 10.5[*].

The comet was again missed at the 1904 return and in 1908 it passed 1.33 AU from Jupiter. Observers in 1910 found a maximum magnitude of only 12 and apparitions thereafter became steadily worse as the comet's perihelion distance steadily increased. The comet was missed in 1917, and an approach to within 0.49 AU of Jupiter in 1920 increased the perihelion distance from 1.27 to 1.36 AU. Although the comet reached 11th magnitude in 1923, it was missed in 1930 and 1937, due to its being too faint and unfavorably situated.

By 1943, the perihelion distance had further increased to 1.38 AU and, after being recovered at a magnitude of 12.5, the comet rapidly faded as it quickly moved away from both the sun and Earth. Although the 1950 return was slightly less favorable, observers reported an outburst in brightness about three weeks after passing perihelion which caused the total magnitude to reach 10.5. The comet was again missed in 1957, but it was recovered by Roemer (Flagstaff, Arizona) in 1963. Her magnitude estimates were of the nuclear region and reached a maximum of about 18. This would indicate a total magnitude of near 14.5, based on previous apparitions.

In 1968, the comet encountered Jupiter again (0.42 AU), which caused the perihelion distance to drop to 1.17 AU--equal to the discovery apparition. Subsequently, observers of the favorable 1970 return reported a maximum brightness of 11. The smaller perihelion distance held through the 1976 return, when it happened that the comet reached perihelion and perigee on the same day. This placed the comet at its closest possible distance from Earth (0.15 AU) and caused the total magnitude to reach 5.0. As in the past, the presence of Jupiter again changed certain orbital parameters when the comet passed 0.30 AU away in 1979. The value of the perihelion distance was then increased to 1.30 AU and future close approaches to Jupiter in 1990, 2034 and 2050 will prevent this value from becoming much smaller.

	T	q	P	Max. Mag.	Max. Tail
1851 II	Jul. 9	1.17	6.39	10.0[*]	none
1857 VII	Nov. 29	1.17	6.38	9.5[*]	none
1870 III	Sep. 23	1.28	6.57	8.5[*]	none
1877 IV	May 11	1.32	6.66	10.0[*]	none
1890 V	Sep. 18	1.32	6.68	9.5[*]	none
1897 II	May 24	1.33	6.69	10.5[*]	none

	T	q	P	Max. Mag.	Max. Tail
1910 III	Sep. 17	1.27	6.54	12.0	5 arcmin
1923 II	Sep. 16	1.36	6.64	11.0	none
1943 III	Sep. 22	1.39	6.72	12.5	"short"
1950 II	Jun. 6	1.38	6.70	10.5	3 arcmin
1963 VII	Oct. 23	1.37	6.67	14.5	none
1970 VII	May 18	1.17	6.23	11.0	none
1976 XI	Aug. 13	1.16	6.23	4.9	1 degree
1982 VII	Sep. 14	1.29	6.38	8.1	none

Denning

Discovered: March 26, 1894   Discovery Magnitude: 10.0

William F. Denning (Bristol, England) discovered this comet as it moved southeastward through Leo Minor on March 26, 1894. It was described as faint, with a coma diameter of 1 arcmin, and possessed a star-like nucleus of magnitude 11 or 12. A faint tail was also present, which measured 2 arcmin long and 1 arcmin wide.

At discovery, the comet was located 0.37 AU from Earth, but both this distance and the distance from the sun increased during the following weeks and the brightness rapidly faded. At the beginning of April, the magnitude was near 10.5 and by the beginning of May, it had dropped to about 11.5[*]. A tail was frequently observed from the moment of discovery until late April, with the longest length being about 3 arcmin. Observations continued throughout May. The comet was last seen on June 5, when the total magnitude had dropped to 13 or 14[*].

Orbital calculations indicated a period of 7.4 years and a study by Schulhof revealed that the comet had passed within 0.2 AU of Jupiter in 1889. The comet was badly placed for observation in 1901, and in 1903, Gast made a detailed study which gave the 1894 orbital period as 7.418 years and predicted the next perihelion date as December 1908. However, searches made around the indicated time proved fruitless and no trace of the comet has ever been found. Recent calculations by Buckley (1976) and Belyaev (1978) confirm the correctness of Gast's computations and show that the probable error amounted to only 10 days.

	T	q	P	Max. Mag.	Max. Tail
1894 I	Feb. 10	1.15	7.42	10.0[*]	3 arcmin

Denning-Fujikawa

Discovered: October 4, 1881   Discovery Magnitude: 7.5

During a routine sweep for comets, Denning discovered this object in Leo on October 4, 1881. The comet was then described as a small, bright nebula with a central condensation. Unfortunately, the comet was nearly one month past perihelion and, with its distances from the sun and Earth increasing, it rapidly faded. Observers shortly after mid-October indicated a magnitude near 8.5 and when the comet was last detected on November 25, the total magnitude was near 11.5[*]. Interestingly, computations revealed that this comet had passed 0.12 AU from Earth on August 4, 1881, but was then located in Southern Hemisphere skies at a time when no regular sweeps for comets were being made. The total magnitude should then have been near 5[*].

Orbital calculations revealed a period between 8.5 and 9.0 years and the comet was next expected in 1890. During 1889, Plummer and Matthiessen independently calculated periods of 8.856 and 8.687 years, respectively. This dif-

ference of two months in the period indicated that a large portion of the sky
would have to be searched and since the 1890 apparition was not particularly
favorable, the comet was expected to be faint. Subsequently, no trace was
found.

The comet remained lost until it was accidentally rediscovered on October
9, 1978, by Shigehisa Fujikawa (Onohara, Japan). Then near magnitude 11, the
comet was seven days past perihelion (0.78 AU) and perigee (0.25 AU). This
magnitude may have been an underestimate, since visual observers in the next
few days reported it to be near 10. Thereafter, the comet faded rapidly with
the total magnitude reaching 13 on October 17. Photos exposed at Harvard's
Agassiz Station on November 1 and 2 revealed fairly weak images of nuclear
magnitude 18, which displayed a short tail 1 to 1.5 arcmin long.

Brian Marsden linked both apparitions and found the comet's period to have
been 8.71 years in 1881 and 9.01 years in 1978. This comet's orbit is especially
interesting in that it can experience very close approaches to Jupiter, Mars,
Earth and Venus.

	T	q	P	Max. Mag.	Max. Tail
1881 V	Sep. 14	0.73	8.71	7.5	1 arcmin
1978 XIX	Oct. 2	0.78	9.01	10.0	2 arcmin

de Vico

Discovered: February 20, 1846   Discovery Magnitude: 7.0

This comet was discovered on February 20, 1846, by Francesco de Vico,
director of the Collegio Romano Observatory (Rome, Italy). It was described as
fairly bright, with a substantial condensation and a tail. Then located in
Cetus, the comet moved rapidly northward. An independent discovery was made by
William C. Bond (Harvard Observatory, Massachusetts) on February 26, but was
not immediately reported.

Although the comet was moving away from Earth, it brightened fairly rapidly
as it approached perihelion, and at the beginning of March, observers indicated
a magnitude near 6.5. The comet passed perihelion on March 6, and continued to
brighten in the next week or so to near magnitude 5*. Thereafter, the comet
faded slowly and was seen for the last time on May 20.

Orbital calculations during the remainder of 1846 indicated the comet was
periodic, but the exact period was not known with any great degree of accuracy,
with G. P. Bond obtaining a result of 69.70 years, and Benjamin Peirce arriving
at a value of 73.72 years. The first in-depth study of this comet's orbit was
made in 1887, when von Hepperger calculated the period as 75.72 years. This
indicated that the next perihelion date would come during the latter half of
1921; however, von Hepperger added that a probable uncertainty of 3 years in
the period would allow the comet to reappear anytime between 1919 and 1925.
Despite searches using prepared ephemerides, Comet de Vico was not recovered at
that return and it should be noted that had the comet arrived at perihelion
during the latter half of 1921, it would have experienced a close approach to
Jupiter in late 1923. In 1976, Buckley obtained a period of 76.30 years for
the 1846 apparition which would have placed the perihelion date in early 1922.
The margin of error for this calculation amounted to only 2 years.

	T	q	P	Max. Mag.	Max. Tail
1846 IV	Mar. 6	0.66	76.30	5.0	short

de Vico-Swift

Discovered: August 23, 1844   Discovery Magnitude: 7.0

During a routine sweep for comets, de Vico discovered this object in
Aquarius on August 23, 1844, and described it as a bright telescopic object.
The comet arrived at perigee (0.19 AU) on September 1, and was at perihelion
(1.19 AU) on the 2nd. At that time, the total magnitude was near 5[*]and indepen-
dent discoveries were made by Melhop (Hamburg, Germany) on the 6th and Hamilton
L. Smith (Cleveland, Ohio) on the 10th. The comet's distances from the sun and
Earth increased slowly during the next couple of months and its fading was slow.
By the end of October, the total magnitude was near 8 and by mid-November it
had dropped to about 9.5[*]. The comet was last detected on December 31.

The comet was recognized as being of short period as early as late September 1844, when Hervé Faye calculated an elliptical orbit with a period of 5.28
years. During the next two years further calculations were made by Nicolai
(5.459 years), Hind (5.421 years), Goldschmidt (5.488 years) and Faye (5.121
years). The comet's first predicted return came in February 1850, but that unfavorable apparition went by unobserved. The 1855 return was favorable due to
an approach to within 0.58 AU of Earth during August; however, cloudy skies
hampered most observers and nothing was found. (Goldschmidt did find a faint
comet on May 17, which was only 1.5 degrees from the expected position, but
later calculations showed that Comet de Vico could not have been in that position and the object was not seen again.) After unsuccessful searches in 1860
and 1866, Comet de Vico was finally given up as lost.

On November 21, 1894, Edward Swift (Echo Mountain, California) discovered
a comet in Aquarius, which he described as very faint, with a small nucleus and
a faint, short tail. During the next few days, Barnard (Lick Observatory)
estimated the total magnitude as near 13. Based on the comet's location in the
sky and direction of motion, A. Berberich immediately suggested that Comet
Swift was the same as Comet de Vico. Shortly after several accurate positions
had been obtained, Schulhof calculated a short-period orbit which bore a striking resemblance to the 1844 comet. The differences in the orbits could even be
explained by a close approach (0.60 AU) to Jupiter in 1885. With perihelion
having occurred on October 12, Swift's comet faded after discovery and was
followed until late January 1895, when near magnitude 14. In 1899, F. H. Seares
calculated a definitive orbit for Comet Swift which gave a period of 5.855 years.
He also indicated that a close approach (0.44 AU) to Jupiter in 1897 had increased the period to 6.40 years, which made subsequent returns neither particularly favorable nor unfavorable. Subsequently, the comet's return in 1901 was
very unfavorable and no trace was found. At the more favorable 1907 return,
numerous searches were made, the highlight of which was a 3.5-hour photographic
exposure by A. Kopff, but no one detected the comet. It was again lost.

In 1963, Brian Marsden decided to investigate the problem of whether comets
de Vico and Swift were the same. Using an electronic computer he was able to
link the 1844 and 1894 apparitions with remarkable precision, and he then proceeded to calculate the comet's next favorable return. He found that an approach
of 1.52 AU from Jupiter in 1957 had slightly reduced the orbital period so that
the 1965 apparition would be very favorable. Marsden requested Joachim Schubart
(Heidelberg) to undertake an independent calculation, which resulted in a difference in the perihelion date of only 11 days. Using an ephemeris supplied
by Schubart, Dr. Arnold Klemola (Yale-Columbia Southern Observatory, Argentina)

recovered Comet de Vico-Swift on June 30, 1965, at a magnitude of 17. The comet
was on a line connecting Marsden and Schubart's predictions, though slightly
nearer to the latter. The comet passed perihelion on August 23, and was then
near magnitude 16. It continued to slowly brighten during the next month, as it
neared Earth, and when at perigee (0.68 AU) in late September, it was near magnitude 15. The comet was lost in the sun's glare after October 15.

The comet passed very close (0.16 AU) to Jupiter in 1968, which acted to
increase the perihelion distance from 1.62 AU to 2.26 AU and the period from 6.3
years to 7.5 years. In the present orbit, the comet can not become brighter
than magnitude 18. The comet was impossibly placed for recovery in 1973, but
Marsden believes that even favorable returns in the future may go by unobserved
"unless some dramatic improvement in observing techniques occurs."

It should be noted that since 1844, this comet was believed to have been
identical to Comet La Hire of 1678, and as recently as 1963, approximate orbits
were being calculated for it based on the assumption of its identity to Comet
de Vico-Swift. However, in 1967, E. I. Kazimirchak-Polonskaya investigated the
orbit of Comet de Vico-Swift and found that it had been captured by Jupiter in
1671, so that in 1678, it was actually "located in the opposite part of its
orbit" from Comet La Hire. Thus, the comets "are different objects, with
materially different orbits."

	T	q	P	Max. Mag.	Max. Tail
1844 I	Sep. 3	1.19	5.46	5.0*	small
1894 IV	Oct. 13	1.39	5.86	13.0	short
1965 VII	Aug. 23	1.62	6.31	15.0	trace

Dubiago

Discovered: April 24, 1921   Discovery Magnitude: 10.0

This comet was discovered by A. D. Dubiago (Kazan, Russia) near the
Auriga-Lynx border on April 24, 1921. It was described as a diffuse 10th-
magnitude nebula moving toward the southeast.

Though the comet neared perihelion after discovery, it slowly faded due to
an increasing distance from Earth. During the first week of May, it reached
magnitude 11 and on May 14, Schorr (Hamburg Observatory) described it as near
magnitude 11.5, with a coma diameter of 3 arcmin. By month's end, the comet
was nearing magnitude 12 and when last detected on June 11, it was near 12.5.
On the latter date, the comet was becoming a difficult object in the evening
twilight; however, there is a possibility that it was detected one last time on
July 7, though this sighting is presently considered uncertain.

Despite a duration of visibility of only 49 days, this comet was found to
possess a period of between 60 and 70 years. The first study of the orbit was
conducted in 1936, when Hirose obtained a value of 67.01 years; however, in
more recent times, values of 61.01 years and 62.35 years were obtained by I.
Muraveva (1978) and B. G. Marsden (1979), respectively. These latter values
are considered more accurate than Hirose's since the probable error amounts to
only about 2 years.

	T	q	P	Max. Mag.	Max. Tail
1921 I	May 5	1.11	62.35	10.0	none

du Toit

Discovered: May 16, 1944   Discovery Magnitude: 10.0

D. du Toit (Boyden Observatory, South Africa) discovered this 10th-magnitude comet as it moved rapidly southeastward through Pavo on May 16, 1944. Although the comet was then nearing a June 17 perihelion, its distances from the sun and Earth varied so little during May, June and July that the total magnitude remained between 10 and 11. Thereafter, the comet began moving northward and steadily faded. By the time van Biesbroeck (Yerkes Observatory) obtained the first Northern Hemisphere observation on September 26 the total magnitude had declined to nearly 17. Further photos by van Biesbroeck between October 19 and 22 revealed a total magnitude of 18 and when last photographed by him on November 14 and 20 faint images of magnitude 19.5 were detected.

The comet was first recognized as being of short period on July 24, 1944, when F. J. Bobone (Cordoba) published an orbit giving the period as 14.0 years. Shortly after the comet was last detected, Bobone revised the orbital period to 14.87 years and in 1955 he published a major study which gave the orbital period as 14.79 years. On the latter date a prediction was made that the comet would return to perihelion on April 10, 1959; however, between December 1958 and February 1959 Elizabeth Roemer (Flagstaff) exposed several photographs of the comet's predicted position, but without success. Several predictions were made for the comet's next return in 1974 with perihelion dates ranging from April 4 to 6. Subsequently, photographs were exposed by C. Kowal (Palomar Observatory) on January 26 and 27, C. Torres (Cerro El Roble) in March and April, and Roemer (Steward Observatory) between June 15 and 17, but without immediate success. However, during January 1975 Torres was reexamining his plates and detected a 19th-magnitude image near the edge of the March 22, 1974 exposure. Further examination revealed another image on the April 22 photo and accurate measurement left no doubt that the diffuse images were of Comet du Toit. It was concluded that the comet had been too faint to register on Kowal's photos in January and was outside of the region covered by Roemer's photos in June.

	T	q	P	Max. Mag.	Max. Tail
1944 III	Jun. 17	1.28	14.78	10.0	none
1974 IV	Apr. 1	1.29	14.97	18.5	none

du Toit-Hartley

Discovered: April 9, 1945   Discovery Magnitude: 10.0

This comet was discovered by du Toit on the evening of April 9, 1945 as it moved southeastward near the Leo-Sextans border. Then located 0.27 AU from Earth, the total magnitude was estimated as near 10.

Although the comet was well placed for observation in the Northern Hemisphere, patrol plates taken at Harvard Observatory stations in Oak Ridge and Cambridge failed to reveal it. Therefore, instead of making a formal announcement of the discovery, Harvard Observatory sent a letter to its southern station requesting it to confirm the report. Nearly two months elapsed before a reply was received on June 9 from John Paraskevopoulos (director of Boyden Observatory) which informed Harvard that the comet had been continuously observed in South Africa at both the Boyden and Union observatories until June 6, when it had become too faint for their cameras. The radiogram gave a rough position for the comet on June 1 but since no orbit was communicated, attempts by the larger Northern Hemisphere telescopes were futile when they began searches on June 12.

Later in 1945, precise positions of Comet du Toit reached the Northern Hemisphere, along with a parabolic orbit computed by C. Jackson (Union Observatory). Leland E. Cunningham recomputed the orbit and found it to be elliptical with a period of only 4.56 years. Since du Toit had already found a periodic comet in 1944, this comet became known as du Toit 2. Few if any searches were made for this comet at its return in 1949-50, and in 1952 K. Hurukawa computed a definitive orbit for the 1945 apparition which revealed a longer period of 5.27 years. He predicted the comet would return in early 1955; however, photographs exposed at McDonald Observatory during April of that year failed to reveal the comet. The chances for recovery seemed more favorable in 1961, since, earlier in the year, new observations had become available from the 1945 apparition. M. P. Candy utilized these to recompute the 1945 orbit and found the orbital period to be 5.28 years. Subsequently, he predicted a favorable return in 1961, but searches at Lowell and Perth observatories were unsuccessful. The comet's next apparitions in 1966, 1971 and 1976 were also predicted in various publications; however, few, if any, searches were carried out.

In 1982, the interest of the astronomical world was suddenly drawn to two comets discovered by Malcolm Hartley on photographic plates made with the 120-cm Schmidt telescope at Siding Spring on February 5 and 6. The comets were separated by only 43 arcmin and the similarity of their motions immediately brought suggestions of their having recently been one object. On February 11 the International Astronomical Union published a _Circular_ which gave both parabolic and elliptical orbits for the two objects. Interestingly, on this same date, _Yamamoto Circular_ no. 1970 was also published and it gave a prediction by S. Nakano (Sumoto, Japan) for the return of du Toit 2. When the IAU _Circular_ reached Japan a few days later, Nakano immediately noted the resemblance between the orbits of each Comet Hartley and the comet du Toit 2-- with only perihelion dates differing by 48 days. After he relayed his suggestion to the United States the comets became known as du Toit-Hartley.

The two comets received the designations 1982 b and 1982 c and possessed discovery magnitudes of 14 and 17, respectively. Based on the preliminary orbits computed by Marsden, Z. Sekanina strongly believed that the comets had been one object up until 1976 (a time which later proved to have been the comet's previous visit to perihelion), but later in that year 1982 b separated from 1982 c. In order to lend support to the theory, Sekanina remarked that 1982 b would have to begin fading as the comets approached both the sun and Earth, which it did, with estimates dropping to 17 on February 19 and 19.5 by May 1 as the comet became more diffuse. During the same time 1982 c brightened as predicted with estimates of 16 on March 2 and 14.5 on May 1. The comet had been closest to Earth (0.28 AU) around April 6--nearly one week after perihelion. Thereafter, 1982 c faded as its distances from the sun and Earth increased.

	T	q	P	Max. Mag.	Max. Tail
1945 II	Apr. 19	1.25	5.28	10.0	none
1982 II	Mar. 30	1.19	5.20	14.5	trace

du Toit-Neujmin-Delporte

Discovered: July 18, 1941   Discovery Magnitude: 9.5

D. du Toit (Boyden Observatory) discovered this comet on July 18, 1941 as it moved northeastward through Aquila. He estimated the total magnitude as 10 and cabled his discovery information to Harvard College Observatory. The delays caused by wartime conditions prevented the notice from arriving before July 27, and, when received, Harvard decided to hold the information until further confirmation. Meanwhile, G. N. Neujmin (Simeis, Crimea) had independently found the comet on July 25 during the examination of an asteroid plate. He estimated the total magnitude as 9 and further observed the comet on July 29 and August 1. Details of those observations were radiogrammed from Moscow to Harvard Observatory. Again delays were inevitable, and, on this occasion, it took nearly 20 days for the message to reach its destination. Meanwhile, a third independent discovery had been made on August 19, this time by E. Delporte (Uccle Observatory, Belgium), who estimated the total magnitude as 9.5. A general announcement of the comet's discovery was finally relayed by Harvard to other observatories on August 22.

The comet passed perihelion on July 21, and was then being estimated as between magnitudes 9 and 10. A few days later it passed perigee (0.3 AU). Thereafter, as the comet's distances from the sun and Earth increased, the brightness declined. By the end of August it had dropped to magnitude 11 and at the end of September it was near 13.5. The comet was last detected on October 20 when the total magnitude had faded to 15. Curiously, the decline in brightness occurred at a faster rate than predicted and some astronomers believe the comet had undergone an outburst in brightness shortly before discovery.

Orbital calculations over the next few years gave periods ranging from 5.45 to 5.52 years, with the latter seeming closer to the truth; however, searches in 1946 and 1952 were not successful partly due to the fact that the discovery outburst in 1941 made it difficult to predict what the actual brightness would be. In 1954 the comet passed 0.66 AU from Jupiter, which acted to increase the period to 5.9 years. This caused the 1958 return to be fairly favorable and five observatories conducted searches at various times between May and October. These covered magnitudes as faint as 17, but nothing cometary was found.

The 1964 apparition was virtually ignored, primarily due to the comet's having been missed at three returns, but also because it was badly placed for observations. In 1966 another close approach to Jupiter increased the period to 6.31 years and when the effects of this approach were coupled with those of the 1954 approach, astronomers noted a curious development. The 1941 orbit possessed an uncertainty of 10 days in the period of this comet and with five revolutions having occurred, the 1970 return would normally possess an uncertainty of 50 days; however, a focusing effect was caused by the Jupiter approaches which decreased the error to only 5 days for the 1970 apparition. On July 6 and 7, 1970, Charles Kowal (Palomar Observatory) recovered the comet very close to the predicted position. The comet was then described as diffuse with a total magnitude of 19. This recovery also indicated that the 1958 ephemerides were 4 days off and that the comet should have brightened to magnitude 16 or 17. The comet was impossibly placed for recovery at the 1976-77 return.

	T	q	P	Max. Mag.	Max. Tail
1941 VII	Jul. 21	1.31	5.55	9.5	3 arcmin
1970 XIII	Oct. 8	1.68	6.31	18.5	none

Encke

Discovered: January 17, 1786   Discovery Magnitude: 5.0

Comet Encke has been one of the most studied comets in history. Possessing

the shortest period (3.3 years) of any known comet, it has been observed at 52
perihelion passages--nearly three times more than any other comet, except
Halley's--since first detected in the 18th century. Named for the German
mathematician Johann F. Encke after he firmly established its periodic nature
in 1819, the comet had been discovered at four previous apparitions before an
elliptical orbit was finally calculated.

The comet was first discovered on January 17, 1786, when Pierre Mechain
found it at 5th magnitude[*] while sweeping for comets in the Aquarius region.
Then situated in twilight, the comet possessed a bright nucleus, but no tail.
It was not observed after the 19th, and no orbit was calculated. The second
discovery came on November 7, 1795, when Caroline Herschel found it near magnitude 5.5[*]. Several accurate positions were obtained during the 23 days of visibility; however, a parabolic orbit made no sense to the mathematicians and the
comet was again lost. The third discovery was made on October 20, 1805, when
Pons (Marseilles), Huth (Frankfurt-on-Oder) and Bouvard (Paris) independently
found the comet within hours of each other. Then near magnitude 5.5[*], the comet
brightened and developed a tail which extended 3 degrees by November 1. Thereafter, the tail gradually shrank as the comet entered twilight and when the
comet was last detected at magnitude 4[*] on November 20, no tail was visible.
On this occasion Encke was one of several mathematicians to acquire the accurate
positions, but while the others struggled with parabolic orbits, Encke found
that the 32-day arc was best represented by an elliptical orbit with a period
of 12.12 years. Unfortunately, no further work was carried out and the comet
was virtually forgotten. The comet passed through the next three perihelion
passages unobserved, but on November 26, 1818, Pons again discovered it. This
time the total magnitude was near 8[*] and observations continued for 48 days.
Encke again set out to compute the orbit, but immediately discarded a parabolic
solution. Then, by using a new computing technique developed by K. F. Gauss, he
found the comet travelled in an ellipse with a period of 3.3 years. Encke
immediately noted a similarity in the paths of comets seen in 1786, 1795 and
1805 and proceeded to calculate the orbit of Pons' comet of 1819 backward in
time, taking planetary perturbations into account. After six weeks of work he
verified that all four comets were actually the same. He then turned his calculations to the future and predicted the comet would return to perihelion on
May 24, 1822. On June 2, 1822, C. L. Rumker (New South Wales, Australia)
succeeded in recovering the comet at a magnitude near 4.5[*]. The comet was then
so close to Encke's predicted position that the comet now bears his name. The
comet has since been observed at every return, except that of 1944 when its
unfavorable position made observations difficult at a time when most major
observatories were hampered by wartime conditions.

Using modern computers the orbit of Comet Encke has been studied in great
detail; however, no hint has been found as to how the comet was placed into its
present orbit. The orbit is considered very stable and, from most indications,
it has been for several thousand years, due to the small aphelion distance of
4.09 AU. The only point against the accepted long stability of the orbit is
the fact that the comet has become a naked-eye object on several occasions, but
a search through Chinese historical records for prediscovery sightings has,
thus far, ended in failure. Therefore, the time the comet has spent in the
present orbit is still open to debate.

The orbit of Comet Encke is in an interesting cycle since three of its revolutions of 3.3 years add up to almost exactly 10 years. Thus, every 10 years the comet nearly repeats itself in terms of its path across the sky as seen from Earth. For example, the 1786 apparition was very favorable for Northern Hemisphere observations and the comet basically repeated that path in 1795 and 1805--all discovery apparitions. It is also interesting to note that during a 10-year cycle observers on Earth see three distinct apparitions: a favorable Northern Hemisphere apparition (perihelion then falling between November and February), an intermediate apparition (perihelion in March, April, September or October) and a favorable Southern Hemisphere apparition (perihelion between May and August).

From a study of the comet's physical features several distinct characteristics have been uncovered by astronomers in the past. First of all, the comet has the curious habit of suddenly becoming unusually active two to three weeks prior to its perihelion passage. At that time the total magnitude becomes about two magnitudes greater than predicted and the tail attains its maximum length. Secondly, the tail always appears as a long, narrow feature and, during the autumn months, an antitail has occasionally been seen as a fan of luminous material within the coma. Thirdly, the comet's two centuries of observations have allowed several analyses of its fading from repeated encounters with the sun. Unfortunately this area is still open to much debate with some astronomers assigning a rate of fading of 2 to 3 magnitudes per century, while others indicate it to be slightly less than 1 magnitude per century. Proponents of the former value believe the comet is nearing its death, which could occur as early as 1996; however, it should be noted that in 1971 and 1980 the comet attained a maximum magnitude of 6, thus indicating that the comet can still become a prominent object comparable to the past.

Comet Encke was afforded the honor of being the first comet observed throughout its orbit when photographed near aphelion in 1913. The idea of it being a possible annual comet was first suggested in 1912 by Professor E. E. Barnard, who expressed the opinion that the 152-cm reflector at Mount Wilson Observatory might be able to detect the comet at aphelion. Subsequently, Mr. F. E. Seagrave prepared an ephemeris to aid a photographic search in 1913, and on September 1 of that year, a photo made at Mount Wilson revealed the comet. Despite this success, searches at the comet's next aphelion in 1916 were unsuccessful--with an object reported by Professor Max Wolf (Heidelberg, Germany) in September later being shown to be too far from the predicted position for a positive identification with Comet Encke. The comet was not again detected at aphelion until August 15, 1972, when Elizabeth Roemer (Steward Observatory) photographed it at a nuclear magnitude of 20.5. The comet has remained an annual comet ever since that date.

	T	q	P	Max. Mag.	Max. Tail
1786 I	Jan. 31	0.34	3.30	5.0*	"short"
1795	Dec. 22	0.34	3.30	5.0*	none
1805	Nov. 22	0.34	3.31	4.0*	3 degrees
1819 I	Jan. 28	0.34	3.29	6.0*	trace
1822 II	May 24	0.35	3.32	4.5*	none
1825 III	Sep. 17	0.34	3.31	5.5*	none
1829	Jan. 10	0.35	3.32	3.5*	18 arcmin
1832 I	May 4	0.34	3.31	6.5*	none

	T	q	P	Max. Mag.	Max. Tail
1835 II	Aug. 27	0.34	3.31	6.5*	none
1838	Dec. 20	0.34	3.31	5.0*	none
1842 I	Apr. 13	0.34	3.31	5.5*	none
1845 IV	Aug. 10	0.34	3.30	7.5*	none
1848 II	Nov. 27	0.34	3.30	4.5*	90 arcmin
1852 I	Mar. 15	0.34	3.30	4.5*	45 arcmin
1855 III	Jul. 2	0.34	3.30	6.5*	none
1858 VIII	Oct. 19	0.34	3.30	6.0*	trace
1862 I	Feb. 7	0.34	3.30	5.5*	10 arcmin
1865 II	May 28	0.34	3.30	6.5*	none
1868 III	Sep. 15	0.33	3.29	6.0*	"short"
1871 V	Dec. 29	0.33	3.29	4.5*	2 degrees
1875 II	Apr. 13	0.33	3.29	5.5*	2 arcmin
1878 II	Jul. 27	0.33	3.29	6.5*	none
1881 VII	Nov. 16	0.34	3.31	5.0*	7 arcmin
1885 I	Mar. 8	0.34	3.31	5.0*	18 arcmin
1888 II	Jun. 28	0.34	3.31	6.0*	none
1891 III	Oct. 18	0.34	3.30	4.5*	15 arcmin
1895 I	Feb. 5	0.34	3.30	5.5	90 arcmin
1898 III	May 27	0.34	3.30	6.0	none
1901 II	Sep. 16	0.34	3.31	6.0	trace
1905 I	Jan. 12	0.34	3.30	5.0	5 arcmin
1908 I	May 1	0.34	3.30	8.0	none
1911 III	Aug. 20	0.34	3.30	7.5	none
1914 VI	Dec. 5	0.34	3.30	6.5	70 arcmin
1918 I	Mar. 25	0.34	3.30	6.5	none
1921 IV	Jul. 14	0.34	3.30	8.5	none
1924 III	Nov. 1	0.34	3.30	7.0	20 arcmin
1928 II	Feb. 20	0.33	3.29	6.0	5 arcmin
1931 II	Jun. 3	0.33	3.28	7.0	none
1934 III	Sep. 15	0.33	3.28	6.5	3 arcmin
1937 VI	Dec. 28	0.33	3.29	5.0	1 arcmin
1941 V	Apr. 17	0.34	3.31	13.0	none
1947 XI	Nov. 26	0.34	3.30	5.0	1 arcmin
1951 III	Mar. 16	0.34	3.30	7.0	15 arcmin
1954 IX	Jul. 3	0.34	3.30	9.6	none
1957 VIII	Oct. 20	0.34	3.30	6.5	1 degree
1961 I	Feb. 6	0.34	3.30	6.0	2 degrees
1964 IV	Jun. 3	0.34	3.30	5.0	none
1967 XIII	Sep. 22	0.34	3.30	8.5	none
1971 II	Jan. 10	0.34	3.30	6.0	2 arcmin
1974 V	Apr. 29	0.34	3.30	7.5	trace
1977 XI	Aug. 17	0.34	3.31	8.8	2 arcmin
1980 XI	Dec. 7	0.34	3.30	6.2	30 arcmin

Faye

Discovered: November 23, 1843   Discovery Magnitude: 6.0

This comet was discovered on November 23, 1843 by the French amateur
astronomer Hervé Faye (Paris), who described it as fairly bright, with a short

tail and a prominent nucleus. Having passed perihelion one month earlier, the
comet was nearing perigee (0.7 AU) and brightened slightly in the next few days,
until barely visible to the naked eye at the end of November. Thereafter, it
faded, with the total magnitude dropping to near 6.5* in mid-December, 7* in mid-
January and 8* by mid-February. The comet was last seen on April 10, 1844 when
the magnitude must have been fainter than 10*.

Shortly after the comet's discovery, calculations revealed Comet Faye to be
a new short-period comet and in the next two years several astronomers and
mathematicians found orbital periods ranging from 7.25 to 7.39 years. As the
comet's next perihelion neared it was realized that a more precise calculation
would be needed if the comet was expected to be recovered. This task was undertaken by Urbain Jean Joseph Leverrier, whose calculations revealed that the
comet had passed 0.25 AU from Jupiter in 1841. This had decreased the perihelion distance from 1.81 AU to 1.69 AU. Leverrier then calculated an orbital
period of 7.44 years and predicted the next perihelion date as early April 1851.
On November 28, 1850 James Challis (Cambridge, England) recovered the comet
very near the predicted position at a magnitude between 10 and 11*. This apparition was an unfavorable one with the comet being 2.6 AU from Earth at the time
of its perihelion passage and it brightened to only magnitude 9.5*.

The comet was recovered at every apparition following 1851 until 1903 when
the orbit had been altered slightly by a somewhat close approach to Jupiter.
Predictions for the 1910 apparition indicated a perihelion passage sometime
during the latter half of October, but searches proved unsuccessful. Then, on
November 8, 1910 Professor Cerulli (Teramo, Italy) discovered a comet which was
described as of magnitude 10.2. The comet was kept under easy observation for
the next two weeks before orbital calculations in late November revealed it to
be none other than Comet Faye. Observations continued until March 31, 1911.
Comet Faye was missed in 1918, again due to perturbations by Jupiter, but after
its recovery in 1925 it has never been missed again.

Several studies have been conducted on this comet during the past two de-
cades which cover both the comet's brightness and orbital evolution. The former
topic was examined in 1971 by S. K. Vsekhsvyatskij and N. I. Il'ichishina and
revealed a decrease in the comet's absolute magnitude of nearly 8 magnitudes.
Although this estimate is high compared to similar studies in the recent past,
it still illustrates that the comet's brightness is dropping rapidly. This has
led to several predictions of the comet's "death" date ranging from 1967 to 1985.
The comet was last observed during the 1977 apparition when the total magnitude
reached 12.5. It should be noted that at the 1969 return, the total magnitude
reached 10 and the maximum visual tail length reached 15 arcmin--a good indication that the comet is quite active.

The orbit of this comet has been fairly stable, except for the Jupiter
perturbations in 1841. Studies in both the United States and Russia have shown
that for at least 300 years prior to 1841 the perihelion fluctuated from 1.7 to
1.9 AU. Since 1841, the perihelion has fluctuated between 1.61 and 1.74 AU.

	T	q	P	Max. Mag.	Max. Tail
1843 III	Oct. 18	1.69	7.44	5.5*	small
1851 I	Apr. 2	1.70	7.46	9.5*	short
1858 V	Sep. 13	1.69	7.45	10.5*	none
1866 II	Feb. 14	1.68	7.41	9.5*	none
1873 III	Jul. 19	1.68	7.41	11.5*	none

	T	q	P	Max. Mag.	Max. Tail
1881 I	Jan. 23	1.74	7.56	10.5*	none
1888 IV	Aug. 20	1.75	7.59	10.0*	trace
1896 II	Mar. 20	1.74	7.56	11.5	none
1910 V	Nov. 2	1.66	7.42	9.5	10 arcmin
1925 V	Aug. 8	1.62	7.32	13.0	1 degree
1932 IX	Dec. 6	1.62	7.32	10.0	10 arcmin
1940 II	Apr. 25	1.65	7.42	15.0	none
1947 IX	Sep. 28	1.66	7.44	12.0	4 arcmin
1955 II	Mar. 5	1.65	7.41	14.0	45 arcsec
1962 VII	May 15	1.61	7.38	13.0	short
1969 VI	Oct. 8	1.62	7.41	10.0	15 arcmin
1977 IV	Feb. 28	1.61	7.39	12.5	16 arcsec

Finlay

Discovered: September 26, 1886   Discovery Magnitude: 11.0

Mr. W. H. Finlay (Cape of Good Hope, South Africa) discovered this comet on September 26, 1886 as it slowly moved eastward through Scorpius. He described it as faint and circular, with a diameter of 1 arcmin. The comet brightened slowly during the next three months, with magnitude estimates reaching 9* as it passed perihelion (r= 1.00 AU) on November 22 and 8.5* as it reached perigee (0.8 AU) shortly after mid-December. Thereafter, it faded to magnitude 10* in late January and 12* when last seen on April 12.

Orbital calculations a few months after discovery stirred interest in this comet's possible identity with the lost Comet de Vico (1844 I). The angular aspects of the orbit were strikingly close, but Comet Finlay's orbit was more elliptical and possessed a slightly smaller perihelion distance. Investigations were made by Boss, Krueger, Holetschek and Oppenheim to decide whether perturbations had caused the differences between the orbits of both comets, but it was finally decided that the comets were not the same.

The comet was recovered at its next apparition in 1893 by its discoverer and, as in 1886, it appeared as a circular, 11th-magnitude object 1 arcmin across. The comet reached a maximum magnitude of 8.5 in late June, as it passed perigee (1.15 AU), and, thereafter, it faded until last seen on September 22.

The next apparition was missed due to the comet's bad placement with respect to the sun, but on July 15, 1906 it was recovered near its predicted position and was described as near magnitude 9.5. This apparition was exceptionally favorable due to a close approach to Earth (0.27 AU) on August 16 at which time the total magnitude was near 8.5 and the coma was 12 arcmin across. The comet continued to brighten during the latter half of August and by month's end it was near 6th magnitude. Thereafter, it faded and when last seen on January 9, 1907 it was near magnitude 16.

During June 1910 the comet passed 0.46 AU from Jupiter which acted to increase the orbital period from 6.54 years to 6.69 years. Subsequently, the comet arrived at perihelion in 1913 when in conjunction with the sun and no observations were made. Predictions for the comet's return in 1919 were widely publicized in astronomy and science magazines; however, searches proved fruitless. The comet was finally located purely by accident as T. Sasaki (Kyoto Observatory, Japan) found what he thought was a new comet on October 25, 1919.

Then near magnitude 9, the comet steadily faded until last seen on March 9, 1920.
The comet was recovered at magnitude 11.5 at its next return in 1926; however,
the next three apparitions went by unobserved since the comet was badly placed
with respect to the sun. It was finally recovered with the aid of photography
on December 4, 1953 at a magnitude of 13.5. The comet has been observed at
every apparition since that of 1953.

	T	q	P	Max. Mag.	Max. Tail
1886 VII	Nov. 23	1.00	6.65	8.5*	none
1893 III	Jul. 13	0.99	6.62	8.5	none
1906 V	Sep. 9	0.96	6.54	6.0	none
1919 II	Oct. 16	1.01	6.69	8.5	none
1926 V	Aug. 8	1.06	6.84	11.0	none
1953 VII	Dec. 26	1.05	6.81	10.5	3 arcmin
1960 VIII	Sep. 1	1.08	6.89	11.0	none
1967 IX	Jul. 28	1.08	6.90	14.0	none
1974 X	Jul. 4	1.10	6.95	13.5	"short"
1981 XII	Jun. 20	1.10	6.97	15.0	none

Forbes

Discovered: August 1, 1929   Discovery Magnitude: 10.0

A. F. I. Forbes (Rosebank, South Africa) discovered this comet on August 1,
1929 while searching for comets in the clear South African skies. He described
it as near 10th magnitude, with a northeastern motion through Microscopium.
The comet steadily faded thereafter, as its distances from the sun and Earth
increased, and by the first week of October the visual magnitude had declined
to 14. Due to an error in the transmission of the comet's discovery positions,
searches in the Northern Hemisphere proved unsuccessful. It was not until
October 1 that an accurate orbit calculated by B. Dawson was received at Yerkes
Observatory and that evening van Biesbroeck succeeded in finding the comet. He
remained the only observer from that moment on and last detected the comet on
November 22 at a magnitude of 16.

Refined orbital calculations gave a period of 6.38 years and indicated the
comet would be unfavorably situated during its return in 1935. Subsequently,
searches revealed nothing and observers looked to recover the comet in 1942.
In that year perihelion was due in April, but the comet failed to reappear until
June 15, when van Biesbroeck found it as a 14th-magnitude object near the position predicted by F. R. Cripps. Likewise, Cripps' prediction enabled Jeffers
(Lick Observatory) to recover the comet on May 14, 1948 when near magnitude 17.

The comet was again missed in 1955 despite searches at Yerkes and Lick
observatories in June and July, but Elizabeth Roemer (Flagstaff) managed to
recover it near magnitude 20 at the next return in 1961. In that year the comet
was under observation from January 15 to December 5 and brightened to magnitude
10 when near perihelion in July--thus making it the most favorable apparition
since that of 1929. The comet was again missed at the unfavorable return of
1967 despite searches in that year and the next at Yerkes and Tokyo observatories; however, Roemer and L. M. Vaughn (Catalina Observatory) managed to recover it on January 19, 1974 when near magnitude 19.5. In that year the maximum
magnitude reached 12.5 when near perigee in July. The comet was next recovered
on March 12, 1980 by Hans-Emil Schuster (European Southern Observatory) near
magnitude 19.5. During the summer months of that year the total magnitude
reached 13.

Studies of the past orbital history indicate Comet Forbes has had no significant change in its perihelion distance for at least 200 years, despite several close encounters with Jupiter. Prior to discovery, the comet made its most significant approaches to Jupiter in 1894 (0.54 AU), 1906 (0.37 AU) and 1918 (1.12 AU). The latter approach acted to increase the orbital period from 6.18 to 6.38 years.

	T	q	P	Max. Mag.	Max. Tail
1929 II	Jun. 26	1.53	6.38	10.0	1 arcmin
1942 III	Apr. 16	1.55	6.43	14.5	1 arcmin
1948 VIII	Sep. 16	1.54	6.42	14.5	3 arcmin
1961 VI	Jul. 25	1.54	6.42	10.0	3 arcmin
1974 IX	May 20	1.53	6.40	12.5	6 arcmin
1980 VI	Sep. 25	1.48	6.27	13.2	"short"

Gale

Discovered: June 7, 1927   Discovery Magnitude: 8.0

During one of his many routine comet-hunting sessions, Walter F. Gale (Sydney, Australia) came across a diffuse, circular object 3 arcmin in diameter on June 7, 1927. The total magnitude was then near 8 and the comet was moving slowly eastward through Pisces Austrinus. The comet was found near its perihelion date and, thereafter, it slowly faded until between magnitude 12 and 13 when last seen on September 2.

Orbital calculations indicated an elliptical orbit with a period near 11 years, but the comet's short observational history of three months left this open to some error. Predictions for the comet's next apparition in 1938 gave a perihelion date of April 19; however, searches were unsuccessful and Mr. L. E. Cunningham (Cambridge, Massachusetts) decided to tackle the problem in a different way. First, he recalculated the orbit and found a perihelion date of May 16 as more probable; then he established a series of ephemerides by varying the perihelion date by 32 days before and after that date. Cunningham then proceeded to photograph a large area of the sky with the 20-cm photographic refractor at the Harvard Oak Ridge Station and on May 1 he found the comet. The position indicated a perihelion date of June 18 and the total magnitude of 10 indicated the brightness would reach 8 during June. Observations were numerous, but the comet failed to live up to predictions as it barely reached 9th magnitude; however, as the comet faded a sudden outburst in brightness occurred in mid-July, which E. L. Johnson (Johannesburg, South Africa) reported to have peaked at magnitude 8.5 on the 20th. Thereafter, fading was rapid and observations ceased after July 29.

The comet has never been seen since, including in 1949, when extensive searches were conducted by a half dozen observatories. Astronomers said their photos indicated the comet was either far off course or much fainter than predicted, although mathematicians disagreed with the former belief. It should be noted that Johnson reexamined photos taken in 1927 and 1938 and found that the sharp nucleus present in the former year was absent in 1938. He added that the comet was much more diffuse in 1938 and possessed a central condensation on only one occasion.

This comet is remarkable for being the only known comet to have passed through Saturn's sphere of influence (radius 0.365 AU). The encounter occurred

on February 14, 1798 at a distance of only 0.17 AU. In 1967, E. I. Kazimirchak-
Polonskaya examined the orbit for the period 1660 to 2060 and found nine close
encounters with Jupiter and eight with Saturn. None of the encounters changed
the orbit to any large degree, including that of 1798, when the period decreased
from 13.05 years to 12.44 years and the perihelion distance decreased from 1.46
to 1.30 AU. The angular elements changed by only a degree or two due to the
sun-Saturn-comet configuration at the time of the approach.

	T	q	P	Max. Mag.	Max. Tail
1927 VI	Jun. 15	1.21	11.28	8.0	none
1938	Jun. 18	1.18	10.99	8.5	short

Gehrels 1

Discovered: October 11, 1972   Discovery Magnitude: 19.0

Dr. Tom Gehrels (Lunar and Planetary Laboratory) discovered this comet on
plates exposed with the 122-cm Palomar Schmidt on October 11, 1972 during the
course of a regular minor planet survey. He described it as diffuse, with some
slight condensation, and estimated the total magnitude as 19. A further image
was also identified on a plate exposed October 14, but moonlight prevented
additional observations during the remainder of October. Marsden prepared a
set of search ephemerides for early November and Gehrels recovered the comet on
the 10th. With this position at hand, Marsden computed an elliptical orbit with
a period near 15 years.

During the next two months the comet brightened slowly as the decreasing
distance from the sun was nearly countered by an increasing distance from Earth.
When last seen on January 29, 1973, before entrance into the sun's glare, the
comet was photographed as a sharp nuclear condensation of magnitude 18.4. The
comet was recovered on September 23 not long after it had exited the sun's glare;
however, further attempts to photograph the comet at Kitt Peak on October 21 and
December 31 were unsuccessful. Refined orbital calculations reveal the comet
to possess a period of 14.5 years and a perihelion distance of 2.94 AU.

	T	q	P	Max. Mag.	Max. Tail
1973 I	Jan. 25	2.94	14.52	18.0	none

Gehrels 2

Discovered: September 28, 1973   Discovery Magnitude: 15.5

Gehrels discovered this comet on plates exposed during a minor planet
survey on September 29, 30 and October 4, 1973. He described it as diffuse,
with a total magnitude between 15 and 16 and a tail length of 2 arcmin.
Gehrels did not announce his discovery until late October, but it was immediately confirmed by E. Helin (California Institute of Technology) on October 26
(magnitude 15) and C. Y. Shao (Harvard College Observatory, Agassiz Station) on
October 28 (magnitude 16). Shortly after the comet's official announcement on
October 31, a preliminary orbit by Marsden enabled Helin to identify a prediscovery image on a 46-cm Palomar Schmidt plate obtained on September 28, and
Paul Wild (Astronomical Institute, Berne) located an image on supernova-search
plates made at Zimmerwald on September 30.

The comet was nearest Earth (1.4 AU) at the end of October and remained
near magnitude 15 and 16 through the end of November as it passed through perihelion. Thereafter, the large distance from the sun caused fading to be slow
with total magnitude estimates still being near 16 at the end of the year.

Before entering the sun's glare, the comet was photographed on February 25, 1974 by Roemer (Steward Observatory), who estimated the nuclear magnitude to be 19.0. The total magnitude must then have been near 17. After conjunction with the sun, an attempt by Shao to recover the comet on November 16 failed; however, Roemer obtained very weak images of magnitude 21.5 on two plates exposed on December 20. She also recorded the comet for the final time on March 7 when a small weak spot of magnitude 21.0 to 21.5 was found at the expected position.

Orbital calculations by Marsden in 1975 produced a period of 7.94 years and indicated that the comet had passed 0.9 AU from Jupiter in 1971, which acted to decrease the perihelion distance from 2.6 AU to 2.4 AU and the period from 8.5 years to 7.9 years. In 1980 he predicted the comet would next arrive at perihelion on November 18.67, 1981. Based on this prediction and using Marsden's prepared ephemeris, W. and A. Cochran (McDonald Observatory) recovered the comet on June 8, 1981 at a total magnitude of 19.5. The recovery indicated Marsden's prediction was off by only 0.15 day.

	T	q	P	Max. Mag.	Max. Tail
1973 XI	Dec. 2	2.35	7.94	15.0	2 arcmin
1981 XVII	Nov. 19	2.36	7.98	16.0	"short"

Gehrels 3

Discovered: October 27, 1975   Discovery Magnitude: 17.0

Gehrels discovered this comet on 122-cm Palomar Schmidt plates exposed during a survey for unusual solar system objects on October 27, 28 and 30, 1975. The object was then so nearly stellar in appearance that some doubt existed as to whether it was a comet or an asteroid, but photos exposed at Harvard's Agassiz Station on November 7 and 9 confirmed Gehrels' belief that some diffuseness was present. The plates also showed the comet to be of short period with a perihelion date still 1.5 years away. It also became obvious that the comet's small eccentricity of 0.15 would enable it to be observed throughout its orbit.

Since the comet was just past opposition it slowly faded to magnitude 18 by year's end. Fading continued during the first two months of 1976 as the comet neared the sun's glare, and when last detected on February 26 it was between magnitude 19.5 and 20. After conjunction with the sun the comet was recovered near magnitude 19 at Agassiz and was photographed once a month until last detected before entering the sun's glare on March 22, 1977. The comet had reached magnitude 18 when closest to Earth in January (2.45 AU), but had already dropped to 19 when last seen.

The comet was again recovered after conjunction by observers at Agassiz on November 18, 1977 and observations continued there until the end of the year. At the comet's opposition in March 1978 a flurry of observations came from Japanese observatories which gave total magnitudes of the comet varying from 16.5 to 17. On the 5th H. Kosai (Tokyo Observatory's Kiso Station) photographed a tail 10 arcmin long and by the 31st A. Mrkos (Klet Observatory, Czechoslovakia) estimated the total magnitude as 17.5. The comet was followed until May 9, 1978 when it again became lost in the sun's glare. The comet has been recovered at opposition at each year since 1978, including 1981 when near aphelion.

Orbital calculations by Marsden in 1975 had indicated the comet had passed near Jupiter in 1972; however, a more elaborate study based on a more precise orbit was conducted by A. Carusi and G. B. Valsecchi (Rome, Italy) in 1980 and

indicated that the comet had actually been temporarily captured by Jupiter on
three occasions. The first time was between 1783.9 and 1786.6, the second time
occurred between 1833.6 and 1835.3 and the final time occurred between 1967.2
and 1974.8.

	T	q	P	Max. Mag.	Max. Tail
1977 VII	Apr. 23	3.42	8.11	16.5	10 arcmin

Giacobini

Discovered: September 7, 1896   Discovery Magnitude: 11.5

M. Giacobini (Nice, France) discovered this comet in Serpens on September
4, 1896 as a faint, circular object 1 arcmin across. He described its motion in
the next few days as eastward. On September 5 Villiger (Munich, Germany) observed the comet with a 26.7-cm refractor and estimated the total magnitude as
11.3. He also estimated the coma diameter as 1 arcmin and commented on the
lack of a sharp nucleus.

Beginning on September 26, the first of a series of observations occurred
which indicated that the comet had split. On that date, and the following two
days, Perrotin (Nice Observatory) detected an extremely faint companion very
near the main nucleus while using a 76-cm equatorial. Although these observations were the only direct references to a second nucleus, cautious reports of a
second condensation were made on September 30 and October 1 by Perrine and
Hussey (Lick Observatory) while using a 91-cm refractor. In addition, on October
10 Sy (Algiers Observatory) commented that the nucleus appeared elongated in
his 32-cm refractor. This latter observation had occasionally been left out of
past studies of this comet's separation, because of an apparent discrepancy
in the position angle: Sy gave it as 160 degrees, while Perrotin measured it as
225 degrees. However, calculations by Zdenek Sekanina in 1978 showed that in
this comet's case "the position angle of the companion should have changed
rapidly with time." Sekanina also gave the date of splitting as April 24, 1896.

Although the comet steadily approached the sun after discovery, its
brightness changed very little as its distance from Earth increased. After the
October 28 perihelion, fading occurred at nearly one-half magnitude per month
and when last detected on January 5, 1897, the comet must have been near magnitude 13.*

The periodic nature of this comet's orbit was detected not long after discovery. Immediately after the final observations, calculations of the period
varied from a value of 6.52 years to 8.99 years, though the former was later
found to be closer to the truth. At the comet's expected return in 1903, Ebell
recalculated the orbit and found a period of 6.64 years. He predicted the perihelion would come in June, but extensive searches by Wolf and Campbell revealed
nothing. The next perihelion was expected in February 1910, but at the end of
1909 Giacobini recalculated the orbit and corrected it for perturbations by
Jupiter. The result was a perihelion date of December 19, 1909--making the
comet impossibly placed for observation. The last major search for this comet
was made in 1929, when van Biesbroeck (Yerkes), as well as other observers,
conducted a photographic search during the summer months. These searches
proved unsuccessful and the comet remains lost.

	T	q	P	Max. Mag.	Max. Tail
1896 V	Oct. 29	1.45	6.65	11.5	none

Giacobini-Zinner

Discovered: December 20, 1900   Discovery Magnitude: 10.5

Giacobini (Nice Observatory) discovered this comet about 1 degree south of Upsilon Aquarii on December 20, 1900. The comet was described as a faint telescopic object with a diameter of 1 arcmin. With perihelion having been passed one month earlier, the comet faded as its distances from the sun and Earth increased. By the end of December the total magnitude had dropped to 11 and by mid-January 1901 it was 13. The comet was last detected on February 16 when near magnitude 15.5[*]. With an observed arc of 59 days, the comet was recognized as a new short-period comet with a period of 6.758 years.

The comet's predicted return in 1907 was unfavorable for observation and astronomers were expecting the 1914 return to be almost impossibly placed. However, on October 23, 1913 Zinner (Bamberg, Germany) accidentally found the comet while observing variable stars near Beta Scuti. He described it as near magnitude 10, with a coma 3 arcmin across and a tail 30 arcmin long. Observations of Comet Zinner were made for a week before identity with Comet Giacobini was first suggested and shortly after mid-November identity was firmly established by Professor M. Ebell. These calculations showed that the actual period in 1900 had been 6.46 years, thus causing the comet to arrive at perihelion 6 months earlier than expected at the 1913-14 return. The comet managed to reach 9th magnitude at the end of October--shortly before passing perihelion--and slowly faded thereafter as it moved away from both the sun and Earth. The comet was last seen on December 27 near magnitude 13.

Comet Giacobini-Zinner was missed at its unfavorable 1920 return, but on October 16, 1926 it was successfully recovered by A. Schwassmann (Bergedorf, Germany) at a total magnitude of 14. The measured position indicated that Cripps' prediction was only 5 days off and the comet reached a maximum brightness of magnitude 11.

Since 1926 Comet Giacobini-Zinner has been missed at only the 1953 return (due to very unfavorable conditions). The comet's most favorable apparition to date was that of 1946 when it passed within 0.26 AU of Earth in late September. At that time the comet was near magnitude 7, but an apparent outburst in brightness during the first days of October caused it to increase to 6.1. Interestingly, the comet again experienced sudden changes in brightness at its next observed return in 1959. These outbursts amounted to increases of one-half magnitude which occurred on August 31, September 23 and October 24.

Comet Giacobini-Zinner is especially noteworthy for its system of meteors, which was first detected at its 1926 apparition. In October of that year an unexpected shower occurred when Earth crossed the comet's orbit 70 days before the comet. The shower became known as the October Draconids and the Giacobinids, but it was absent until the comet next appeared in 1933. In that year Earth crossed the orbit 80 days after the comet on October 9 and a shower of 300 to 1,000 meteors per minute was noted at one time.

In 1939, Earth crossed the comet's orbit 136 days ahead of it, but no shower appeared; however, in 1946, when Earth crossed the comet's orbit only 15 days after it, a spectacular shower appeared with observed rates varying from 3,000 to 32,000 per hour. After that year no further shower was observed until 1972, due to the perturbing effects of Jupiter, but an approach to within 0.58 AU of that planet in 1969 brought the orbit very close to Earth's again. In October 1972, Japanese radar stations detected a peak of 84 meteors during one ten-minute period.

	T	q	P	Max. Mag.	Max. Tail
1900 III	Nov. 28	0.93	6.46	10.5	3 arcmin
1913 V	Nov. 3	0.98	6.53	9.0	30 arcmin
1926 VI	Dec. 12	0.99	6.58	11.0	2 arcmin
1933 III	Jul. 15	1.00	6.60	11.5	5 arcmin
1940 I	Feb. 17	1.00	6.59	15.0	none
1946 V	Sep. 18	1.00	6.59	6.0	25 arcmin
1959 VIII	Oct. 27	0.94	6.42	7.0	1 degree
1966 I	Mar. 28	0.93	6.41	20.0	none
1972 VI	Aug. 5	0.99	6.52	9.0	22 arcmin
1979 III	Feb. 13	1.00	6.52	18.0	none

Giclas

Discovered: September 3, 1978   Discovery Magnitude: 15.6

Henry L. Giclas (Lowell Observatory) discovered this comet on September 8, 1978 slightly west of Iota Ceti. It was described as diffuse, with a condensation, and possessed a total magnitude of 15.6. Around mid-October, a prediscovery image was found on a photo that had been taken on September 3.

Despite a steadily decreasing distance from the sun, the comet changed little in brightness after discovery since it reached perigee (0.81 AU) at the end of September and thereafter steadily moved away from Earth. On October 23 T. Seki (Geisei, Japan) estimated the total magnitude as 16, but after perihelion on November 21, fading was very slow with Giclas estimating the total magnitude as 17.0 on December 27. The comet was last seen on March 28, 1979 near magnitude 19, with further observations being affected by the sun's glare.

The comet was recognized as a new short-period object on September 14, when Brian G. Marsden computed an elliptical orbit with a period of 6.74 years. After a longer duration of visibility had been achieved, this value was revised to 6.68 years.

	T	q	P	Max. Mag.	Max. Tail
1978 XXII	Nov. 21	1.73	6.68	15.5	none

Grigg-Skjellerup

Discovered: July 23, 1902   Discovery Magnitude: 9.5

John Grigg (Thames, New Zealand) was sweeping for comets in the Leo-Virgo region with his 9-cm Wray telescope on July 23, 1902 when he came across a suspicious-looking object. With the great number of galaxies in the region, Grigg checked several charts and tables for possible identification, but it soon became apparent that the object was a new comet. The comet was observed a total of 14 times between July 23 and August 3, with observations after the latter date being impossible due to moonlight. Delayed notice of the discovery allowed no other observations by other observers and only a parabolic orbit could be calculated.

On May 17, 1922 J. F. Skjellerup (Cape of Good Hope, South Africa) discovered a 12th-magnitude comet near the Gemini-Cancer border which was moving northeastward. Then located 0.39 AU from Earth, the comet was 5 arcmin across and brightened to magnitude 10.5 when it passed perigee (0.25 AU) in late May. Thereafter, the comet faded rapidly with total magnitude estimates reaching 15 on July 23 and 16 when last detected on August 19.

Professors Crawford and Meyer (University of California) computed an elliptical orbit for Comet Skjellerup during July 1922 and were the first to call

attention to the similarity between this orbit and the parabolic orbit of
Comet Grigg. Though the period of Comet Skjellerup was then thought to be 5.52
years, a later study by G. Merton found it to be only 5 years and also firmly
established the identity of the two comets. Merton's prediction for the 1927
return indicated a perihelion date of May 10 and using his own ephemeris he and
F. J. Hargreaves managed to recover the comet on March 27 at a total magnitude
of 12. The indicated correction to the predicted perihelion date was less than
5 hours. On this occasion, the comet passed 0.20 AU from Earth on June 4 and
was then near 9th magnitude with a coma 10 arcmin across.

The comet has been observed at every apparition since its 1922 rediscovery,
although problems arose in 1947 when the predictions were 5 days off--thus delay-
ing recovery by at least a month. It is characterized by its short durations
of visibility (not exceeding 130 days until 1972, when larger telescopes were
used to recover comets) primarily caused by the rapid development of the coma
as it approaches perihelion and the rapid exhaustion of the coma after perihelion.

Orbital calculations show that since 1725 the comet has undergone several
perturbations from Jupiter, which are highlighted by the steady increase in the
comet's perihelion distance. In 1725 this distance was 0.77 AU and in 1922 it
was 0.89. As of 1977, the perihelion distance was 0.99 AU and the orbit is   .
presently situated so that a meteor shower can be observed around April 23. The
shower was discovered in 1977 when it peaked at a rate of 40 meteors per hour.
From the lack of observations in the following years, it seems this shower may
be similar to that of the comet Giacobini-Zinner in that it will appear only
shortly before or after the comet passes the intersection point.

	T	q	P	Max. Mag.	Max. Tail
1902 II	Jul. 4	0.75	4.83	9.5	none
1922 I	May 16	0.89	4.98	10.5	trace
1927 V	May 10	0.89	4.99	9.0	"short"
1932 II	May 13	0.91	5.02	9.5	"short"
1937 III	May 23	0.91	5.02	12.0	none
1942 V	May 23	0.86	4.90	9.0	1 arcmin
1947 II	Apr. 18	0.85	4.90	9.0	none
1952 IV	Mar. 11	0.86	4.90	11.0	trace
1957 I	Feb. 3	0.86	4.90	14.0	none
1961 IX	Dec. 31	0.86	4.91	16.0	none
1967 I	Jan. 16	1.00	5.12	16.0	none
1972 II	Mar. 3	1.00	5.12	17.0	none
1977 VI	Apr. 11	0.99	5.10	9.0	5 arcmin
1982 IV	May 15	0.99	5.09	9.5	trace

Grischow

Discovered: February 10, 1743   Discovery Magnitude: 3.0

Grischow (Berlin, Germany) discovered this comet on February 10, 1743 and
on the next evening he described it as 18 arcmin in diameter with a short tail.
Independent discoveries were made on February 11 by Franz (Vienna), and February
12 by Zanotti (Bologna) and J. D. Maraldi (Paris).

The comet had passed only 0.028 AU from Earth on February 9 and, since
perihelion had occurred one month earlier, the comet steadily faded after its
discovery. When the comet was last detected on February 28, the total magnitude
must have been near 7.[*]

The comet was poorly observed at this apparition with only 4 good positions being available for orbital calculations. These give a parabolic orbit with an inclination of only 2 degrees. An elliptical orbit with a period of 5.44 years was calculated in 1833 by Clausen, but this was based on an incorrect assumption that the comet was identical to Comet Blanpain (1819 IV). It should however be noted that the low inclination, as well as the low latitude of perihelion (1 degree), has led several astronomers to believe the comet belongs to the short-period comet class.

	T	q	P	Max. Mag.	Max. Tail
1743 I	Jan. 11	0.84	5.46	3.0[*]	short

Gunn

Discovered: August 8, 1954   Discovery Magnitude: 19.0

While examining a plate of the galaxy cluster Abell 194 which had been exposed on October 27, 1970 with the 122-cm Palomar Schmidt, James E. Gunn detected the trailed image of a 16th-magnitude comet. The comet was described as possessing a central condensation and a small tail, and was moving slowly southwestward. Gunn requested confirmation from J. W. Young (Table Mountain Observatory, California), but photos taken with the 61-cm reflector on November 6 only showed faint, threshold images near magnitude 15. More definite confirmation had to await additional photos with the 122-cm Palomar Schmidt, which were made by J. N. Bahcall on November 22 and 23. These plates indicated the comet was near magnitude 16 and an announcement was finally made to the Central Bureau on December 8.

After the telegram was received at the Central Bureau, Brian Marsden was immediately able to utilize the observations in an elliptical orbit with a period of 6.7 years. The comet's small eccentricity of 0.32 caused fading to be very slow and by January 2, 1971, the total magnitude had dropped to only 17. A revised orbit a short time later not only indicated a slightly longer period, but also implied that the comet had passed within 0.35 AU from Jupiter in September 1965. This close approach decreased the perihelion distance from 3.39 AU to 2.44 AU, and also decreased the period from 8.5 years to the present value of 6.8 years.

Observations of this comet continued throughout 1971, and in 1972 it was observed by Roemer in September when at aphelion. The magnitude was then near 19.5. The comet has been observed at every opposition up to the present time. At perihelion in 1976, the total magnitude reached a maximum value of 14.

Interestingly, in early November 1980 images of an unknown comet were found on two plates exposed on August 8, 1954 during the Palomar Sky Survey. The weak image was described as near magnitude 19, with a coma about 5 arcsec across and a faint, narrow tail about 1 arcmin long. Nearly 5 months later, T. Nomura (Waseda University, Tokyo) identified this unknown comet as a prediscovery image of Comet Gunn. This identification extended the orbital record of this comet by 16 years.

	T	q	P	Max. Mag.	Max. Tail
1969 II	Apr. 19	2.44	6.80	15.0	small
1976 III	Feb. 11	2.44	6.80	14.0	3 arcmin

Halley

Discovered: May -239   Discovery Magnitude: ?

Halley's Comet has long fascinated man. With a recorded history that
spans at least 23 centuries, the comet has been the subject of fear and delight,
and of poems and books. However, despite the attention it received at each
return, no one ever considered the possibility that one comet could be returning
to the sun's vicinity once every generation--at least not until 1682. In that
year the comet became more than a mere visual sensation to one man--28-year-old
Edmond Halley.

Halley's memory of the comet of 1682 blossomed over the years into an overall interest in comets and after accepting the Chair of Geometry at Oxford in
1704, he began an orbital study of 24 comets seen between 1337 and 1698, using
the celestial mechanics developed by his friend Isaac Newton. Upon completion
of the project, Halley immediately noted a similarity in the parabolic orbits
of comets observed in 1531, 1607 and 1682. After further noting that the
differences in the dates amounted to about 76 years, Halley utilized Newton's
formulas for calculating ellipses and predicted that the comet would return in
1758.

Halley did not live to see his prediction come true. Soon after his
death, three French astronomers--Joseph Jerome de Lalande, Alexis Claude
Clairaut and Madame Nicole Lepaute--independently recalculated the comet's path
in greater detail. The outcome of their calculations provided astronomers with
ephemerides to aid their searches for the comet in 1758. On December 25 of
that year, the German astronomer Johann Georg Palitzsch recovered the comet
and calculations based on the new positions indicated Clairaut had missed the
actual perihelion date by only 32 days.

After the 1759 apparition, astronomers began to calculate when the comet
would next return; however, there was an aura of mystery as to why the 1759
predictions had been so far off--even when perturbations from the six known
planets were considered. The problem was solved on March 13, 1781, when Sir
William Herschel accidentally discovered the planet Uranus. Thereafter, predictions for the 1835 perihelion date were made by M. C. T. Damoiseau (November 4), Count M. G. de Pontecoulant (November 13) and O. A. Rosenberger
(November 26).

Searches along the comet's orbital path began as early as December 1834,
but these early sweeps were unsuccessful and the first glimpse of the comet
did not come until August 6, 1835, when Father M. Dumouchel (Rome) detected it
in a position which indicated a perihelion date of November 16--only three days
later than Pontecoulant's prediction.

After the 1835 return, work again began on the comet's orbit, but this time
astronomers calculated past perihelion dates in order to detect further apparitions. Such work was not new since Halley himself had identified the comet of
1456 as a return of his comet and J. K. Burckhardt linked the comet with one
that appeared in AD 989; however, two men took these past investigations to new
limits. The first was P. A. F. Laugier, an astronomer at Paris Observatory,
who succeeded in linking Halley's Comet to comets which appeared in 451, 760 and
1378. The second investigator was John Russell Hind.

Although Hind had only been 12 years old in 1835, his interest grew over
the years until one day he began the most elaborate search for previous appearances of this famous comet. By using the Chinese annals, he identified 15

"probable" apparitions. These probabilities were not so much based on mathematics as they were on the comet's stated appearance and its path across the sky, but Hind managed to extend the history of Halley's Comet back to 12 BC--the year when the comet's appearance over Rome terrified the citizens.

Hind's list remained definitive until the studies of Philip H. Cowell and A. C. D. Crommelin (Royal Observatory, Greenwich) commenced at the beginning of the 20th century. These men were out to compute the most accurate orbit to date and one point in their favor was the use of perturbations from Neptune, which had been discovered in 1846. Their work began by calculating past perihelion dates and when finished with this aspect they had proven 12 of Hind's previous identifications to be correct and had added several of their own going back to 240 BC. Cowell and Crommelin then turned to the future and predicted that the comet would return to perihelion on April 16.6, 1910.

The first searches for Halley's Comet began in the winter of 1908-9, with astronomers utilizing the photographic plate for the first time with this comet. O. J. Lee (Yerkes Observatory) exposed one of the earliest plates on December 22, 1908, but the first to announce the comet's recovery was Max Wolf (Heidelberg) who obtained a small, weak image on a plate exposed on September 11, 1909. Despite the care taken by Cowell and Crommelin, the precise photographic positions obtained by Wolf and other astronomers indicated their orbit was still in error by three days--the same error encountered in 1835. Today, further investigations into the orbit of Halley's Comet, as well as other comets, indicate a force called nongravitational effect may be responsible for discrepancies between the predicted and actual orbits. This jet-like effect causes either a steady acceleration or deceleration in active comets. Based on Halley's Comet's past apparitions a value for the nongravitational force has been obtained and this has been utilized in several predictions for the 1986 perihelion date. The next perihelion is expected on February 9, rather than the earlier calculated February 5.

Although Halley's Comet has 29 definite apparitions on record, most of these occurred long before the invention of the telescope and virtually the only available data on the comet's physical appearance are estimates of the tail length. These estimates come from Chinese annals and are always given in terms of chi (feet). In 1972, T. Kiang estimated that one chi was equal to about 1.5 degrees and this conversion allows the tail length to be put into terms more compatible with today's estimates. In looking over the various ancient observations it is seen that the comet's maximum tail length usually falls between 10 and 20 degrees, except on two occasions: In AD 374 and 837 the tail was at least 100 degrees in length, with the maximum occurring when the comet was within 0.1 AU of Earth.

The 1835 return was the first "modern" observation of the comet. In that year the maximum magnitude reached 1[*] and the tail grew to a length of 30 degrees. During October, as the comet passed within 0.4 AU of Earth, observers noted a protrusion from the nucleus, which shifted rapidly.

The 1910 apparition was well observed by virtually every observatory in the world. In comparison with past apparitions the maximum brightness reached 0 and the tail reached a length of nearly 100 degrees as the comet passed 0.15 AU from Earth on May 20. As in 1835, observers noted jets protruding from the nucleus, but a new phenomenon was also noted as large telescopes observed the comet between April 16 and June 4, 1910--secondary nuclei were being detected.

The extra nuclei were never observed for longer than 1 or 2 days although one
secondary nucleus did seem to last 3 days beginning on May 31. It is interesting
to note that most of those secondary nuclei were detected between May 13 and
June 4, when the comet was within 0.45 AU of Earth.

Throughout the 20th century astronomers have succeeded in tracing Halley's
Comet back to 240 BC, but no definite identifications have been made prior to
that date.   In 1981, D. K. Yeomans and T. Kiang published one of the most ambi-
tious orbital studies of Halley's Comet to date. They calculated the perihelion
dates back to 1404 BC and remarked on an interesting circumstance that may explain why the comet was not seen before 240 BC. They found that the 29 established apparitions included 14 in "which the Earth-comet distance became less
than 0.25 AU." However, the 16 apparitions between 1404 BC and 315 BC included
only two--those of 1266 BC and 1404 BC. Thus, the comet was more unfavorably
placed for discovery.

T	q	P	Max. Mag.[†]	Max. Tail[††]
-239	May 25	0.59	76.75	---	---
-163	Nov. 13	0.58	76.88	---	---
-86	Aug. 6	0.59	77.12	---	---
-11	Oct. 11	0.59	76.33	---	---
66	Jan. 26	0.59	76.55	---	12 degrees
141	Mar. 22	0.58	77.28	---	10 degrees
218	May 18	0.58	77.37	---	---
295	Apr. 20	0.58	79.13	---	---
374	Feb. 16	0.58	78.76	-3.0[*]	100 degrees
451	Jun. 28	0.57	79.29	-3.0[*]	---
530	Sep. 27	0.58	78.90	---	---
607	Mar. 15	0.58	77.47	---	---
684	Oct. 3	0.58	77.62	---	15 degrees
760	May 21	0.58	77.00	0.0[*]	8 degrees
837	Feb. 28	0.58	76.90	-2.0[*]	100 degrees
912	Jul. 19	0.58	77.45	---	---
989	Sep. 6	0.58	77.14	0.0[*]	long
1066	Mar. 21	0.57	79.26	-4.0[*]	long
1145	Apr. 19	0.57	79.02	-3.0[*]	60 degrees
1222	Sep. 29	0.57	79.12	---	30 degrees
1301	Oct. 26	0.57	79.14	0.0[*]	15 degrees
1378	Nov. 11	0.58	77.76	-1.0[*]	15 degrees
1456	Jun. 10	0.58	77.10	0.0[*]	22 degrees
1531	Aug. 26	0.58	76.50	0.0[*]	15 degrees
1607	Oct. 28	0.58	76.06	0.0[*]	10 degrees
1682	Sep. 15	0.58	77.41	0.0[*]	30 degrees
1759 I	Mar. 13	0.58	76.89	0.0[*]	25 degrees
1835 III	Nov. 16	0.59	76.27	1.0[*]	30 degrees
1910 II	Apr. 20	0.59	76.08	0.0	90 degrees

[†] The maximum magnitudes prior to 1835 are considered very questionable and
    when sufficient observations are available values were adopted from works
    by Holetschek, S. K. Vsekhsvyatsky and H. Mucke.

[††] Prior to 1682, tail lengths are acquired from Chinese annals and con-
    verted from chi to degrees using T. Kiang's equation of 1 chi equals 1.5
    degrees.

Haneda-Campos

Discovered: August 9, 1978   Discovery Magnitude: 11.0

On September 1, 1978 Toshio Haneda (Haranomachi, Japan) and Jose da Silva Campos (Durban, South Africa) independently discovered this comet within 8 hours of each other about 2 degrees south of Omega Capricorni. Then between magnitude 9 and 10, the comet was described as diffuse, with a condensation, but no tail. Less than a week after discovery, V. Verveer (Perth Observatory) found an 11th-magnitude prediscovery image on a photo exposed on August 11. Shortly thereafter, another prediscovery image was found, this time at Palomar Observatory, where E. Helin and S. J. Bus found a 13th- to 14th-magnitude image on a photo exposed on August 10. They added that "exposures with the 122-cm Schmidt (limiting magnitude about 19.5) during July 10-14 show no trace of the comet." Finally, at the end of September, G. Pizarro (European Southern Observatory) found a prediscovery image on a photo exposed on August 9. The total magnitude was estimated as 11.

Immediately following discovery, observers were estimating the total magnitude as 10 and the coma diameter as 1.5 arcmin. The comet moved southeastward and was predicted to slowly brighten as it neared both perihelion (1.10 AU) and perigee (0.15 AU), which were expected to occur on October 9; however, on that date the total magnitude was still estimated as 10--indicating the comet was at least one-half magnitude fainter then expected. This was only one of three occasions which indicated the comet was experiencing unexpected changes in brightness. The first occurred at discovery when several observers indicated a slow fading during the first week which amounted to one magnitude. Following the unexpected fading shortly before perihelion, observers indicated a rapid fading in late October. On the 29th, T. Seki (Geisei, Japan) estimated a total magnitude of 14.5. Two days later it was at 15 and on November 8 it was near 17. The comet was last seen on November 29.

Orbital calculations shortly after discovery indicated a period between 4.9 and 5.8 years--the latter of which indicated a close approach to Jupiter in 1969. However, a few days later a new orbit gave a period of 5.4 years and indicated no approach to Jupiter since 1950. The most recent orbital calculation is based on 24 precise positions and gives a period of 5.97 years. This indicates a moderate approach to Jupiter in 1969 with a closer encounter in 1957.

	T	q	P	Max. Mag.	Max. Tail
1978 XX	Oct. 9	1.10	5.97	9.5	none

Harrington

Discovered: August 5, 1953   Discovery Magnitude: 15.5

Robert G. Harrington discovered this comet on a sky survey plate exposed on August 14, 1953 with the 122-cm Palomar Schmidt. The comet was then in Aquarius and was described as a diffuse, 15th-magnitude object with a central condensation and a short tail. E. L. Johnson (Johannesburg, South Africa) later found a prediscovery image on a photo exposed on August 5 and estimated the total magnitude as 15.5.

The comet changed little in brightness during the next month and when observed by van Biesbroeck (Yerkes Observatory) between September 15 and 17 the total magnitude was estimated as 15.2. Van Biesbroeck also measured the coma diameter as 20 arcsec and the tail length as 3 arcmin. Perihelion came on September 22 and by October 1 the total magnitude was near 16. The comet

was last seen on December 10, by Roemer (Lick Observatory), who described it
as nearly stellar with a magnitude of 18.8.

Orbital calculations by Joseph L. Brady towards the end of 1953 gave an
elliptical orbit with a period of 6.90 years. In 1957, Brian Marsden computed
a definitive orbit for the 1953 apparition which gave a period of 6.98 years
and this was supported in 1954 by C. Dinwoodie. Marsden predicted the comet
would next arrive at perihelion on June 28.3, 1960.

The comet was recovered on August 3, 1960 on a plate exposed at Flagstaff
by Elizabeth Roemer. The measured position indicated a correction of -0.6 day
in Marsden's prediction. Then near magnitude 19, the comet changed little in
brightness during the next two months and was last seen at magnitude 19.8 on
October 26.

The comet was not detected at its returns in 1967 and 1974, since perihelion
came when the comet was in conjunction with the sun; however, on September 4,
1980 P. Jekabsons (Perth Observatory) recovered the comet at a magnitude near
18.5. The comet was observed only until October 6.

	T	q	P	Max. Mag.	Max. Tail
1953 VI	Sep. 22	1.69	6.97	15.0	3 arcmin
1960 VII	Jun. 28	1.58	6.80	19.0	short
1980 XIV	Dec. 24	1.60	6.86	18.5	none

Harrington-Abell

Discovered: March 22, 1955   Discovery Magnitude: 17.0

R. G. Harrington and George O. Abell discovered this 17th-magnitude comet
on routine sky survey plates exposed with the 122-cm Palomar Schmidt on March
22, 1955. The comet was diffuse, with a central condensation and a short tail.
Two more Palomar photographs during the next 8 days allowed L. E. Cunningham to
compute an elliptical orbit which gave a period of 7.01 years and showed the
comet to be 3 months past perihelion. This allowed van Biesbroeck to recover
the comet on April 24 when the nuclear magnitude was near 19.5. The comet was
last detected on May 18, when Roemer obtained a 120-minute exposure to reveal
a nuclear magnitude of 19.2.

In 1959, T. Seki computed a definitive orbit which gave a period of 7.22
years. In 1960, I. Hasegawa confirmed the correctness of Seki's orbit, but
gave a slightly shorter period of 7.20 years. Upon recovery of the comet by
Alan McClure on January 26, 1962, it was found that Hasegawa's calculations
were closest. A few days later, Roemer estimated the nuclear magnitude as 17.8
and the comet changed little during the next 2 months as it passed through perihelion in late February. By April 3 the nucleus had dropped to 19.0 and when
last detected on May 26 it was near 20.2.

The comet was next recovered on November 23, 1968 by Roemer at a magnitude
of 19.4. Shortly thereafter, pre-recovery images exposed on October 27 were
also identified by Roemer and were described as near magnitude 20.5. The comet
was observed for 176 days at this return--more than 50 days longer than at the
previous appearance--and reached a maximum magnitude of 19 before becoming lost
in evening twilight after April 20.

Nearly 2 years after the comet had passed its aphelion in 1972, it passed
0.037 AU from Jupiter in April 1974. The angular elements of its orbit were
altered significantly; however, the perihelion distance was hardly changed and
the period was increased by less than 5 months. Predictions for the 1976 return

were not expected to be very accurate, but on October 6, 1975 Roemer and M. A.
Daniel (Steward Observatory) recovered the comet only 30 arcsec from the prediction computed by Marsden. The comet was then near magnitude 20.4 and brightened another magnitude by December 5. A slow fading set in thereafter and when
last detected on March 25, 1976 the magnitude was near 20.5

It should be noted that a study published by Marsden in 1967 indicated that
since 1725 Comet Harrington-Abell had experienced only minor perturbations by
Jupiter prior to its encounter with that planet in 1974.

	T	q	P	Max. Mag.	Max. Tail
1954 XIII	Dec. 13	1.77	7.20	17.0	short
1962 II	Feb. 24	1.78	7.22	17.8	none
1969 III	May 11	1.77	7.19	19.0	none
1976 VIII	Apr. 22	1.78	7.59	19.5	none

Harrington-Wilson

Discovered: January 30, 1952   Discovery Magnitude: 15.0

While examining a sky survey plate taken with the 122-cm Palomar Schmidt
on January 30, 1952, R. G. Harrington and Albert G. Wilson found a comet near
the Coma Berenices-Virgo border. The comet was described as of the 15th magnitude and possessed some condensation and a short tail. Having already passed
perihelion on October 30, 1951, the comet slowly faded, with magnitude estimates reaching 15.5 to 16 by late February. When last seen by the discoverers
on April 16 and 19, the total magnitude had dropped to near 19.5.

Orbital calculations revealed the comet to be of short period with
Cunningham obtaining a period of 6.38 years. A prediction for the 1958 return
was calculated by B. O. Wheel, B. G. Marsden and W. H. Julian and indicated a
a perihelion date of March 21; however, searches at Flagstaff in late 1957 and
February 1958 proved unsuccessful, as did a search at Mt. Palomar on April 12-
13.

The comet should have passed 0.62 AU from Jupiter in February 1961 and was
next expected at perihelion on October 2, 1964, according to predictions by P.
Egerton and B. O. Wheel (an independent calculation by Marsden gave September
29). Few searches were made, however, due to the comet being somewhat unfavorably placed. The comet is now considered lost.

	T	q	P	Max. Mag.	Max. Tail
1951 IX	Oct. 30	1.66	6.36	15.0	short

Helfenzrieder

Discovered: April 1, 1766   Discovery Magnitude: 3.0

Helfenzrieder (Dillingen, Germany) discovered this comet with his naked eye
on April 1, 1766 at a total magnitude of 3*. As the comet approached perihelion
in the following days it slowly brightened and developed a tail. On April 8
Charles Messier (Paris) independently found the comet and described it as bright
with a tail over 4 degrees long. The nucleus was said to resemble a 3rd-magni-
tude star. With a total magnitude then near 2*, the comet changed little in
brightness during the next few days as the increasing distance from Earth countered the effects of a nearing perihelion. The comet was lost in evening twilight after April 11, but was recovered in Southern Hemisphere skies after perihelion and was observed from April 29 until May 13.

Orbital calculations by Burckhardt at the end of the 18th century indicated
this comet traveled in an elliptical orbit with a period of 5.01 years. His
orbit also indicated that the comet had passed very close to Jupiter in 1764,
but there was no indication as to why the comet was not spotted at its next
returns. In 1915, Wirtz recalculated the orbit and found a period of 4.51 years,
although he added that a period of 3.9 years was not out of the question. Again
a close approach to Jupiter in 1764 was indicated with the longer period giving
a minimum approach of 0.03 AU, thus providing a definite capture date. Several
astronomers have indicated that the comet must have been abnormally bright in
1766 and was subsequently missed at succeeding returns due to faintness. A
later ejection by Jupiter into a new orbit is also a strong theory, though the
uncertainty in the period of nearly one year makes the ejection date very difficult to obtain.

	T	q	P	Max. Mag.	Max. Tail
1766 II	Apr. 28	0.41	4.35	2.0[*]	7 degrees

Herschel-Rigollet

Discovered: December 21, 1788   Discovery Magnitude: 7.5

Caroline Herschel (Slough) discovered this comet a little more than one
degree south of Beta Lyrae on December 21, 1788. That same evening her brother
William described it as "a considerably bright nebula, of an irregular form,
very gradually brighter in the middle, and about five or six minutes in diameter." The comet was well observed as it traveled northward through Draco and
was kept under observation until February 5, 1789. No tail was ever observed
and the magnitude when last detected must have been near 10[*].

Orbital calculations gave only a parabolic orbit and no return of this
comet was expected; however, on July 28, 1939 Roger Rigollet (Lagny, France)
found an 8th-magnitude comet in Taurus. When preliminary orbits began being
calculated on August 1, L. E. Cunningham suggested that it was identical to
Herschel's comet. This theory was later proven correct.

The comet was brightest in early August when near magnitude 7.3, but
after perihelion on the 9th, fading was slow and steady as the comet moved
away from both the sun and Earth. By September 5, van Biesbroeck gave the
total magnitude as 8.4. Some of his further estimates were 11 on October 11,
13 on November 6 and 16 on December 22. Visually, a tail was frequently observed extending 2 arcmin during September and October, but photographs on
August 15 had showed it stretching 1 degree. The comet was last seen on January 16, 1940, when Jeffers (Lick Observatory) estimated the nuclear magnitude
as 19.

Orbital calculations by Dr. Maxwell and Mrs. Kaster in September 1939 indicated the comet possessed a period of 150 years. In 1940, with further positions at hand, Maxwell and Kaster computed a period of 156 years. In 1974,
Marsden approximately linked the two apparitions and computed periods of 162
years and 155 years for 1788 and 1939, respectively.

	T	q	P	Max. Mag.	Max. Tail
1788 II	Nov. 21	0.75	162.34	7.5[*]	none
1939 VI	Aug. 9	0.75	154.91	7.3	1 degree

Holmes

Discovered: November 7, 1892   Discovery Magnitude: 4.0

While engaged in observing the Andromeda Galaxy on November 7, 1892, Mr. Edwin Holmes (London, England) discovered this comet and described it as a round, nebulous mass 5 arcmin in diameter. He added that it exceeded the Andromeda Galaxy in brightness. Independent discoveries were made on November 9 by Anderson (Edinburgh, Scotland) and Davidson (Mackay, Australia).

The comet was at first suspected to be P/ Biela, but, although observations within the next weeks did show it to be of short period, it was not that famous comet. With both perihelion and perigee already having been passed, the comet was expected to rapidly fade; however, as November progressed the brightness changed little. In addition, the comet developed a tail shortly after discovery which extended one-half degree on photographs exposed by Barnard on November 11. Barnard's photos also revealed a diffuse nebulous object one-half degree beyond the tail which moved with the comet.

The unexpected continued to occur in late November. First, Deslandres photographed an apparent double nucleus on the 21st, which was absent on later dates, and, second, the coma expanded until its diameter reached 30 arcmin at month's end. During December the comet rapidly grew faint and by early January 1893 it was between magnitude 9 and 10. On January 16 a remarkable outburst occurred which brought the total magnitude to 5 or brighter. Thereafter, the brightness declined, with the total magnitude being near 10 during the first week of February and about 12 by March 10. The comet was last seen on April 6.

With 5 months of observations to work with, astronomers calculated the the orbital period as 6.90 years. They also found that the comet and Jupiter could come within 0.4 AU of one another as had apparently happened in 1861. The comet was next expected at perihelion in 1899 and on June 11 Professor C. D. Perrine (Lick Observatory) recovered it very near the expected position. The total magnitude was then near 16. Although the comet was expected to become fairly bright with respect to its previous appearance it failed to surpass magnitude 13--5 to 6 magnitudes fainter than expected.

The comet was next detected on August 29, 1906 when Wolf photographed it as an object of magnitude 15.5. The comet failed to become brighter than 15--at least one magnitude fainter than predicted--and was last seen on December 7. As the comet continued to move away from the sun it encountered Jupiter in 1908 at a minimum distance of 0.54 AU. This acted to increase the period by 0.47 years. Predictions for the 1912 return failed to take these perturbations into account and the comet was not seen. In fact, it was not until 1926 that this close approach was identified by J. F. Polak. Both his predictions and those of the British Astronomical Association were utilized at the 1928 apparition, but conditions were unfavorable. However, unsuccessful searches were also made at the more favorable returns of 1935, 1942 and 1950.

Although the comet had finally been given up as lost, Brian Marsden reinvestigated its orbit in 1963. He found that the orbits computed by Polak and other astronomers were actually quite accurate and expressed his opinion that "its extraordinary appearance in 1892" indicated that the comet was physically unstable, thus leading to "an alternative reason for its disappearance." Nevertheless, Marsden brought the orbit up to date and produced a search ephemeris for the 1964-65 apparition. On July 16, 1964 Elizabeth Roemer (U. S. Naval Observatory) recovered the comet as a well-condensed object of magnitude 19.5.

The indicated correction to Marsden's predicted perihelion time was only +0.7
day. The comet remained several magnitudes fainter than predicted during this
apparition and never became brighter than magnitude 18.5.

During 1968 the comet passed 1.03 AU from Jupiter which acted to decrease
the period by 0.3 year and the perihelion distance by 0.2 AU. Nevertheless,
the comet was again recovered in 1971 when Roemer and Alice H. Ferguson located
images of magnitude 20 on a photo exposed on June 20. It never became brighter
than 19 as was also the case when seen in 1979-80.

	T	q	P	Max. Mag.	Max. Tail
1892 III	Jun. 14	2.14	6.90	4.0	30 arcmin
1899 II	Apr. 29	2.13	6.87	13.0	none
1906 III	Mar. 15	2.12	6.86	15.0	none
1964 X	Nov. 16	2.35	7.35	18.5	none
1972 I	Jan. 31	2.16	7.05	19.0	20 arcsec
1979 IV	Feb. 23	2.16	7.06	19.5	none

Honda-Mrkos-Pajdusakova

Discovered: December 3, 1948   Discovery Magnitude: 9.0

Shortly before dawn on December 3, 1948 Minora Honda (Kurashiki, Japan)
discovered this comet while searching for comets in eastern Hydra. He described
it as diffuse, without a nucleus or tail, and of magnitude 9. On December 6
Ludmilla Pajdusakova (Skalnate Pleso Observatory) was conducting a routine comet
search with 25x100 binoculars when she independently found the object; however,
twilight prevented a definite identification and she thought there was a chance
she was actually seeing the galaxy M83. The following morning her colleague
Antonin Mrkos--then unaware of the possible comet--became the third person to
independently discover it.

A faint trace of tail was detected by van Biesbroeck (Yerkes) on December
10 and his magnitude estimate of 10 indicated that the comet had begun to fade
rapidly. By December 14 it had dropped to 11 and on January 8 it was near 14.
Also on the 8th, van Biesbroeck and Jeffers independently obtained nuclear
magnitude estimates of 17 and 17.5, respectively. The comet was not seen after
January 10, 1949.

A. Schmitt (Algiers) was the first astronomer to establish that the comet's
orbit was elliptical and, shortly after it was last seen, he utilized the 5 most
precise positions to obtain a period of 5.31 years. This indicated a very close
approach to Jupiter in August 1935. Shortly thereafter, Cunningham used 19 positions and obtained a period of 5.00 years. In 1953 G. Merton computed a definitive orbit which gave a period of 5.22 years. His prediction for the perihelion date in 1954 was February 6. Using Merton's prepared ephemeris, the
comet was independently recovered by T. Mitani (Kwasan Observatory, Japan) on
January 28, and van Biesbroeck and Miss D. Jehoulet (McDonald Observatory) on
February 4. The total magnitude was then near 9 and a narrow tail extended 12
arcmin. By February 20 photos at Lick Observatory indicated the total magnitude had dropped to 11.5 and on March 29 the nucleus was estimated as 17. Van
Biesbroeck obtained total magnitude estimates of 15 on March 27 and 16 on April
1. The comet was not seen thereafter.

The comet was missed in 1959 and although badly placed for recovery in
1964 Roemer (U. S. Naval Observatory) still managed to photograph it on June 14
and 15 when near magnitude 16. The 1969 apparition was more favorable with the

maximum magnitude reaching 8.1 on September 21 and at the comet's next return
in 1974-75 it was described as magnitude 7.7 between December 31 and January 2.
On February 5, 1975 the comet passed 0.23 AU from Earth and came within 1 degree
of the south celestial pole. The 1980 apparition was again very unfavorable
with observations being obtained only between May 1 and 14. The maximum magnitude was then near 14.

	T	q	P	Max. Mag.	Max. Tail
1948 XII	Nov. 18	0.56	5.22	9.0	trace
1954 III	Feb. 5	0.56	5.21	9.0	12 arcmin
1964 VII	Jul. 7	0.56	5.21	16.0	none
1969 V	Sep. 23	0.56	5.22	7.6	25 arcmin
1974 XVI	Dec. 28	0.58	5.28	7.6	13 arcmin
1980 I	Apr. 11	0.58	5.28	14.0	40 arcsec

Howell

Discovered: August 29, 1981   Discovery Magnitude: 15.0

This comet was found by Ellen Howell (California Institute of Technology) on photographic plates taken on August 29 and 30 with the 46-cm
Palomar Schmidt telescope. On the 31st, Charles Kowal confirmed the discovery with the 122-cm Palomar Schmidt and described the comet as diffuse
and tailless, with some evidence of condensation.

Precise positions were obtained by A. C. Gilmore and P. M. Kilmartin
(Mt. John University Observatory, Australia) on September 1, 2 and 6--the
nuclear magnitude being estimated as 15 on each occasion--and Brian G.
Marsden (Harvard-Smithsonian Center for Astrophysics) was able to compute
both parabolic and elliptical orbits on September 8, of which he said the
latter was preferred. By September 11, Marsden was able to confirm the
elliptical orbit and showed the period to be only 5.94 years and the perihelion distance to be 1.6 AU.

The comet was found to have "a hint of tail to the southwest" on a
photo taken by Kowal on September 4, and this observer added that the total
magnitude was still near 15. However, the comet's steadily increasing distances from the sun and Earth caused it to steadily fade. By December 21,
observers at Oak Ridge Observatory detected the comet as a "weak image" on
photos. The comet was last observed before entering the sun's glare on
January 15 and 16, 1982, when J. Gibson used the 122-cm Palomar Schmidt to
obtain images. The total magnitude was then probably between 18 and 19[*].

Computations by Marsden indicate the comet passed 0.6 AU from Jupiter
in 1978, with the comet orbit having been larger before then.

	T	q	P	Max. Mag.	Max. Tail
1981 X	May 4	1.62	5.94	15.0	trace

Jackson-Neujmin

Discovered: September 9, 1936   Discovery Magnitude: 12.0

This comet was discovered by Cyril Jackson (Union Observatory, South
Africa) on September 20 while examining photographic plates exposed for minor
planets on September 15. He described it as faint and diffuse, with a total
magnitude of 12. On September 21, G. Neujmin (Simeis, Russia) independently
found the comet and a short time later, Rigaux (Uccle Observatory, Belgium)
found a prediscovery image on a plate exposed on September 9.

Then located in Aquarius, the comet passed perigee (0.47 AU) on September 21 and perihelion (1.46 AU) on October 3. With the Earth-comet distance increasing after the discovery, the comet steadily faded, with van Biesbroeck giving magnitude estimates of 13 on September 22, 14 on October 12 and 15 on October 16. The comet was last detected on November 5 when the photographic magnitude was near 17.5.

Based on the 5-week arc, both Jackson and L. E. Cunningham computed elliptical orbits with periods of 8.06 years and 8.53 years, respectively. The comet was overlooked in 1945, due to its having been unfavorably situated for observations, but predictions were made for the 1953 return. Between September 3 and 6, Elizabeth Roemer (Lick Observatory) used a 51-cm astrograph to photographically search for the comet. She covered variations in the perihelion date amounting to -20 to +40 days from the prediction, but despite a limiting magnitude of between 16 and 17, nothing was found.

In 1959, B. G. Marsden investigated the orbit of this comet and found the period in 1936 to equal 8.57 years. His orbit indicated that the predictions in 1953 may have been off by nearly one month and he found a close approach (0.8 AU) to Jupiter in 1956. In 1960 Marsden recomputed the orbit using 29 observations obtained in 1936 and again arrived at a period of 8.57 years. He then predicted the comet would again arrive at perihelion on March 23, 1962, but added that it would be very badly situated. No searches were made.

In 1968 Marsden again revised the 1936 orbit--this time using 33 positions --and used this as a basis for his prediction for the 1970 return, for which a perihelion date of August 6 was given. On September 6 and 7, 1970 Charles T. Kowal (Palomar Observatory) recovered the comet 6 degrees from the central prediction--indicating a correction of -7 days. He described it as of magnitude 14, with a central condensation and a short tail. It steadily faded thereafter and was last seen on November 25 when near magnitude 19.5. The 1978 return was particularly unfavorable; however, Kowal managed to photograph the comet on November 28 and 29 near magnitude 19.5. No further observations were obtained.

	T	q	P	Max. Mag.	Max. Tail
1936 IV	Oct. 3	1.46	8.57	12.0	none
1970 IX	Aug. 6	1.43	8.39	14.0	short
1978 XXVI	Dec. 25	1.43	8.37	19.5	none

Johnson

Discovered: August 15, 1949   Discovery Magnitude: 13.7

Ernest Leonard Johnson (Union Observatory, South Africa) discovered this comet on a plate exposed August 25, 1949 for the minor planet 1949 OG. Then in Capricornus, the comet was described as small and diffuse, with no condensation, and a magnitude of 13.7. Shortly thereafter, Johnson found prediscovery images on plates exposed on August 15 and 20.

The comet slowly faded in the following weeks as it moved away from Earth and when at perihelion in mid-September it was still near magnitude 14. By late October it had dimmed to magnitude 15 and when last detected on November 19, A. W. J. Cousins (Radcliffe Observatory, South Africa) estimated the photographic magnitude as 16.

Orbital calculations revealed an elliptical orbit with a period of 6.85 years and the comet was found to have made a very close approach to Jupiter in 1931. For the comet's 1956 return a prediction was published by W. H. Julian and B. O. Wheel which gave the perihelion date as July 24. On August 6, 1956 J. A. Bruwer (Union Observatory) recovered the comet at a magnitude of 13.5. The measured position indicated the prediction had been off by only -2.4 days. The comet steadily faded after its recovery and was near magnitude 15 when first detected in the Northern Hemisphere by van Biesbroeck. Observations continued until October 28, by which time the total magnitude had dropped to 17.8.

The comet was next recovered on April 24, 1963 when Roemer estimated the magnitude as 17.5. The nuclear magnitude remained near 18 throughout the summer months as the comet passed perihelion in June and on July 15 Roemer detected a trace of a tail extending 6 arcsec to the northeast. The comet was last detected on January 9 when near magnitude 19.3.

The 1970 apparition was a particularly unfavorable one with the March perihelion coming when the comet was in conjunction with the sun. Several searches were conducted in 1969 by Tomita (March 14, April 13 and July 13) and Roemer (May 17, 25 and June 8) but with no success. Roemer finally recovered the comet on July 5, 1970 at magnitude 18.8. After a reexamination of her plates exposed in 1969, images near magnitude 20 were located on pairs of plates taken on May 17 and 25. The comet was last photographed on October 3, 1970 near magnitude 20.1. During the next apparition in 1976-77 the maximum magnitude only reached 18.5.

	T	q	P	Max. Mag.	Max. Tail
1949 II	Sep. 17	2.25	6.86	13.7	none
1956 V	Jul. 27	2.26	6.88	13.5	none
1963 IV	Jun. 9	2.25	6.86	17.0	6 arcsec
1970 IV	Mar. 30	2.20	6.77	18.8	12 arcsec
1977 I	Jan. 8	2.20	6.76	18.5	none

Kearns-Kwee

Discovered: August 17, 1963   Discovery Magnitude: 12.0

E. Kearns and K. K. Kwee (Palomar Observatory) discovered this comet on a plate exposed August 17, 1963 in a search for the long-lost Periodic Comet Tempel-Swift. Although the total magnitude was estimated as 12 by the discoverers, observations in the following weeks indicated that the comet had either undergone an outburst at discovery or the discovery brightness was overestimated by at least 1 magnitude. On August 24 Roemer described it as diffuse, with a nucleus of magnitude 17.1 and a narrow tail 30 arcsec long.

The comet's identity with P/ Tempel-Swift was disproven immediately after discovery, but Marsden still showed it to be of short period, with preliminary calculations giving a period of 8.5 years. The comet slowly brightened during the next few months as it approached both the sun and Earth, with magnitude estimates being near 12 in mid-November and 11.5 on December 24. The comet was brightest on January 3, 1964 when Beyer gave a total magnitude of 11.1. Thereafter, with the comet moving away from both the sun and Earth, it slowly faded to a total magnitude of 14 by March 8. Thereafter, Roemer became the main observer and provided estimates of the nuclear magnitude on into 1965. On March 9, 1964 she estimated the nuclear magnitude as 16.5. This faded to 17.0 by April 10, 17.9 by May 3 and 19.2 by June 4. The comet was recovered after conjunction with the sun on January 12, 1965 at magnitude 19.6 and when last detected on April 24 it was near 20.

Even before the comet was last seen, it had become obvious to astronomers that this comet had passed very close to Jupiter in 1961. After all of the observations were at hand, investigations were made to determine what kind of orbit the comet had possessed prior to 1961. E. I. Kazimirchak-Polonskaya found the comet to have passed only 0.03 AU from Jupiter on November 13, 1961 and that prior to this it had possessed a period of about 53 years. This period indicated an approach to within 0.04 AU of Jupiter in 1855 which had transformed the comet's orbit from a hyperbola to an ellipse. In his own investigation of the comet's past orbits B. G. Marsden remarked that the orbit was "too imperfectly known for us to say when this (the last close approach to Jupiter) might have been and what the previous nature of the orbit was."

Predictions for the 1972 return gave a perihelion date of November 28.9 and on July 26, 1971 Roemer and L. M. Vaughn recovered the comet at a position which indicated a correction to the predicted perihelion of -0.4 day. The magnitude was estimated as 20 and the comet was brightest in late December 1972 and early January 1973 when John E. Bortle (Stormville, New York) gave the magnitude as 12.8. This estimate, as well as others in 1972 and 1973, consistently ran 1.5 magnitudes fainter than predicted. The comet was observed until April 6, 1973.

In 1976 Kazimirchak-Polonskaya and S. D. Shaporev reexamined the past orbit of P/ Kearns-Kwee and found that prior to the 1961 encounter with Jupiter the orbital period had been 52 years. There was no encounter in 1855; however, an approach to within 0.012 AU of Jupiter on February 16, 1701 did alter the orbit significantly. Interestingly, the orbit prior to 1701 was not a hyperbolic one, but rather it was one of short period where the perihelion distance was only 1.6 AU and the comet circled the sun once every 6.8 years.

	T	q	P	Max. Mag.	Max. Tail
1963 VIII	Dec. 7	2.21	8.95	11.1	4 arcmin
1972 XI	Nov. 28	2.23	9.01	12.8	1 arcmin
1981 XX	Nov. 30	2.22	8.99	13.0	1 arcmin

Klemola

Discovered: October 28, 1965   Discovery Magnitude: 17.0

In mid-November 1965 A. R. Klemola (Yale-Columbia Southern Station, Argentina) found a diffuse object on plates he had taken with a 51-cm double astrograph on October 28, 29, 31, November 1 and 2. The comet was moving slowly to the southeast and was described as slightly nebulous with a coma diameter of 12 arcsec. Marsden immediately calculated a parabolic orbit which was published on November 15 with his statement, "the orbit is quite possibly an ellipse of short period." Dr. A. D. Andrews (Boyden Observatory) photographed the comet on November 18 in a position which confirmed Marsden's belief in its short-period nature. Further observations were obtained by Tomita (Dodaira) on November 21 and 22, and by Klemola on November 27 and December 13. The magnitude on the latter date was between 18 and 19. No further observations were made.

Orbital calculations by Marsden in 1967 gave the period as 11.0 years, with an uncertainty of 2 months. Predictions for the comet's return in 1976 gave a perihelion date of August 20 and, with the help of variational ephemerides extending to 40 days either side of this date, G. Sause (Haute Provence Observatory) recovered it on August 6. Then described as of magnitude 12 with a tail 2 to 3 arcmin long, the precise positions indicated the predictions were only 10 days off--making the perihelion date August 10, 1976.

Observations in the following weeks indicated the magnitude estimate at discovery was slightly exaggerated. Observers found the total magnitude to steadily brighten from a value of 13 in mid-August to about 12 by month's end. The comet changed little in brightness during the first half of September, but at the beginning of October it had dropped to 12.5. One month later A. Mrkos (Klet Observatory) estimated the total magnitude to be 15.5. The comet was last seen on January 21, 1977.

	T	q	P	Max. Mag.	Max. Tail
1965 VI	Aug. 18	1.76	10.94	17.0	none
1976 X	Aug. 10	1.77	10.94	11.0	2 arcmin

Kohoutek

Discovered: February 9, 1975   Discovery Magnitude: 14.0

On February 9, 1975 Lubos Kohoutek (Hamburg-Bergedorf Observatory) obtained a 14-minute exposure of the planetary nebula Baade 1. While examining the plate on February 17 the trailed image of a 14th-magnitude comet was noted, but moonlight prevented an immediate confirmation. Finally, on February 27, he exposed three plates: two to the west-southwest and one to the east-northeast of the February 9 position. Upon examining these he immediately found a 13th-magnitude comet on the former plates and reported his discovery to the Central Bureau. On March 1 and 2 K. Ikemura discovered a comet moving in the opposite direction of Kohoutek's. Further images were then discovered by Kojima, on a plate exposed on February 28, and by McCrosky and Schwartz, on a plate exposed on March 1. Interestingly, these indicated that Ikemura's comet was identical to Kohoutek's February 27th object and that the February 9th object was a different comet. Kohoutek reexamined his plates exposed to the east-northeast on February 27 and soon found a 15th-magnitude object at the plate limit. On March 5, J. H. Bulgar (Harvard Observatory's Agassiz Station) confirmed that this object was Kohoutek's February 9th comet.

Orbital calculations immediately revealed Kohoutek's comet to be moving in an elliptical orbit with a period near 5.5 years. They also revealed a possible encounter with Jupiter in 1972. On March 7, Roemer photographed a 17.2-magnitude nucleus within a coma 1 arcmin across. On the same date, T. Seki estimated the total magnitude as 15 and on March 18 the comet's increasing distances from the sun and Earth had caused it to fade to magnitude 16. On April 3 the comet had suddenly brightened to magnitude 14, only to fade back to 15.5 in just a few hours. On April 29, C. Y. Shao reported the central condensation had brightened to nearly 14.5 and six nights later Roemer estimated the nuclear magnitude as 16.5. Thereafter, the comet gradually faded and was last detected on June 8 shortly before entering the sun's glare. At opposition in 1976, Roemer secured a photograph on April 29 which indicated a nuclear magnitude of 21.5.

Improved orbital elements for this apparition were obtained by several astronomers between 1976 and 1979. These indicated an orbital period of 6.23 years and confirmed the close approach to Jupiter in 1972. That encounter occurred in July and the minimum distance of 0.15 AU acted to reduce the perihelion distance from 2.4 AU to 1.6 AU and the period from 8.2 years to 6.2 years.

Predictions for the 1980 return proved to be only -0.68 day off from the actual perihelion date, according to the recovery position obtained on August 6 by H.-E. Schuster (European Southern Observatory). The total magnitude was then near 19 and the comet was brightest during January 1981, when near magnitude 18.

	T	q	P	Max. Mag.	Max. Tail
1975 III	Jan. 18	1.57	6.23	14.0	none
1981 IX	Apr. 18	1.57	6.24	18.0	none

Kojima

Discovered: December 27, 1970   Discovery Magnitude: 14.0

Nobuhisa Kojima (Ishiki, Japan) discovered this comet on plates he had
exposed with his 31-cm reflector on December 27, while searching for the lost
Periodic Comet Neujmin 2. The new comet was described as diffuse with a condensation. Four days later, both Seki and Tomita confirmed the comet, with the
former observer also detecting a short tail.

During the first days of 1971 the comet was described as near magnitude
13.5 with a tail 5 arcmin long. By early March the total magnitude had dropped
to near 16.5 and on the 27th, Roemer estimated the nuclear magnitude as 17.2.
Roemer continued her observations until June 27, when the nucleus had faded to
19.0. On July 15 Tomita exposed a photo on the region of the comet, but the
failure to detect anything indicated the total magnitude was fainter than 18.

Orbital calculations indicated this comet traveled in an elliptical orbit
with a period of 6.16 years. From these it was found that the comet had passed
0.4 AU from Jupiter in 1962 which acted to decrease the perihelion distance from
2.0 AU to 1.6 AU and the period from 6.9 years to 6.2 years. It was also noted
that an approach of 0.15 AU from Jupiter in 1973 would increase the perihelion
distance to 2.40 AU and the period to 7.85 years.

In 1976 Marsden published his prediction for the comet's return in 1978.
He gave the perihelion date as May 25. On December 9 and 10, 1977 K. Hurukawa
and Y. Kosai (Tokyo Astronomical Observatory) recovered the comet at magnitude
18. The comet was subsequently found on a photo exposed by Kowal on December 8
and the indicated correction to the perihelion date was only -0.18 day. The
comet was observed until March 13, 1978 and never became brighter than 18.

	T	q	P	Max. Mag.	Max. Tail
1970 XII	Oct. 7	1.63	6.16	13.5	5 arcmin
1978 X	May 25	2.40	7.85	18.0	none

Kopff

Discovered: August 20, 1906   Discovery Magnitude: 11.5

This comet was discovered photographically by Kopff (Königstuhl Observatory) on August 22, 1906. It was then described as round with a diameter of
1.5 arcmin and possessed a total magnitude of between 11 and 12. A nucleus of
magnitude 13 was also present. Shortly thereafter a prediscovery image was
found on a plate exposed by Kopff on August 20.

The comet steadily moved away from both the sun and Earth after discovery
and slowly faded to magnitude 12 by mid-September and 13 by the end of October.
It was last detected on December 15 and 16 when Max Wolf estimated the photographic magnitude as 16. The comet was first recognized as periodic in mid-
September when orbital calculations gave periods ranging from 6.6 to 6.7 years.
After the comet's final observations, Ebell recomputed the orbit and obtained a
period of 6.64 years. This revealed that the 1912-13 apparition would be unfavorably situated for observations and no searches were subsequently made.
In 1913 G. Zappa (Naples) calculated definitive elements for the 1906 apparition which gave the period to be 6.58 years. He added that the comet would be

favorably situated for recovery in the summer of 1919. On July 30, 1919 the
comet was recovered by Wolf fairly close to the predicted position. The magnitude was then 10.5 and steadily faded as the comet moved away from both the sun
and Earth. The last observation came on December 11.

The comet has not been missed since 1919. When it returned in 1926 it was
recovered near magnitude 16 by Wolf. A few days later van Biesbroeck (Yerkes
Observatory) said the total magnitude seemed closer to 17, but he indicated the
comet slowly brightened until by September it was near 16. Although the comet
was then moving away from the sun, it brightened as the distance from Earth
decreased. The comet was last seen on November 2. In 1932, the comet was recovered by J. Bobone (Cordoba) on May 25, when near opposition. It was then
near magnitude 12 and proceeded to fade very slowly as the increasing distance
from Earth countered the effects of the decreasing distance from the sun. The
comet's fifth recorded apparition officially began on April 21, 1939 when it
was recovered by van Biesbroeck. Then near magnitude 13.5, the comet steadily
faded to magnitude 14 by June 23, 15.5 by mid-September and 17 by November 17.

Perturbations due to a close approach to Jupiter were responsible for
slightly altering this comet's orbit in 1942-43 and as it neared the 1945 apparition the period had been reduced to 6.2 years and the perihelion distance had
decreased to 1.49 AU. On May 7, 1945 the comet was recovered by H. L. Giclas
(Lowell Observatory) at a magnitude of 13. The comet was then 2 degrees away
from the position predicted by W. E. Beart and W. P. Henderson and was found due
to Giclas' use of a wide-field photographic telescope. The comet brightened
throughout the next three months and reached a maximum magnitude of 8.6 at the
beginning of August. Thereafter, it faded and was last detected on January 2,
1946. The comet's 1951 apparition was relatively unusual. It was recovered on
April 12 and at that time, as well as during the next few months, the comet was
consistently about 3 magnitudes fainter than predicted; however, around the time
of perihelion in late October, the comet suddenly brightened to magnitude 10.5--
an increase of 2 magnitudes--and was still near 11.5 when last detected on November 29.

In 1954 the comet passed very close to Jupiter and although it experienced
only a 0.13-year increase in its period and a 0.2 AU increase in its perihelion
distance, major alterations occurred in the angular elements of its orbit.
Nevertheless, painstaking calculations were made by Kepinski which allowed the
comet to be recovered within 3 arcmin of the predicted position on June 25, 1958.
The comet was not particularly well placed for observations at this return and
the maximum magnitude reached only 18.5.

The 1964 apparition was particularly favorable. Recovery came on December
18, 1963 and the comet steadily brightened from a nuclear magnitude of 18.8 to
14.8 by July 3. During early June the total magnitude attained its maximum
value of 9. The comet's next two appearances in 1970 and 1977 were not particularly notable with the maximum magnitude never exceeding 16.

	T	q	P	Max. Mag.	Max. Tail
1906 IV	May 3	1.70	6.58	11.5	none
1919 I	Jun. 29	1.71	6.60	10.5	none
1926 II	Jan. 28	1.70	6.58	16.0	none
1932 III	Aug. 21	1.69	6.56	12.0	none
1939 II	Mar. 13	1.68	6.54	13.5	none
1945 V	Aug. 11	1.50	6.18	8.0	1 arcmin

	T	q	P	Max. Mag.	Max. Tail
1951 VII	Oct. 20	1.49	6.18	10.5	none
1958 I	Jan. 20	1.52	6.31	18.5	none
1964 III	May 16	1.52	6.31	9.0	5 arcmin
1970 XI	Oct. 2	1.57	6.41	16.8	none
1977 V	Mar. 8	1.57	6.43	17.0	none

Kowal 1

Discovered: April 24, 1977   Discovery Magnitude: 16.5

     Charles T. Kowal (Palomar Observatory) discovered this comet on photographic plates exposed on April 24 and 25, 1977. Then between magnitude 16 and 17, the comet was diffuse with some condensation and had a tail extending 2 arcmin. Kowal confirmed the comet on the 26th, as did E. A. Harlan (Lick Observatory), but it then became inaccessible due to moonlight and was not reobserved until May 17 and 19 when Kowal obtained photographs and estimated the total magnitude as 17. From the available data Marsden was able to calculate an elliptical orbit with a period of 18.6 years. Further positions soon caused the period to drop to 15.1 years. The comet was last seen on June 17 when C.-Y. Shao (Harvard College Observatory's Agassiz Station) estimated the total magnitude as 19. In 1979 Marsden recalculated the orbit and obtained a period of 15.11 years. He considers the uncertainty to be about 2 weeks.

	T	q	P	Max. Mag.	Max. Tail
1977 III	Feb. 23	4.66	15.11	16.5	2 arcmin

Kowal 2

Discovered: January 27, 1979   Discovery Magnitude: 17.0

     Kowal discovered this comet on 122-cm Schmidt plates exposed on January 27, 28 and 29, 1979. He described it as diffuse, with slight condensation, but no tail was visible. The total magnitude was given as 17. By January 31 Marsden already suspected the comet to be of short period and after T. Seki (Kochi Observatory) provided his precise positions obtained on February 1, a period of 10.3 years was estimated. Further observations by Seki on February 16, 27 and 28 allowed Marsden to reduce the period to 7.05 years. Another position was obtained on March 1 by P. Young and on March 23 Kowal photographed it at a magnitude of 18. The comet was last detected on March 28 by observers of Harvard Observatory's Agassiz Station. From these new observations Marsden calculated the orbital period as 6.51 years and a more precise calculation by S. Nakano in 1981 raised this to 6.69 years.

	T	q	P	Max. Mag.	Max. Tail
1979 II	Jan. 14	1.52	6.60	16.5	none

Lexell

Discovered: June 14, 1770   Discovery Magnitude: 17.0

     Charles Messier (Paris) discovered this bright, telescopic object while searching for comets on June 14, 1770. The comet was then located 0.21 AU from Earth and proceeded to brighten rapidly as it neared both the sun and Earth. By June 20 it had become an easy naked-eye object and on the 28th, the coma was estimated as 54 arcmin across. The comet was closest to Earth (0.015 AU) on July 2 and observers then indicated a total magnitude of 2[*] and estimated the coma diameter as 2.4 degrees. The comet was lost in the sun's glare after July

4. Alexander Guy Pingre took Messier's positions and computed a parabolic orbit.
A subsequent ephemeris allowed the comet to be recovered on August 3 by Messier.
Then described as faintly visible to the naked eye, the comet faded to below
naked-eye visibility by August 27. Although no tail had been reported up until
mid-August, observers between August 19 and 27 could trace the tail for a length
of 1 degree in telescopes. The comet was last seen on October 3.

     After the final observation attempts were made to compute a parabolic orbit
from the available positions. Pingre was the first to try, but no parabola
would represent the observations. Soon afterward the German physicist Johann
Heinrich Lambert also attempted to calculate a parabolic orbit, but his attempts
were also unsuccessful and both men stated their belief that the orbit might
be elliptical. Finally, in 1779, Anders Johann Lexell succeeded in computing an
elliptical orbit and demonstrated that the period was only 5.6 years. It immediately became apparent that the comet had again been at perihelion in 1776,
but Lexell showed that it was then unfavorably situated for observations.
The question also arose as to why the comet had not been seen prior to 1770 and
Lexell's further calculations showed that the comet had passed very close to
Jupiter in 1767, which he surmised had caused drastic changes in the orbit. He
also found that the comet had been in Jupiter's vicinity in 1779 and predicted
that the comet was then placed into a totally different orbit. In 1806,
Burckhardt verified Lexell's calculations.

     The most recent study of this comet was published in 1967 by E. I.
Kazimirchak-Polonskaya. It was shown that the comet had passed only 0.02 AU
from Jupiter on March 27, 1767 which decreased the period from 10 years to 5.6
years and the perihelion distance from 3.3 AU to 0.67 AU. On July 27, 1779 the
comet passed 0.0015 AU from Jupiter which increased the period to 260 years and
the perihelion distance to 5.2 AU. There is a chance that the orbit could have
been transferred into a hyperbolic one. Thus, the comet is hopelessly lost.

	T	q	P	Max. Mag.	Max. Tail
1770 I	Aug. 14	0.67	5.60	2.0[*]	1 degree

Longmore

Discovered: June 10, 1975   Discovery Magnitude: 17.0

     Andrew Jonathan Longmore (Siding Spring Observatory, Australia) discovered
this comet on a sky survey plate exposed by P. R. Standen with the 122-cm
Schmidt telescope on June 10, 1975. The comet was then described as near magnitude 17, with some condensation, and possessed a faint tail 15 arcsec long. On
June 11 Longmore and P. Wallace confirmed the comet and described its motion as
slow to the southwest. During the next month no further observations were reported, first due to confusion concerning the comet's magnitude and, second,
because of the uncertainty in its position. Subsequently, photographs by
Gilmore (Wellington) on June 17 and 19 were unsuccessful; however, on July 9
M. E. Sim (Siding Spring) obtained a further photograph, though an approximate
position was not reported until July 25. At that time, Marsden was able to
calculate a rough orbit which indicated the comet might be of short period, but
the existing uncertainty still made it difficult to predict the comet's exact
position. Subsequently, Gilmore was unable to photograph the comet on four 60-
minute exposures made on July 30 and August 1.

     In early August precise positions were given for all available observa-
tions and Marsden calculated the orbital period as 7.05 years. On August 6, T.

Hawarden (Siding Spring) obtained a further position which confirmed the elliptical orbit and indicated a period of 6.98 years. The total magnitude was then estimated as 19 and a further observation by Richard M. West (European Southern Observatory) on August 11 gave the magnitude as slightly brighter than 20. Further photographs exposed on August 13 and 16 by Torres (Cerro El Roble) and on August 25 by Gilmore were not successful, probably due to the comet's faintness. Although no further observations were expected, the comet was again found on October 4 by V. M. and B. M. Blanco (Cerro Tololo Observatory) and was described as near magnitude 19. A more precise orbit by Marsden confirmed the 6.98-year period and indicated that the present orbit arose due to a close approach to Jupiter in 1963.

As the comet approached its next perihelion, S. Nakano computed an orbit which gave the perihelion date as October 21, 1981. An ephemeris published in late 1980 allowed T. Seki (Geisei Station) to recover the comet on January 2, 1981 when near magnitude 18. The correction to the orbit amounted to only 0.26 day. The comet brightened slightly by June as its distance from the sun diminished, but observations ended on June 6 as it entered the sun's glare.

	T	q	P	Max. Mag.	Max. Tail
1974 XIV	Nov. 4	2.40	6.98	17.0	15 arcsec
1981 XVI	Oct. 22	2.40	6.98	17.0	none

Lovas

Discovered: December 5, 1980   Discovery Magnitude: 17.0

On a photographic plate exposed by Miklos Lovas (Konkoly Observatory, Hungary) on December 5, was found a probable comet which was simply described as diffuse, with condensation. It was then near the Lynx-Cancer border and was moving to the southeast. No further word was received until shortly after mid-December, when Lovas sent a rough position obtained on December 9, which confirmed the direction of motion. At about this same time, Charles Kowal (Palomar Mountain Observatory, California) reported that he had found an unknown comet on December 14 that was moving to the west. Observations by Kowal on the 15th and T. Seki (Geisei, Japan) on the 9th indicated the comet was the same as Lovas', with the motion having nearly reversed between the 9th and 14th. All observers agreed the magnitude was 17.

On January 2, 1981, Brian Marsden (Harvard-Smithsonian Center for Astrophysics) published a preliminary ephemeris based on a rough parabolic orbit computed from 3 precise positions provided by Kowal and Seki. He said the comet "may be a short-period one" and added that observations were "urgently needed." Precise positions by S. J. Bus (Palomar) on January 6 and G. Schwartz and C.-Y. Shao (Agassiz station of Harvard College Observatory) for January 9 allowed a more precise parabolic orbit to be computed, but it was not until February 2 that Marsden's periodic-comet assumption could be confirmed. This was made possible by precise positions obtained by Gibson (Palomar) on January 26, and J. Bulger (Agassiz) on the 31st. The perihelion date turned out to be September 3, thus confirming that the comet was moving away from the sun and fading.

The Agassiz plate of January 31 showed the comet's magnitude to be 17.5 and when the comet was last detected on April 3 (also at Agassiz) it appeared as only "a weak image" of nuclear magnitude 19.5. A 40-minute exposure one month later failed to detect the comet.

	T	q	P	Max. Mag.	Max. Tail
1980 V	Sep. 3	1.68	9.06	17.0	none

Mellish

Discovered: March 20, 1917   Discovery Magnitude: 7.0

    John E. Mellish (Leetonia, Ohio) discovered this comet on March 20, 1917 while searching for comets. It was then located in Aries and was described as diffuse and near magnitude 8. Observers during the next few days indicated a total magnitude closer to 7.

    The comet brightened rapidly as it approached both the sun and Earth and by the end of March the magnitude was being estimated as about 5.5 and the tail was extending 1 degree. Around April 4 and 5 the comet was independently discovered by several observers in the Northern Hemisphere as a naked-eye object of magnitude 2, which possessed a short bright tail. The comet's low altitude shortly after sunset made photographs and positional measures impossible and it was lost in the sun's glare after April 5.

    Perihelion came on April 11 (r= 0.19 AU) and on April 15 the first of several independent discoveries were made by observers in the Southern Hemisphere as the comet exited the sun's glare. The total magnitude was then between 1 and 2 and visual estimates of the tail length were as high as 20 degrees. Fading was rapid as the comet moved away from both the sun and Earth, with magnitude estimates being near 7.5 after the first week of May and 9.5 by the end of May. The comet was last detected on June 25 as an object of magnitude 11.5.

    This comet was first recognized as one of short period in late May 1917, when A. C. D. Crommelin obtained elliptical elements which gave the orbital period as 141.70 years. In 1932 Asklof computed a definitive orbit which gave the period as 145.37 years. He added that the uncertainty seems to be less than 8 months.

	T	q	P	Max. Mag.	Max. Tail
1917 I	Apr. 11	0.19	145.37	1.0	20 degrees

Metcalf

Discovered: November 15, 1906   Discovery Magnitude: 11.5

    The Reverend Joel H. Metcalf (Taunton, Massachusetts) discovered this comet photographically as it moved slowly to the southwest in Eridanus on November 15, 1906. He described it as between magnitude 11 and 12, with a round coma 2 arcmin across and a distinct central condensation.

    Having already passed perihelion on October 10, 1906 the comet slowly faded after discovery as its distances from the sun and Earth increased. By December 10 the total magnitude had dropped to 12 and on January 1 it was near 12.5. The comet was last seen on January 16, 1907 near magnitude 13. It should be noted that M. E. Esclangon (Bordeaux Observatory) observed two nebulous objects near the comet on November 22 which he described as easily visible. The first object was elongated with a length of 30 arcsec, while the second object was circular, with a diameter of 20 arcsec. Esclangon was unable to detect the objects on November 23 and concluded that water had been present between the objective lenses; however, Professor Kreutz suggested they were actual companions to the comet.

    The comet was first recognized as a member of the short-period family in

early December 1906, when Crawford (Berkeley, California) calculated an elliptical orbit with a period of 6.89 years. By late December Crawford recomputed the orbit with more recent observations and obtained a period of 8.23
years. The most accurate orbit came a short time after Comet Metcalf was last
seen. The computer was Ebell and he found the orbital period to be 7.59 years.
This orbit was used to supply a prediction for the comet's return in 1914; however, although the comet was expected at perihelion in June, searches throughout most of the year proved unsuccessful. A report did come from Harvard
Observatory in February 1915 which said that Miss Leavitt had recovered the
comet on February 9, but the object found by Miss Leavitt proved to be a minor
planet.

     The comet was next expected in 1922 and in that year Bianchi computed an
orbit which fit the observations better than any other orbit. Bianchi found
the orbital period in 1906 to be 7.77 years; however, searches in his predicted positions proved fruitless. The comet was again expected in 1929 and
G. Merton recomputed the 1906 orbit and found a period of 7.73 years. His predicted perihelion date for 1929 was November 23, but, again, all searches
revealed no trace of the comet. Few searches were made at the following returns as the comet was given up as lost. In 1974 N. A. Belyaev, N. Yu.
Goryajnova and U. V. Emel'yanenko undertook the most extensive calculations to
date. Using 73 observations from the 1906-7 apparition they connected Bianchi's
orbit and arrived at a period of 7.78 years. They then advanced the comet and
applied the perturbations of the planets Venus to Saturn from 1906 until 1976.
Their calculations revealed close encounters with Jupiter in September 1911
(0.86 AU), August 1935 (1.17 AU) and August 1969 (1.05 AU) and they predicted
the comet would arrive at perihelion on June 21, 1975. Unfortunately, the
return was not particularly favorable and no searches were made.

	T	q	P	Max. Mag.	Max. Tail
1906 VI	Oct. 10	1.63	7.78	11.5	none

Neujmin 1

Discovered: September 4, 1913   Discovery Magnitude: 10.0

     This comet was discovered by G. N. Neujmin (Simeis, Crimea) on a photo-
graphic plate exposed September 4, 1913 during an asteroid survey. The 10th-
magnitude object was distinctly asteroidal in appearance and was not identified
as a comet until photos on September 7 by Dr. K. Graff (Bergedorf, Germany)
showed a short tail and an 11th-magnitude nucleus. At that time, Graff thought
he had discovered a new comet while searching for Neujmin's asteroid.

     Neujmin 1 was at its closest possible distance from Earth at discovery
(0.54 AU) and, therefore, was at its brightest possible magnitude of 10.0.
Shortly after discovery, a tail extending up to 4 arcmin was occasionally
photographed on long exposures. The comet faded slowly after discovery and due
to the frequent lack of a coma, most magnitude estimates were made of the stellar
nucleus: Graff estimated it as 11.0 on September 6 and George van Biesbroeck
(Uccle, Belgium) found it to be 12.2 on October 20. The same two observers, as
well as others in Italy and Russia, reported brightness variations during September and October. The comet was last observed at Helwan Observatory (Egypt)
and Hamburg Observatory (Germany) on December 31.

     The comet was identified as a new short-period comet in late September 1913
when Drs. Einarrson and Seth B. Nicholson computed elliptical elements with

a period of 17.4 years. By late October they recomputed the orbit and revised the period to 17.83 years. Shortly after the comet was last seen F. E. Seagrave set out to compute a definitive orbit, which he completed in March 1914. His value for the orbital period was 17.56 years.

The comet was next expected in 1931 and van Biesbroeck (Yerkes Observatory) recomputed the comet's orbit to obtain the best prediction possible for the comet's return. He found the 1913 period to be 17.75 years and predicted it would be reduced to 17.68 years by the time it passed perihelion on May 9, 1931. Numerous unsuccessful searches were made, but the comet was finally recovered on September 17, 1931 by S. B. Nicholson (Mount Wilson Observatory). The precise position indicated the prediction was 9 days too late. Although Nicholson estimated the total magnitude as 15, this seems to have been an overestimate since observers in early October found it to be 17. The comet brightened slightly in October as its distance from Earth decreased and it was near 16th magnitude by mid-month. Thereafter, it slowly faded and was near 17.5 when last seen on January 10, 1932. This apparition is especially noteworthy in that no tail was ever observed and the comet rarely exhibited a coma.

The comet was next recovered on May 6, 1948, nearly 3.5 days ahead of the prediction made by van Biesbroeck. Nicholson was again the observer and he estimated the magnitude as 17.5. Observations continued until December 3, with the comet reaching magnitude 16 between September and November. As before, photographs rarely showed any nebulosity around the comet's nucleus. The comet was next recovered on May 16, 1966. On that date A. D. Andrews (Boyden Observatory, South Africa) photographed the comet in a position just 0.68 day ahead of the prediction made by Dr. H. Raudsaar. The magnitude was then estimated as 17 and the comet was diffuse in appearance. When photographed by Z. M. Pereyra (Cordoba Observatory) on June 23 and 24 it was completely stellar and near magnitude 16. The comet was last detected on August 7, 1967.

The primary influences on this comet are Jupiter and Saturn, but according to studies by Soviet and American astronomers, the comet has not been in Jupiter's vicinity for at least 1,200 years. Although in the vicinity of Saturn six times between 1660 and 2060, the approaches have had only minor effects. Thus, during the 400 years studied, the period has remained primarily between 17.93 years and 18.38 years, with similar minor variations in the other orbital elements.

	T	q	P	Max. Mag.	Max. Tail
1913 III	Aug. 17	1.53	17.76	10.0	4 arcmin
1931 I	Apr. 30	1.53	17.69	16.0	none
1948 XIII	Dec. 16	1.55	17.97	16.0	none
1966 VI	Dec. 9	1.54	17.93	16.0	none

Neujmin 2

Discovered: February 24, 1916   Discovery Magnitude: 11.0

G. N. Neujmin (Simeis, Crimea) discovered this comet on routine minor planet survey photographs made on February 24, 1916. The comet was then described as near magnitude 11, with a slow southward movement through Cancer. On February 27 the comet passed only 0.38 AU from Earth--just 0.4 AU greater than its closest possible approach.

Although the comet passed perihelion on March 12, observers indicated that it continued to slowly brighten until well into April as magnitude estimates of between 10 and 10.5 were frequently reported. Thereafter, the comet slowly

faded with observations ceasing after June 5, 1916. The comet was shown to be
moving in an elliptical orbit with a period of 5.4 years and was next expected
in mid-1921. In that year the comet was expected to be very unfavorably situated for observations, but searches were begun in late 1920. On November 16,
Neujmin succeeded in photographing a moving object not brighter than magnitude
15 or 16, which he suspected was the comet, but no further observations were
obtained before the comet entered the sun's glare and identity with Comet
Neujmin 2 seemed doubtful. Nevertheless, for the 1927 return Neujmin computed
two orbits: one being attained by applying perturbations to the 1916 orbit and
the other being attained by applying perturbations to an orbit linking the 1920
object with the 1916 apparition. On November 5, 1926 Neujmin recovered his
comet midway between his two predictions and estimated the magnitude as 14.5.
The comet brightened as it approached both the sun and Earth, with magnitude
estimates being near 13 in mid-December and 12 shortly before the comet's perihelion passage on January 16, 1927. As in 1916 the comet continued to brighten
for the next month after passing perihelion and reached a maximum magnitude of
between 11 and 11.5 during the first days of February. Thereafter, the comet
faded rapidly with magnitude estimates being near 13 by mid-February. The
comet was last detected on March 9 as a very faint, diffuse image near magnitude 14.

Orbital calculations by Neujmin gave the 1927 period as 5.42 years and
cast serious doubt on the comet's identity with the object photographed in 1920.
Predictions for the 1932 return were supplied by both Neujmin and A. C. D.
Crommelin with each giving a perihelion date near June 19; however, searches
carried out at both Yerkes Observatory and Bamberg Observatory revealed nothing.
Although unsuccessful searches were also made in 1937, astronomers considered
the comet to be too faint and unfavorably placed for recovery and chances seemed
to be better for the 1943 return. A perihelion date of April 27, 1943 was predicted and the comet was expected to be recovered well before that date; however,
several photographic searches by van Biesbroeck (Yerkes Observatory), which
should have detected the comet if brighter than 16th magnitude, revealed nothing.

The unfavorable 1948 apparition was virtually ignored, but in 1953 E. A.
Mitrofanova provided ephemerides for the rather favorable 1954 return, but, once
again, all searches were unsuccessful. In 1963 B. G. Marsden studied the past
apparitions of the comet to try and provide a very accurate orbit to enable a
future recovery. He found that the comet had passed 1.00 AU from Jupiter in
1950 and 0.33 AU from Jupiter in 1962. Curiously, his orbit for the 1954 apparition was very close to that of Mitrofanova's, thus hinting that an alternative
other than inaccurate ephemerides might have to be sought to explain the comet's
disappearance. Marsden found the 1963 return would be unfavorable, but his
calculations for the 1971 return indicated an apparition comparable to the very
favorable 1927 apparition. Between October 5, 1970 and March 1971 observers at
5 major observatories conducted exhaustive photographic searches for the comet,
but with no result. The comet thus seems no longer to exist.

	T	q	P	Max. Mag.	Max. Tail
1916 II	Mar. 12	1.34	5.43	10.0	none
1927 I	Jan. 16	1.34	5.43	11.0	none

Neujmin 3

Discovered: August 2, 1929   Discovery Magnitude: 13.0

G. N. Neujmin (Simeis, Crimea) discovered this comet on routine minor
planet survey plates exposed on August 2.9, 1929. He described it as near magnitude 13 and called its southwestern motion "planetary." The comet was then near
opposition and with perihelion having already occurred on June 29, 1929 it
rapidly faded as it moved away from both the sun and Earth. On August 11 van
Biesbroeck (Yerkes Observatory) estimated the total magnitude as 14 and measured
the coma diameter as 42 arcsec. By August 31 Max Wolf (Heidelberg) estimated
the total magnitude as 15 and when the comet was last detected on September 9
it was described as a 15th-magnitude diffuse object without a clear nucleus.

After the comet had been observed for two weeks astronomers already recognized that parabolic elements did not satisfy the observations and in mid-August
M. Ebell (Kiel) calculated the first elliptical orbit and found a period of
11.98 years. Shortly after the comet was last seen, E. C. Bower and H. C. Willis
(Berkeley, California) computed an elliptical orbit using all available observations and arrived at a period of 10.90 years. Ebell also arrived at this value
a short time later.

For the comet's expected return in 1940 H. Q. Rasmusen (Copenhagen) applied
planetary perturbations to Ebell's orbit and predicted the comet would arrive
at perihelion on May 8. Maximum brightness was expected in June and July, but
all searches proved unsuccessful. For the comet's next return in 1951 W. H.
Julian took the perturbations deduced by Rasmusen for the period 1929 to 1940
and applied them to a definitive 1929 orbit computed by Itaru Imai in 1938.
Calculating his own set of pertubations from 1940 to 1951, Julian predicted the
perihelion date would be May 28, 1951. Searches began in early 1951 and on May
4 L. E. Cunningham recovered the comet only 0.5 degree from the predicted posi-
tion--indicating perihelion came 1.5 days earlier than expected. The comet
was then near magnitude 17 and for most of its apparition it remained near 17
and never possessed a coma larger than a few seconds of arc. It was last seen
on November 26 at a magnitude of 19.5.

The comet was again missed in 1961 despite several photographic searches,
but on April 17, 1972 Roemer and McCallister (Steward Observatory) recovered
the comet in a position which indicated a correction of just -0.03 day to the
perihelion date predicted by Marsden in 1971. Then estimated as of magnitude
18.9, the comet slowly brightened and by August it was at magnitude 17.7. The
final observation came on October 3.

A study by Russian mathematician E. I. Kazimirchak- Polonskaya of the comet's
orbital evolution between 1660 and 2060 revealed nine encounters with Jupiter
and six with Saturn. Basically, the comet has traveled in two distinct orbits
during this period. The first orbit slowly increased in period from 13.9 to 15.3
years between 1660 and 1845, while the perihelion distance increased from 2.4 to
2.8 AU. The aphelion lay close to Saturn so that during this time most of the
shaping resulted from this planet. Jupiter took over as the sole influence in
1850, when the comet passed 0.12 AU away. The orbital period was reduced to
11.6 years and the perihelion distance decreased to 2.2 AU. From then until
2060 the period would decline to about 10.6 years and the perihelion would decrease to 2.0 AU.

	T	q	P	Max. Mag.	Max. Tail
1929 III	Jun. 29	2.04	10.90	13.0	none

	T	q	P	Max. Mag.	Max. Tail
1951 V	May 27	2.03	10.95	16.5	none
1972 IV	May 17	1.98	10.57	17.7	none

Olbers

Discovered: March 6, 1815   Discovery Magnitude: 7.5

Dr. Heinrich Wilhelm Olbers (Bremen, Germany) discovered this comet on March 6, 1815 and described it as a small, faint object with a diffuse nucleus. Then located in Camelopardalis, the comet remained in high declination throughout its period of visibility.

The comet slowly approached both the sun and Earth following discovery, which not only caused the brightness to slowly increase to magnitude 5[*] by late April, but also caused the tail to develop to a maximum length of 1 degree. After the April 26th perihelion, the comet steadily faded and when last seen on August 25 it was near 9th magnitude[*]. Numerous orbits were computed for this comet and prior to the final observations it had already become apparent that it was traveling in an elliptical orbit. The period was consistently given as between 72 and 77 years and Bessel's calculation of a definitive orbit gave a value of 74.10 years.

Bessel predicted the comet would return to perihelion on February 9, 1887, but Dr. Ginzel rediscussed the orbit and found the most probable date to be December 17, 1886--with an uncertainty of 1.6 years. Searches in late 1886 and early 1887 continually proved fruitless, but on August 25, 1887 William R. Brooks (Phelps, New York) discovered a comet which was soon identified as Comet Olbers. Then between magnitude 8 and 9[*], the comet slowly brightened as it approached both the sun and Earth with the actual perihelion date falling on October 9. By September the comet was at its brightest with a total magnitude near 7 and after holding this brightness for the next month, it finally began to fade. By December 23 it was estimated as 10[*]. The comet was followed until July 6, 1888 when it had faded to near magnitude 12 or 13[*].

It was not until 1948 that a definitive orbit for the comet was calculated. In that year Hans Q. Rasmusen (Copenhagen) obtained October 8.48 for the perihelion date of 1887 and predicted the comet would next come to perihelion on June 15.9, 1956. Searches for the comet began in mid-1955, but no trace was found until January 4, 1956, when Antonin Mrkos (Lomnicky Stit) recovered it near magnitude 16, less than 1 degree from Rasmusen's prediction. This indicated a correction of +5.5 days to the time of perihelion. Pre-recovery images were soon identified on two plates exposed at McDonald Observatory on November 12 and on a Tokyo plate exposed on January 2. The comet steadily brightened in the following months as it approached the sun, and during June and July observers estimated the total magnitude as between 6.5 and 7.5 and reported a tail extending up to 1 degree. The comet faded thereafter and was last seen on September 25, near magnitude 10, shortly before entering the sun's glare.

In 1967 E. I. Kazimirchak-Polonskaya examined the orbit of Comet Olbers during the period 1660-2060 and found only one minor encounter with Jupiter. This approach came in January 1889 (1.5 AU) and acted to decrease the orbital period by nearly 3 years.

	T	q	P	Max. Mag.	Max. Tail
1815	Apr. 26	1.21	74.86	5.0[*]	1 degree

	T	q	P	Max. Mag.	Max. Tail
1887 V	Oct. 9	1.20	72.40	7.0*	10 arcmin
1956 IV	Jun. 19	1.18	69.57	6.5	1 degree

Oterma

Discovered: February 17, 1942   Discovery Magnitude: 15.0

Dr. Liisi Oterma (Turku, Finland) discovered this comet on photographs exposed on April 8, 1943. She described it as stellar and near magnitude 15, with
a slow northwestward movement. Several years later a prediscovery image was
found on a photograph exposed on February 17, 1942.

With perihelion having been passed in August 1942, the comet faded very
slowly during the next several months following discovery and when last detected
in early July before becoming lost in the sun's glare, observers were estimating
the total magnitude as 16. The comet's large perihelion distance of 3.4 AU made
it difficult for astronomers to determine the nature of the orbit; however, in
late June, P. Herget computed a rough elliptical orbit which gave the period as
7 years and indicated the comet's near-circular orbit was completely confined
between the orbits of Mars and Jupiter. Shortly after the comet's final observations, several independent orbital calculations confirmed Herget's conclusion,
though the period was found to be nearly 8 years. It was also stated that the
small eccentricity of 0.14 could allow the comet to be observed every year. This
prediction was justified by observations made by van Biesbroeck and Herbig in
1944 in which the comet was observed between January and June at a magnitude of
16. When recovered in April 1945 the comet had faded to magnitude 17 and was
near 17.5 when last seen in August. Thereafter, observations were mainly made
around opposition and the comet's maximum magnitude reached 17.5 in both 1946
and 1947.

As the comet approached its 1950 perihelion it reached a maximum brightness
of 17 in 1948, 16 in 1949 and 15 in 1950. Thereafter, it slowly faded each year
and continued this pattern through its next perihelion date of 1958.

Orbital calculations of this comet were extensive and not long after its
discovery astronomers realized its orbit had been shaped by a close approach
to Jupiter in 1937. In 1958 L. Oterma carried out the most extensive orbital
study of this comet up to that time, which indicated the comet had traveled an
orbit with a period of 18 years and a perihelion distance of 5.65 AU prior to
1937, when it passed 0.17 AU from Jupiter. Oterma also traced the orbit into
the future and found that circumstances would repeat themselves in 1963, when
a close approach (0.10 AU) to Jupiter would increase the period to 19 years and
the perihelion distance to 5.4 AU. The orbit began to be affected in late 1961
and on August 7, 1962 the comet was detected for the last time. Further observations at the comet's future perihelion dates are considered very doubtful.

	T	q	P	Max. Mag.	Max. Tail
1942 VIII	Aug. 21	3.39	7.89	15.0	2 arcmin
1950 III	Jul. 16	3.40	7.92	15.0	3 arcmin
1958 IV	Jun. 10	3.39	7.88	17.0	1 arcmin

Perrine-Mrkos

Discovered: December 9, 1896   Discovery Magnitude: 8.0

Charles Dillon Perrine (Lick Observatory) discovered this comet on December
9, 1896 and described it as near 8th magnitude, with a well-shaped nucleus and

a tail 30 arcmin long. The comet was then located 0.27 AU from Earth and proceeded to fade fairly rapidly as this distance, as well as the distance from the
sun, increased thereafter. By the end of December the total magnitude was near
9th magnitude and by mid-January it was near 10.5. The comet was followed until
March 4, 1897, when the total magnitude had dropped to 12.5.

The comet was recognized as a new short-period comet and astronomers computed the orbital period as slightly less than 6.5 years. Conditions for recovery in 1902 were very unfavorable and observers had to wait until 1909 to reobserve the comet. Ristenpart computed a definitive orbit to aid the coming
searches and found the orbital period of the 1896 apparition to equal 6.44
years. On August 12, 1909 A. Kopff (Königstuhl, Germany) recovered the comet
with the aid of Ristenpart's ephemeris. Then near magnitude 15, the comet
slowly brightened during the next three months as it approached both the sun and
Earth and reached a maximum brightness of magnitude 13 in October. Observations
ceased after November 21 as the comet entered the sun's glare.

The 1916 apparition was overlooked since perihelion came while the comet
was in conjunction with the sun, but the 1922 return was considered more favorable. Numerous searches were conducted during the latter part of the year, but
no comet was found. In late December word came from Japan that the comet had
been recovered by K. Nakamura (Kyoto University Observatory) on November 29.
Observations of the comet extended over only three days and its total magnitude
was estimated as 13.5; however, F. R. Cripps took the approximate positions and
determined that Nakamura's object was not Comet Perrine. The object remains
unidentified. At the 1929 return several unsuccessful photographic searches
were conducted by van Biesbroeck and the comet was considered lost.

In 1955, I. Hasegawa decided to make one last effort at trying to recover
Comet Perrine and published three ephemerides representing the range of possible
perihelion dates for a return in that year. Some searches were made, but the
comet seemed to again be escaping detection--until October 19. On that date
Antonin Mrkos (Lomnicky Stit, Czechoslovakia) discovered a 9th-magnitude comet
which was recognized in the following days as moving in the same general direction as Comet Perrine. Astronomers were cautious in making a definite decision
on the identity of Comet Mrkos with the long-lost periodic comet for two reasons:
its location was outside the range predicted by Hasegawa and its total magnitude
was five magnitudes brighter than expected. The comet faded more rapidly than
expected during the first month of observation and was near magnitude 16 by November 14. This led several astronomers to believe the comet had undergone an
outburst in brightness shortly before discovery. Thereafter, it faded to 17.5
by mid-December and 18 by January 9, 1956. No further observations were reported
and, a short time later, new orbital calculations convinced astronomers that this
was a return of Comet Perrine.

Comet Perrine-Mrkos was again observed during its returns in 1961-62 and
1968-69, with the maximum magnitude reaching 17.5 and 13, respectively. Predictions for the 1975 return gave a perihelion date of August 3 and indicated a
moderately well-placed apparition with the total magnitude expected to reach 15--
based on the previous two apparitions--in early August. On August 8, Roemer
conducted the first search by taking a 30-minute exposure covering a variation
in the perihelion date ranging from -0.42 to +0.55 day. The failure to locate
the comet prompted her to make two 40-minute exposures on September 12 covering
a perihelion date variation of -1.3 to +1.5 day, but again nothing was found

even though the nuclear magnitude limit was between 19.0 and 19.5. On October
2, Kowal exposed a photo which covered a range of -5 to +8 days either side of
the expected perihelion date; however, a magnitude limit of 19 was not enough
to reveal the comet. It should be noted that Seki reported an accurate position
of an uncondensed image he photographed on August 13. The total magnitude was
given as 17 and the position indicated a correction of -0.04 day to the predicted perihelion. No further positions were obtained and although such an
object should have been detected in the searches by Roemer and Kowal, B. G.
Marsden wrote in 1978 that the possibility of a brief outburst in brightness
similar to that of 1955 "cannot be completely dismissed." Nevertheless, the
comet's rapid loss of light since its discovery in 1896 suggests that it may
no longer be detectable.

	T	q	P	Max. Mag.	Max. Tail
1896 VII	Nov. 25	1.11	6.44	8.0	30 arcmin
1909 III	Nov. 1	1.17	6.53	13.0	none
1955 VII	Sep. 27	1.15	6.46	9.0	3 arcmin
1962 I	Feb. 13	1.27	6.70	17.5	none
1968 VIII	Nov. 2	1.27	6.72	13.0	none

Peters-Hartley

Discovered: June 26, 1846   Discovery Magnitude: 9.0

Dr. C. Peters (Clinton, New York) discovered this very small and faint comet
on June 26, 1846. He announced his finding only after obtaining some additional
observations and included a preliminary parabolic orbit. Unfortunately, the
orbit was in error and attempts to observe the comet elsewhere failed. Peters
continued observations until July 21, at which time it had faded to near magnitude 10. Soon afterward it was learned that an independent discovery was made
on July 2 by Francesco de Vico (Rome, Italy).

From the observed arc of 26 days several astronomers computed parabolic
orbits, but Heinrich L. d'Arrest soon found that the positions could be very
well represented by a short-period orbit. Soon afterward Peters set out to
calculate a definitive elliptical orbit for his comet and arrived at a period of
12.85 years. Unfortunately, Peters determined the probable error to be 1.6 years
and few attempts were made to recover the comet during the late 1850s. Nevertheless, as a member of the fairly small group of lost periodic comets, Comet
Peters was an occasional target for possible refined orbital calculations. In
1887 Berberich calculated a period of 13.4 years (with a probable error of one
year) and in 1976 Buckley found it to be 12.71 years (with a probable error of
three years). The latter orbit was considered the most accurate, but after 130
years, Buckley's probable error indicated the comet could be anywhere in its
orbit and no searches were made. This strengthened the long-standing opinion of
astronomers that if the comet was again seen it would be due to a complete accident. On July 11, 1982 that accident occurred.

On that date Malcolm Hartley (Siding Spring, Australia) was examining a
plate he had exposed with a 122-cm Schmidt telescope only a few hours earlier
when he detected the trail of a 15th-magnitude comet. Confirmation was made by
M. P. Candy (Perth Observatory) on the 13th and by the 18th this astronomer had
accumulated enough positions to compute an elliptical orbit with a period of
6.90 years. Immediately after seeing the orbit I. Hasegawa and S. Nakano
notified the Central Bureau that the orbit bore a striking resemblance to the

orbit of Comet Peters. Based on the range of the orbital period proposed by
Buckley in 1976 (9 to 15 years), Marsden found no way to link the apparitions--
unless the period had actually been less than 9 years in 1846. After July 20,
Candy redetermined the period of Comet Hartley as 8.12 years and Marsden inte-
grated the elements backward and arrived at a perihelion date differing by only
10 days from that already computed for Comet Peters. Marsden then made a minor
adjustment and fit the orbit to the positions obtained for Comet Peters and the
resulting period was only 7.88 years. Further refinements followed, but con-
sidering the low accuracy of the positions that had been obtained for Comet
Peters, Marsden concluded that the orbit represented these positions well enough
to link the two comets together.

	T	q	P	Max. Mag.	Max. Tail
1846 VI	Jun. 4	1.50	7.88	7.0[*]	none
1982 III	May 9	1.63	8.12	15.0	none

Pigott

Discovered: November 19, 1783   Discovery Magnitude: 7.0

Edward Pigott (York, England) discovered this comet on November 19, 1783,
at a magnitude of about 7[*]. The coma was estimated as 3 arcmin across, but no
tail was visible. The comet was independently discovered on November 26 by
Pierre F. A. Mechain (Paris, France).

Pigott remarked on the comet's invisibility to the naked eye and opera
glasses, though it was easily seen in a small telescope. The comet was nearest
the sun when discovered, as well as being at its closest possible distance to
the Earth (0.50 AU). In the days and weeks following discovery the comet faded
rapidly as the comet-Earth distance increased. Most observations ceased during
the first 10 days of December because of moonlight, but observers at Paris
managed to reobserve the comet after December 10 and kept it in view until the
21st--scarcely one month after discovery.

The uncertainty in the period of this comet amounts to several months; however, all indications point to a close approach to Jupiter in 1780. This encounter placed the comet in a 2:1 resonance with Jupiter, which brought about a
second close approach in 1792. Further investigation into the history of this
comet's orbit is impossible due to the uncertainty in the period.

	T	q	P	Max. Mag.	Max. Tail
1783	Nov. 20	1.46	5.89	7.0[*]	none

Pons-Brooks

Discovered: July 21, 1812   Discovery Magnitude: 6.5

Jean Louis Pons (Marseilles, France) discovered this comet near the
Camelopardalis-Lynx border on July 21, 1812. He described it as a shapeless
object, with no apparent tail. Independent discoveries were made by Wisniewsky
(Novocherkassk, Russia) on July 31 and Bouvard (Paris, France) on August 1.

The comet brightened rapidly as it approached both the sun and Earth and it
first reached naked-eye visibility around August 13. By the end of August the
total magnitude was near 4.5[*] and a tail was extending nearly two degrees. As
September progressed, the comet moved rapidly southward. It attained its maximum brightness of 4[*] around the 15th and also possessed a split tail, with each
branch extending 3 degrees. Observations ended after September 28, due to the
increasing southern declination.

Orbital calculations indicated this comet had a period between 65 and 75 years, but Johann Encke computed a definitive orbit and arrived at a period of 70.68 years. Using Encke's orbit astronomers predicted the comet would return to perihelion in early 1883, but searches proved fruitless. The recovery of Comet Pons was finally made on September 2, 1883 when William R. Brooks accidentally found it while searching for comets. The comet was then described as small, with a total magnitude near 10, and it was considered a new comet until the first orbital calculations showed it to be identical to Comet Pons.

The comet brightened rapidly as September progressed and observers continuously described it as a small, tailless nebulosity; however, a change occurred on September 22 and when the comet was next seen by various observers on the 23rd it had become a stellar object of magnitude 7 or 8. Thereafter, the coma slowly reappeared, with the addition of a short tail. At the beginning of November the comet was near magnitude 6.5 and it first reached naked-eye visibility on the 20th. By December 1 the comet was described as a 5th-magnitude object possessing a coma 6 arcmin across and a tail 1 degree long and at the beginning of January 1884 it was a 3rd-magnitude object with a coma 10 arcmin across and a narrow tail 5 degrees long. On the 1st Muller (Potsdam, Germany) detected a nuclear brightening of 0.7 magnitude during 1.75 hours of observation. Nothing further occurred as the comet passed through perigee (0.63 AU) on January 10, but on January 19 the comet suddenly changed. First, it was at least one magnitude brighter than expected and, second, its appearance had completely changed. Trepied (Algiers Observatory) indicated three distinct zones within the coma in a rather schematic drawing. He showed a central, sharply defined nucleus with two jets emanating from it. The nucleus was then surrounded by a circular halo. The comet passed through perihelion (r= 0.78 AU) on January 26 and then possessed a tail which one observer traced for 20 degrees in binoculars. The comet faded steadily thereafter with observers giving magnitude estimates of 4 at the beginning of February, and 6 at the beginning of March. The final observation came on June 2 when near magnitude 9.5.

Several orbital calculations were made after the 1883-84 apparition and several predictions were made for the comet's next return in 1954. The most precise was a set of elements by P. Herget and P. Musen published in 1953. This definitive work predicted the perihelion date to be May 27, 1954. Using an ephemeris from this work Elizabeth Roemer (Lick Observatory) recovered the comet on June 20, 1953--only 25 arcmin from the predicted position. The comet was then estimated as near magnitude 17.5, but, as in 1883-84, it was prone to sudden outbursts in brightness, the first of which came around July 1, when van Biesbroeck (Yerkes Observatory) photographed the comet near magnitude 13. Thereafter, the comet faded back to its original brightness with estimates being near 15 on July 8, 16 on July 12 and 18 by July 16. The comet's normal brightening occurred during the next two months and on September 15 it had reached 16th magnitude; however, on the 28th van Biesbroeck found it at magnitude 12 and, as before, the comet rapidly faded after the outburst, with magnitude estimates falling to 15 by October 3 and 16 by October 11. The comet slowly brightened to magnitude 15.5 by December 2, but on the 7th another outburst occurred with the maximum magnitude reaching 11. After fading to magnitude 13.5 by December 24 a slow, steady brightening occurred as the comet approached both the sun and Earth. By April 23, 1884, when at low altitude, the total magnitude was estimated as 6.4 by van Biesbroeck. Observations ceased soon afterward as the comet approached conjunc-

tion with the sun and perihelion. It reappeared in the southern sky in early
June and gradually faded until last detectea on September 4.

	T	q	P	Max. Mag.	Max. Tail
1812 I	Sep. 16	0.78	72.57	4.0*	3 degrees
1884 I	Jan. 26	0.78	71.68	3.0	20 degrees
1954 VII	May 23	0.77	70.92	6.0	30 degrees

Pons-Gambart

Discovered: June 21, 1827   Discovery Magnitude: 5.5

Jean L. Pons (Florence, Italy) and Adolphe Gambart (Marseilles, France)
independently discovered this comet on June 21, 1827 and indicated it was near
magnitude 5.5*. The comet was then 14 days past perihelion, and it slowly faded
as its distances from the sun and Earth increased. It was last detected on July
21 when Pons indicated a total magnitude near 8*.

Orbital calculations were strictly parabolic in nature, due to the short
observational arc, but in 1917 Ogura computed two elliptical orbits with periods
of 46.0 and 63.8 years--the latter being considered as the more precise. In
1978 S. Nakano recomputed the orbit to more precisely represent the observations
and found the period to be 57.5 years. Nevertheless, the apparent error of 10
years in the period has made this comet hopelessly lost. It should be noted that
in 1979 I. Hasegawa computed a number of orbits from ancient Chinese observations
prior to 1600. One of these, representing a comet seen in 1110, bears a striking
resemblance to the orbit of Comet Pons-Gambart.

	T	q	P	Max. Mag.	Max. Tail
1827 II	Jun. 8	0.81	57.46	5.5*	none

Pons-Winnecke

Discovered: June 12, 1819   Discovery Magnitude: 8.0

Pons (Marseilles, France) discovered this comet in Leo on June 12, 1819
and described it as small, with a central condensation, but no tail. The comet
brightened as it approached both the sun and Earth and at the end of June it was
near magnitude 7*. As July progressed the comet became more and more difficult
to observe due to evening twilight and when last detected on July 22 the total
magnitude was between 5 and 6*.

Johann Encke took special interest in the calculation of this comet's
orbit--primarily because of his earliest calculation of an orbital period of
just 2.3 years. He set out to calculate a definitive orbit to enable a future
recovery and then obtained a period of 5.62 years. Despite Encke's thoroughness,
his orbit did not allow a recovery at later returns and Comet Pons was considered lost.

The comet was accidentally rediscovered on March 9, 1858 by F. A. Winnecke
(Bonn, Germany). Then located in Ophiuchus, the comet was described as diffuse,
with a coma diameter of 3 arcmin, and possessed a total magnitude near 7.5*. The
comet slowly brightened during the next month as it approached both the sun and
Earth and reached its maximum magnitude of about 6* on April 15. Thereafter,
twilight began affecting observations and the comet was lost after April 24.
It was recovered on May 26 in the Southern Hemisphere and was kept under observation until last detected on June 23, when near magnitude 9*.

Orbital calculations not long after the comet's discovery indicated a possible relationship with Comet Pons, but by the time the comet was last observed

the comets were proved to be identical. Not long afterward, Winnecke computed
the 1858 orbit to possess a period of 5.55 years. Predictions for the 1863 return revealed that the comet would arrive at perihelion when in conjunction
with the sun, but conditions for recovery in 1869 were much more favorable and
on April 10, 1869 the comet was recovered by Winnecke very close to the predicted positions. At this particular apparition the comet passed only 0.25 AU
from Earth on July 8 and was described as near magnitude 6.5*with a nuclear
magnitude of 8. Also at that time the coma was estimated as 10 arcmin across
and a tail was clearly visible.

Since 1869, Comet Pons-Winnecke has been missed at only three returns
(1880, 1904 and 1957); however, perturbations by Jupiter have steadily increased
both the perihelion distance and the period to the point of making future recoveries doubtful--especially since the comet has lost much of its initial
brightness over the last century and a half. Since before its discovery, the
comet has been locked into a 2:1 resonance with Jupiter, which has brought about
a close encounter with that planet every 12 years. These close approaches were,
at first, bringing the comet into a more favorable position for observations
and in 1921, 1927 and 1939, when the perihelion distance was between 1.04 and
1.10 AU, the comet passed Earth at distances of only 0.14 AU, 0.04 AU and 0.11
AU respectively. The maximum magnitude subsequently reached 6.5 in 1921 and
1939, while in 1927 it was near 3.5. After 1939, however, the perihelion distance and period continued to increase and the result was that the comet has not
been brighter than magnitude 11.5. In fact, since 1964, the comet's maximum
magnitude has not been brighter than 17.

	T	q	P	Max. Mag.	Max. Tail
1819 III	Jul. 20	0.77	5.56	6.0*	none
1858 II	May 3	0.77	5.55	6.5*	trace
1869 I	Jun. 30	0.78	5.59	6.5*	short
1875 I	Mar. 13	0.83	5.73	7.5*	none
1886 VI	Sep. 5	0.89	5.82	8.0*	none
1892 IV	Jul. 1	0.89	5.82	6.5*	short
1898 II	Mar. 21	0.92	5.83	10.5	none
1909 II	Oct. 10	0.97	5.89	9.5	none
1915 III	Sep. 3	0.97	5.89	9.3	none
1921 III	Jun. 13	1.04	6.02	6.5	30 arcmin
1927 VII	Jun. 21	1.04	6.01	3.5	1 degree
1933 II	May 19	1.10	6.09	9.5	none
1939 V	Jun. 23	1.10	6.09	6.5	none
1945 IV	Jul. 11	1.16	6.16	11.5	trace
1951 VI	Sep. 9	1.16	6.16	14.0	none
1964 I	Mar. 25	1.23	6.30	17.5	none
1970 VIII	Jul. 21	1.25	6.34	17.0	none
1976 XIV	Nov. 29	1.25	6.36	20.0	none

Reinmuth 1

Discovered: January 26, 1928   Discovery Magnitude: 12.0

During the course of the routine minor planet survey carried on at Heidelberg Observatory, Karl Reinmuth discovered this 12th-magnitude comet on plates
exposed on February 22, 1928. Subsequently, prediscovery images were found on
plates exposed at Barcelona on January 26, Heidelberg on January 29 and Moscow

on February 12.

During the last days of February 1928, observers were describing the comet as 1 arcmin in diameter with a total magnitude between 12 and 12.5 and a short, stubby tail. Thereafter, the comet faded as it moved away from the sun and Earth, with magnitude estimates of 13 in mid-March, 13.5 in early April, 14.5 by mid-April and 16 by the end of April. The comet was last detected on June 15, when near magnitude 17.

Comet Reinmuth 1 was identified as a new short-period comet in March 1928 and the suggestion was made that it might be identical to either the periodic comet Taylor, seen in 1916, or the periodic comet Denning, seen in 1894; however, by April both of these possible identifications were dropped as unlikely. By the end of 1928 L. Berman and F. L. Whipple had calculated the most precise orbit of this comet, which gave the period as 7.24 years. In 1934, J. T. Foxell and A. E. Levin used Berman and Whipple's orbit to derive a prediction for the 1935 return. They concluded that perihelion would fall on May 1 and when the comet was recovered by H. M. Jeffers (Lick Observatory) on November 5, 1934 it was found that the prediction was only 0.4 day off.

Comet Reinmuth was not detected in 1942-43, despite several photographic searches, because of unfavorable geometric conditions. But on November 19, 1949, it was recovered by A. Mrkos (Skalnate Pleso Observatory) very near the predicted position. Since that apparition the comet has been seen at each return; however, due to a close approach to Jupiter in 1937, the comet's increased perihelion distance and period has prevented it from becoming brighter than magnitude 17.

	T	q	P	Max. Mag.	Max. Tail
1928 I	Jan. 31	1.86	7.23	12.0	2 arcmin
1935 II	Apr. 30	1.86	7.23	15.0	2 arcmin
1950 IV	Jul. 23	2.04	7.69	17.5	none
1958 II	Mar. 26	2.03	7.66	17.0	none
1965 V	Aug. 8	1.98	7.60	18.0	none
1973 IV	Mar. 21	1.99	7.63	17.0	none
1980 VIII	Oct. 30	1.98	7.59	17.0	none

Reinmuth 2

Discovered: September 10, 1947   Discovery Magnitude: 12.5

K. Reinmuth (Heidelberg, Germany) discovered this comet during a routine minor planet survey. It was then located in southeastern Pegasus and moved slowly in a northwestern direction. Reinmuth described the comet as near magnitude 12.5 with a nucleus of magnitude 14.

Having already passed both perihelion and perigee, Comet Reinmuth 2 slowly faded during the following weeks with total magnitude estimates reaching 13 in late September and 14 in late October. By mid-November observers were reporting it as near 15 and one month later it was near 16. The comet was last seen on February 1, 1948, when near magnitude 17.

The comet was first recognized as a new short-period one on September 22, 1947 when L. E. Cunningham calculated the period to be 6.40 years; however, in late 1947 both Cunningham and E. K. Rabe independently computed orbits with slightly longer periods of 6.59 and 6.57 years, respectively. The comet was found to have been within 1 AU of Jupiter between the middle of 1943 and the end of 1945 and the suggestion was made that the comet might be identical to the

comet Tuttle-Giacobini, which had been seen in 1858 and 1907. Rabe soon proved
this identification to be wrong and took on the problem of computing a definitive
orbit. His result was a period of 6.59 years and he then turned his sights on
predicting the comet's next perihelion date. The result was March 27, 1954.

The comet was recovered on July 5, 1953 by van Biesbroeck (McDonald Observatory) and was very close to Rabe's predicted position. Van Biesbroeck described it as near magnitude 19 with a coma only 3 arcsec across. The comet was
again observed on July 15, but photographic attempts failed at Mount Wilson
Observatory on September 4 and the comet soon moved into the sun's glare. Perihelion occurred while the comet was in conjunction with the sun, but it was recovered in July 1954 at both Lick and Yerkes observatories. Then near magnitude
17.5, the comet steadily faded to a nuclear magnitude of 18 by September 27.
The final observation came on November 25 when Jeffers estimated the total magnitude as 17.

Comet Reinmuth 2 has been seen at every apparition since its discovery, including those of 1954 and 1969 when perihelion arrived while the comet was in
conjunction with the sun. This was possibly due to the large perihelion distance of 1.9 AU. In 1967, the comet underwent its closest approach to Earth
since it was discovered in 1947 (0.95 AU) and brightened to a total magnitude
of 15 in September 1967. It should be noted that two months after passing
closest to Earth in 1967, both Roemer and van Biesbroeck independently found
the comet near magnitude 13--indicating a sudden outburst in brightness.

	T	q	P	Max. Mag.	Max. Tail
1947 VII	Aug. 20	1.87	6.59	12.5	none
1954 VI	Mar. 27	1.87	6.59	17.0	"short"
1960 IX	Nov. 25	1.93	6.71	17.5	none
1967 XI	Aug. 18	1.94	6.73	13.0	none
1974 VI	May 8	1.94	6.74	15.0	none
1981 III	Jan. 30	1.95	6.74	17.0	none

Russell 1

Discovered: February 27, 1979   Discovery Magnitude: 18.5

Kenneth S. Russell (U. K. Schmidt Telescope Unit, Siding Spring,
Australia) discovered this comet on photos taken by P. R. Standen on June
16 and 24. He estimated the total magnitude as 17 and said the comet was
diffuse, with a condensation, and possessed a possible tail.

Russell's initial positions, as well as C. Pratt's precise position
obtained at Perth Observatory (Australia) on June 29, allowed M. P. Candy,
also of Perth, to compute elliptical elements with a period of 7.43 years.
Candy also showed the comet to be 1 month past perihelion and steadily
fading. The total magnitude estimates made by Pratt on the 29th, as well
as by G. Schwartz (Harvard College Observatory's Agassiz station) 24 hours
earlier, showed the comet to still be at 17.

Russell, assisted by J. Barrow, obtained further precise positions
on July 15, 16, 18 and 23. For each date the total magnitude was given
as 18. Daniel W. E. Green (Harvard-Smithsonian Center for Astrophysics)
computed a more precise orbit in late July which gave the period as 6.13
years and the perihelion distance as 1.61 AU. The comet was last seen on
August 14 near magnitude 18.5[*]. No observations were made after conjunction
with the sun due to the comet's apparent faintness; however, images of the

comet were found on photos exposed nearly four months prior to discovery
and allowed Brian Marsden (Harvard-Smithsonian Center for Astrophysics) to
compute a precise orbit which should allow the comet's recovery at its
next return.

	T	q	P	Max. Mag.	Max. Tail
1979 V	May 27	1.61	6.10	17.0	none

Russell 2

Discovered: August 9, 1980   Discovery Magnitude: 16.0

While examining a photographic plate exposed on September 28, 1980,
by J. Barrow, Kenneth S. Russell (U. K. Schmidt Telescope Unit, Siding
Spring, Australia) noticed a 17th-magnitude, sharply defined, trailed image
"with an extended diffuse patch surrounding it." Confirmation was made by
A. Savage on October 2.

Additional precise photographic positions were obtained by Russell on
October 3 and 6, which allowed Brian G. Marsden (Harvard-Smithsonian Center
for Astrophysics) to determine the orbit to be periodic. The preliminary
elements indicated the perihelion date of May 12, 1980, (q=2.14 AU) and
a period of 7.19 years. An ephemeris based on this orbit allowed Russell
to identify a prediscovery trail on a plate taken by Savage on August 9,
1980. The magnitude was then estimated as 16. This position fully confirmed the earlier computed periodic orbit, with the perihelion date being
revised to May 19 (q=2.16 AU). Unfortunately the only other observation came
on October 7 as the comet faded due to increasing distances from both the
sun and Earth.

	T	q	P	Max. Mag.	Max. Tail
1980 III	May 19	2.16	7.12	16.0	none

Sanguin

Discovered: September 13, 1977   Discovery Magnitude: 16.0

J. G. Sanguin (El Leoncito, Chile) found this comet while examining a
photographic plate exposed with the 51-cm double astrograph on October 15, 1977.
Then near magnitude 16, the comet was confirmed on October 20 by H.-E. Schuster
(European Southern Observatory). Both observers reported the presence of a
faint tail. Prediscovery images were found a short time later on plates exposed at Palomar Observatory on September 13 (magnitude 14.5) and at Cerro El
Roble on October 11 (magnitude 13.5 with a tail 20 arcsec long).

The comet had been closest to Earth (0.85 AU) on September 14--only a few
days prior to the perihelion passage. After its discovery the comet slowly
faded as it moved away from both the sun and Earth, with magnitude estimates
being near 16.5 in November and 17.5 in early December. The comet was last
detected on January 31, 1978 when near magnitude 19.

Orbital calculations already revealed the comet's short-period nature as
early as October 25, 1977 with the calculated period then being near 14 years.
After the comet was last seen, B. G. Marsden computed an orbit using 13 precise
positions and arrived at a period of 12.50 years.

	T	q	P	Max. Mag.	Max. Tail
1977 XII	Sep. 18	1.81	12.50	13.5	20 arcsec

Schaumasse

Discovered: December 1, 1911   Discovery Magnitude: 12.0

M. A. Schaumasse (Nice, France) discovered this 12th-magnitude comet near Sigma Virginis on December 1, 1911. He described it as diffuse, with a coma diameter near 3 arcmin. Although the comet had already passed perihelion, it brightened to magnitude 11 by mid-December as its distance from Earth decreased. Thereafter, it faded, with magnitude estimates dropping to 12.5 by the beginning of January 1912 and 14 by February 16. The comet was last seen on February 19.

Orbital calculations by G. Fayet at the beginning of January 1912 revealed the comet to be traveling in an elliptical orbit with a period of 7.1 years. At a later date Fayet and Schaumasse recomputed the orbit and arrived at a period of 8.0 years. Several ephemerides became available as the comet returned to perihelion in 1919, and the comet was recovered on October 30 in a position differing from the predictions by 6 degrees. Then near magnitude 10.5, the comet steadily faded as it moved away from both the sun and Earth. By November 30 it was near magnitude 12 and when last seen on January 2, 1920 it was near magnitude 13.

Comet Schaumasse was again seen at its return in 1927 and reached a maximum magnitude of 12 during October. The comet was not detected in 1935 and, although this apparition was not as favorable as the past returns, several searches were conducted at times when the comet should have been found. The comet passed 0.37 AU from Jupiter on May 5, 1937 and astronomers made very careful computations in predicting the comet's return in 1943. Searches at several observatories were unsuccessful, but on March 24, 1944 Henry L. Giclas (Lowell Observatory) photographed the comet on large-field plates. These showed the comet to be 7 degrees from the expected position. Much of this difference was attributed to an unknown acceleration, which in the 1960s was finally explained as jet-like effects called nongravitational forces. The comet was near magnitude 15 when found by Giclas and it steadily faded until near 19 when last seen on July 20, 1944.

Observations were again made at the apparitions of 1951-52 and 1959-60 with the maximum brightness reaching 4.9 and 9.5, respectively. It should be noted that the maximum magnitude in 1952 was reached during an apparent outburst in brightness. The comet was not detected in 1968 and 1976, primarily due to unfavorable geometric conditions; however, several intense photographic searches were made at each apparition and the apparent loss of the comet is currently a puzzle to astronomers.

	T	q	P	Max. Mag.	Max. Tail
1911 VII	Nov. 14	1.23	8.01	11.0	none
1919 IV	Oct. 21	1.17	7.94	10.5	none
1927 VIII	Oct. 1	1.17	7.95	12.0	none
1943 V	Nov. 26	1.20	8.21	15.0	none
1952 III	Feb. 11	1.19	8.17	4.9	1 degree
1960 III	Apr. 18	1.20	8.18	9.5	none

Schorr

Discovered: November 23, 1918   Discovery Magnitude: 14.0

Professor R. Schorr (Hamburg Observatory, Germany) discovered this comet in Taurus while examining a photographic plate exposed on November 23, 1918 in the course of a regular minor planet tracking program. The comet was then described as near 14th-magnitude with a coma 30 arcsec across; however, despite the steadily increasing distances from both the sun and Earth, it faded slowly and when last detected on December 31 it was estimated as near magnitude 15.

The comet was first recognized as being of short period in early December when H. M. Jeffers (Lick Observatory) calculated a period of 6.71 years. Six months later, Jeffers recomputed the orbit using three precise positions obtained on November 24, December 10 and December 31 and obtained a period of 6.68 years. The comet was next due at perihelion in 1925 and J. Larink calculated a definitive orbit which gave the 1918 period as 6.71 years and the 1925 perihelion date as May 27.9. The comet was not well placed at this return, since perihelion came while the comet was in conjunction with the sun, and the few searches that were made in August 1925 proved unsuccessful. The comet was expected to be more favorably placed as it approached its next perihelion date of January 6, 1932. Unfortunately searches made during the fall of 1931 revealed no trace of the comet.

Although predictions were occasionally made for later returns, the positions were always expected to be in error by several degrees. In 1974 N. A. Belyaev, V. V. Emel'yanenko and N. Yu. Goryajnova recomputed the orbit and found the 1918 period to be 6.66 years and predicted the comet would next arrive at perihelion on December 18, 1974. Nothing was found. In 1978 B. G. Marsden recomputed the 1918 orbit and also arrived at a period of 6.66 years.

	T	q	P	Max. Mag.	Max. Tail
1918 III	Sep. 30	1.88	6.66	14.0	none

Schuster

Discovered: September 5, 1977   Discovery Magnitude: 17.5

Hans-Emil Schuster (European Southern Observatory) discovered this comet on plates exposed on October 9 and 10, 1977. He estimated the magnitude as 17, but added that, except for some "fuzziness to the northeast," the object seemed to be a minor planet. On October 14, Schuster confirmed the cometary nature when his 40-minute exposure with the 100-cm Schmidt telescope revealed a tail 20 arcsec long to the north-northeast. A few days later computations by B. G. Marsden revealed that the comet was identical to a minor planet discovered by Schuster on plates exposed on September 5, 6 and 7.

At discovery, the comet was located near perigee (1.03 AU), but despite the increasing distance from Earth thereafter, it slowly brightened as it approached perihelion. On December 3, observers estimated the magnitude as 16 and when last seen on January 8, 1978 it was near 15.5. Observations thereafter were stopped due to the sun's glare.

The comet was found to be of short period in mid-October 1977--shortly after it was linked to the "minor planet" found in early September. Marsden then computed the orbital period as 7.46 years and added that the comet should "have made a moderately close approach to Jupiter in 1958." A recomputation of the orbit in 1979 gave a value of 7.47 years.

	T	q	P	Max. Mag.	Max. Tail
1978 I	Jan. 7	1.63	7.47	15.5	1 arcmin

Schwassmann-Wachmann 1

Discovered: March 4, 1902   Discovery Magnitude: 12.0

A. Schwassmann and A. A. Wachmann (Hamburg Observatory, Germany) discovered

this comet on photographs exposed on November 15, 1927. Then near magnitude
13.5, with a coma 2 arcmin across, the comet proceeded to fade very rapidly
during the next few days with estimates reaching 15 on November 28 and 16 by
December 1. Interestingly, as December progressed, the comet brightened to
magnitude 14 by the 14th and then faded back to 16 by the 21st. Astronomers
were puzzled by this apparent fluctuation in the magnitude, especially since
revised orbital calculations near the end of December revealed the comet to be
situated about 6 AU from the sun--a region where comets were expected
to be fairly inactive. Even more interesting was the fact that calculations
revealed the comet to be traveling in an elliptical orbit with a period of
16.4 years.

The comet continued to fade as the new year began and by February 1928 it
was near magnitude 17. Observations ceased after February 24, but revised
orbital calculations indicated an eccentricity which would allow the comet to
be observed near opposition every year. Subsequently, it was recovered on
September 21 and was observed from then until the end of the year at a magnitude between 16 and 17. As the comet approached its third opposition searches
were begun on November 6, 1929 by Walter Baade (Hamburg Observatory) but no
trace was found--indicating a total magnitude fainter than 17. Baade conducted
another photographic search on December 2 and, although the comet should only
have been slightly brighter, it was found with a total magnitude of 13.5.
Fading was rapid thereafter, with van Biesbroeck (Yerkes Observatory) giving a
magnitude of 15 on December 28.

The comet has continued to be observed annually up to the present time and
astronomers estimate it will be around for some time to come, since the orbit
is slowly stabilizing just outside Jupiter's orbit. Since 1927 the orbital period has declined from 16.4 years to 15.0 years and the eccentricity has declined
from a value of 0.15 to 0.11. The normal state of the comet has been to reach
a total magnitude of 17 at perihelion and then slowly fade to between 18 and 19
by the time it reaches aphelion. The coma normally measures less than 1 arcmin
across, while a tail is nonexistent.

Of major importance is the fact that the outbursts in brightness have continued at a rate of over one per year and astronomers have yet to arrive at a
theory which adequately explains what is happening. One major theory is that
solar activity causes the outbursts, but this is in jeopardy since no distinct
correlation exists between the outbursts and the sun's 11-year cycle of minimum
and maximum sunspot activity. In addition, it is currently difficult to explain
how any external force could cause the outbursts since the annual comets Oterma,
Gunn, Gehrels 3 and Smirnova-Chernykh have never been observed to undergo any
unexpected change in brightness.

Most of Comet Schwassmann-Wachmann 1's outbursts are characterized by a
nucleus which brightens five magnitudes within a few days. When at maximum
brightness the comet is completely stellar in appearance, but once fading begins,
a bright ring-tailed coma appears. This coma expands and fades during the following weeks until, one month after the outburst, it appears as a faint, ill-
defined feature nearly 3 arcmin in diameter. There are, of course, variations
in the comet's outburst routine: 1) the outburst occasionally occurs within
hours rather than days, 2) the range of the outburst has on three occasions
covered nearly 9 magnitudes and reached a maximum brightness of 10, 3) on
several occasions the coma was mainly concentrated on one side of the nucleus or

was brighter on one side, 4) jets occasionally appear within the coma as an
outburst subsides.

One other point of interest in the history of this comet was the detection
of prediscovery images on photographic plates exposed on March 4 and 5, 1902.
The images were discovered in the spring of 1931 by K. Reinmuth and were described as of 12th magnitude, with a coma 1.5 arcmin across. Shortly thereafter,
A. C. D. Crommelin tried to link the positions to a known periodic comet, but
failed. However, at the end of June 1931, L. E. Cunningham identified the object as none other than Schwassmann-Wachmann 1, thus pushing the observational
record back 25 years.

	T	q	P	Max. Mag.	Max. Tail
1925 II	May 8	5.47	16.44	12.0	none
1941 VI	Apr. 22	5.52	16.14	9.4	none
1957 IV	May 13	5.54	16.10	10.0	none
1974 II	Feb. 15	5.45	15.03	10.0	none

Schwassmann-Wachmann 2

Discovered: December 8, 1928   Discovery Magnitude: 11.5

During a routine search for minor planets, Professor A. Schwassmann and
A. A. Wachmann (Hamburg Observatory) discovered this comet on plates exposed on
January 17, 1929. The total magnitude was then estimated as 11. Shortly there-
after, prediscovery images were found at Yerkes Observatory (January 4, 7 and 12),
Harvard Observatory (December 19 and January 9) and Tokyo Observatory (December
8 and 19).

The comet had passed perigee (1.21 AU) on December 30, 1928, and, when
discovered, it was approaching perihelion. Observers then estimated the total
magnitude as 11.5 and described the tail as 1 to 2 arcmin long; however, as the
comet's March 23 perihelion drew closer the increasing distance from Earth caused
the brightness to fade. By the end of March the comet was near magnitude 13.
The comet was last detected on June 6 at a total magnitude of 14.5. The solar
glare prevented further observations.

The comet was first recognized as being of short period on January 21,
1929, when George van Biesbroeck and Y. C. Chang computed an orbit based on
positions obtained on January 4, 12 and 20. The result was a period of 6.83
years. In February, A. C. D. Crommelin utilized the Harvard prediscovery
images as well as recent positions to obtain a more accurate estimate of 6.43
years. Later computations by C. H. Smiley and S. Kanda gave similar values for
the period. The comet was next expected at perihelion in 1935 and two major
predictions became available as the comet approached the sun. H. Q. Rasmusen
calculated the perihelion date as August 31.4, while P. J. Harris and J. D.
McNeile found it to be August 24.1.

Astronomers found that perihelion would occur when the comet was in conjunction with the sun and searches began during the last half of 1934.
Wachmann found an object very near the expected position on August 15 and movement seemed to be as expected; however, a few days later L. E. Cunningham identified the object as the minor planet Nysa (44). The comet was finally recovered on December 11, 1934, when van Biesbroeck (Yerkes Observatory) estimated the total magnitude as 16.5. The actual date of perihelion turned out to
be August 28.8--just 2.6 days earlier than Rasmusen's prediction.

Astronomers have never failed to recover the comet since its discovery apparition and, on the average, it is observed for 523 days at each return. In 1973, L. Kresak established that Schwassmann-Wachmann 2 was a potential annual comet. This encouraged searches near the time of aphelion in December 1977, but no trace was found. Occasional searches were made thereafter and the comet was finally recovered on December 14, 1979 by G. Schwartz (Harvard College Observatory's Agassiz Station). Then located 3.5 AU from the sun, the comet was found earlier than ever before; however, with a magnitude of 20.5, it was fainter than expected and made an aphelion observation more difficult than previously predicted.

The present orbit was created by an approach within Jupiter's sphere of influence in March 1926 (0.179 AU). Prior to this the comet had a period of 9.3 years and a perihelion distance of 3.55 AU. N. A. Belyaev has examined the orbital changes between 1660 and 2060 and discovered a very unstable orbit. Two close approaches within Jupiter's sphere of influence, as well as three moderate approaches and three minor ones were found. The period fluctuates from 9.36 years to 6.37 years, while the perihelion distance varies from 3.61 AU to 1.90 AU. Belyaev predicts the second entry into Jupiter's sphere of influence will occur in 1997--increasing the period by 2 years and the perihelion distance by 1.3 AU.

	T	q	P	Max. Mag.	Max. Tail
1929 I	Mar. 23	2.09	6.42	11.0	2 arcmin
1935 III	Aug. 29	2.09	6.42	14.0	3 arcmin
1942 I	Feb. 14	2.14	6.51	11.0	3 arcmin
1948 VII	Aug. 24	2.15	6.53	15.0	3 arcmin
1955 I	Feb. 27	2.15	6.53	13.0	2 arcmin
1961 VII	Sep. 5	2.16	6.54	16.0	1 arcmin
1968 II	Mar. 14	2.15	6.52	13.0	30 arcsec
1974 XIII	Sep. 12	2.14	6.51	13.0	10 arcmin
1981 VI	Mar. 17	2.14	6.50	11.8	"short"

Schwassmann-Wachmann 3

Discovered: April 27, 1930   Discovery Magnitude: 9.5

A. Schwassmann and A. A. Wachmann (Hamburg Observatory) discovered this comet on plates exposed on May 2, 1930 for the regular minor planet survey. The comet was then near Tau Corona Borealis and was described as diffuse, with a total magnitude of 9.5. Shortly thereafter, prediscovery images were found on plates exposed at Babelsberg Observatory (Germany) on April 27 and 29.

Observers during the first days following the discovery announcement described the comet as between magnitude 9 and 10, with a star-like nucleus of 12th magnitude and a tail extending 3 arcmin. Rough orbital calculations revealed the comet was about 0.3 AU from Earth, with this distance and the distance from the sun both decreasing in the following weeks. As May progressed, the comet brightened to magnitude 8.5 on the 24th and 7.5 by the 31st. The comet was closest to Earth on June 1 (0.06 AU) and was described as between magnitude 6 and 7, with a tail length of 30 arcmin. On June 3 the comet was independently discovered by Blathwayt (Johannesburg, South Africa) as it entered Southern Hemisphere skies. Thereafter, the comet slowly faded, with magnitude estimates dropping to 8 by June 8, 9 by July 4 and 10 by August 4. Observations ceased after August 24.

The comet was first recognized as a new short-period object at the end of

May 1930, when A. C. D. Crommelin utilized observations from May 2, 12 and 22
to obtain an orbit with a period of 5.46 years. Although other orbital calculations during the next few months gave values varying by up to 0.3 year,
later definitive orbits agreed with Crommelin's initial estimate. The comet
was recognized as an intrinsically faint object and with unfavorable geometric
conditions in 1935-36, no trace was found in searches that could only be conducted six months or more before and after the perihelion passage. The 1941
apparition was expected to be more favorable, though again it was not as well
placed as in 1930. Subsequently, the comet's faintness was apparently to blame
for the unsuccessful searches at several major observatories--especially Yerkes
Observatory, where van Biesbroeck conducted several photographic searches.

In 1946, F. R. Cripps used 143 observations to compute a definitive orbit
for the 1930 appearance. The resulting period was 5.43 years. Unfortunately,
his prediction for the 1947 return showed the geometric conditions to be very
unfavorable. Though the 1952 apparition was expected to be more favorable,
searches again could not be conducted near the perihelion date, due to the
comet's proximity to the sun in February, March and April, and no trace was
found. In 1957, D. A. Kalnin computed a new orbit and predicted the next
perihelion date to be October 12, 1957. Perturbations by Mercury, Venus, Earth,
Mars, Jupiter and Saturn were taken into account and it was found that the comet
had passed 0.9 AU from Jupiter on October 20, 1953. This resulted in an increase of the orbital period from 5.46 years (1952) to 5.61 years (1957).
Again searches proved unsuccessful.

In 1973, N. A. Belyaev and S. D. Shaporev used 190 observations made in 1930
to correct Kalnin's elements. They found the comet to have passed only 0.25 AU
from Jupiter on November 6, 1965 and found the period to have decreased to 5.40
years. The next perihelion was predicted to fall on March 17, 1974--another
unfavorable apparition. From all indications, however, astronomers noted that
the 1979 return would be very favorable--the best since 1930--and several predictions were made. Several searches were conducted based on an ephemeris by
Brian Marsden, but nothing was found. Suddenly, on August 15 M. P. Candy
(Perth Observatory) reported the discovery of a comet by J. Johnston and M.
Buhagiar on a photograph taken on August 13, while searching the plate for minor
planets. Confirmation came on the 15th and Candy pointed out that the direction
and rate of motion resembled those expected for Schwassmann-Wachmann 3, but
the perihelion date was 34 days later than predicted.

At discovery the total magnitude was estimated as 13 and the comet brightened
to a maximum brightness of 12.5 by mid-September, as it passed 1.04 AU from
Earth--more than twice as far as originally predicted, due to the delayed perihelion.

	T	q	P	Max. Mag.	Max. Tail
1930 VI	Jun. 14	1.01	5.43	6.5	30 arcmin
1979 VIII	Sep. 3	0.94	5.38	11.9	"short"

Shajn-Schaldach

Discovered: August 28, 1949   Discovery Magnitude: 12.5

Dr. Pelageja F. Shajn (Crimean Astrophysical Observatory) discovered this
comet on routine minor planet survey plates exposed on September 18.9, 1949.
She described it as a diffuse 12.8-magnitude object, with a tail 30 arcmin long.
On September 20.3 the comet was accidentally photographed by Robert D. Schaldach
(Lowell Observatory), also during the course of a routine minor planet survey.

He indicated the magnitude was 12. A short time later, prediscovery images were found by Shajn on survey plates exposed on August 28 and September 4.

The comet changed little in brightness as it approached its November 27 perihelion since the distance from Earth steadily increased, and magnitude estimates were continually between 12 and 13. Between October 17 and 19 Max Beyer (Hamburg Observatory) described the magnitude as 11.6--nearly 0.5 magnitude brighter than the days before and after--and there was also a faint jet which was no longer visible after the 19th. Fading was rapid during December and reached 14.5 by month's end--nearly 1 magnitude fainter than expected. The comet was last detected on January 11, 1950, when near magnitude 14. The comet was still well placed for observations in January and February, but no sightings were reported.

The comet was first recognized as being of short period near the end of October 1949, when Miss Amelia White computed an elliptical orbit with a period of 7.76 years--the uncertainty then being nearly 1 year. She also discovered that the comet had passed 0.18 AU from Jupiter in 1946. Based on an observed arc of 82 days, A. D. Dubiago provided an improved orbit in 1950, but surpassed this work with a definitive orbit in 1955, which used 66 observations made during 113 days and gave a period of 7.28 years. Using this orbit, ephemerides were published for the 1956 return, but searches at Yerkes, Lick and Johannesburg observatories provided no results. Predictions for the return of 1964 were provided by several astronomers, but again no observations were obtained.

In 1970, Brian Marsden recomputed the 1949 orbit and confirmed Dubiago's period of 7.28 years. He then predicted the next perihelion date as October 4, 1971 and found that conditions would be the most favorable since 1949, although an uncertainty of nearly 5 days existed in the calculations. Searches began in June 1971, but nothing was found until September 29 and 30, when Charles T. Kowal (Palomar Observatory) photographed it with the 122-cm Schmidt telescope. The comet was then described as near magnitude 16, with a short tail. The position indicated a correction of -2.5 days was necessary to the prediction. Subsequently, pre-recovery images were found on plates exposed at Haute Provence Observatory on September 15, 16 and 18. The comet slowly faded after its recovery with estimates of 16.2 on October 24 and 16.5 by November 14. The final observation was made on January 20, 1972.

The comet was again recovered at its next return when C.-Y. Shao and G. Schwartz (Harvard Observatory's Agassiz Station) photographed it on July 2 and 3, 1978. The nuclear magnitude was then between 20 and 20.5 and the positions were almost exactly as predicted. Observations continued until October 31.

	T	q	P	Max. Mag.	Max. Tail
1949 VI	Nov. 27	2.23	7.27	11.6	30 arcmin
1971 IX	Oct. 2	2.23	7.27	16.0	short
1979 I	Jan. 9	2.22	7.25	19.0	none

Slaughter-Burnham

Discovered: December 10, 1958   Discovery Magnitude: 16.0

Between December 10 and 15, 1958, this faint comet was photographed nine times in the course of the proper motion survey at Lowell Observatory, but it managed to escape brief examinations and was not noticed until January 27, 1959. On that date, C. D. Slaughter was blink examining a plate exposed on December 10 with a much older plate when he suddenly found the 16th-magnitude object. A

search for the object was expected to be fruitless, but within the following
days eight additional plates were found. The nine plates were measured by
H. L. Giclas and the positions were used by Elizabeth Roemer to compute an orbit
and ephemeris. On February 2, Roemer recovered the comet and described it as
possessing a nuclear magnitude of 18.5.

The comet had passed perihelion on September 5, 1958 and steadily faded
after its discovery until it was last observed on April 9, 1959 near a nuclear
magnitude of 20.5. In July 1959, Roemer used 3 precise positions obtained
between December 10 and April 1 and obtained elliptical elements with a period
of 11.64 years. She considered these to be accurate to within 2 days. In
1968 G. Sitarski used 14 observations to obtain a more precise orbit with a
period of 11.61 years. He then provided a prediction for the 1970 return.

Searches for the comet began in the latter half of 1969 and a recovery was
reported by Z. M. Pereyra in early September; however, additional observations
proved this object to be an asteroid and searches continued. The comet was
finally recovered on November 4 when Roemer obtained condensed images of nuclear
magnitude 20.1 on plates exposed with the 229-cm reflector at Steward Observatory.
Sitarski's prediction required a correction of only -1.1 day. The comet slowly
brightened to a nuclear magnitude of 19.8 by October 3, 1970. The total magnitude was then near 17 and a tail was estimated as 12 arcsec long. The comet
was followed until April 21, 1971.

The comet was next recovered on July 9, 1981 by observers at Oak Ridge
Observatory. Then near magnitude 20, the predicted perihelion date of November
19, 1981 found by S. Nakano required a correction of only +0.10 day.

	T	q	P	Max. Mag.	Max. Tail
1958 VI	Sep. 5	2.54	11.61	16.0	trace
1970 V	Apr. 13	2.54	11.62	17.0	12 arcsec
1981 XVIII	Nov. 19	2.54	11.62	17.5	trace

Smirnova-Chernykh

Discovered: March 4, 1975   Discovery Magnitude: 15.0

This comet was first noticed by Tamara Mikhajlovna Smirnova (Institute
for Theoretical Astronomy) in late March 1975 while examining 60-minute exposures obtained at the Crimean Astrophysical Observatory on March 4 and 16.
The object was described as of magnitude 15, but its identity as a comet was
then not definite. On March 30, Nikolaj Stepanovich Chernykh (Crimean Astrophysical Observatory) obtained a further exposure which proved the object was a
comet. Shortly thereafter, Smirnova found another image on a plate exposed on
March 14.

The comet changed little in brightness during April and by the end of the
month G. R. Kastel computed elliptical elements which gave the period as 8.49
years, the perihelion distance as 3.57 AU and the eccentricity as 0.14. She
also found that the present orbit had evolved due to a close encounter (0.4 AU)
with Jupiter in 1963 in which the perihelion distance decreased from 5.7 AU and
the eccentricity increased from 0.08. Of primary interest was the comet's eccentricity and intrinsic brightness, which indicated it could be a new annual comet
--joining the ranks of P/Schwassmann-Wachmann 1 and P/Gunn. Roemer began obtaining nuclear magnitude estimates on April 17 when the comet was near 18.0. By
May 4, this had brightened to 17.0. Observations ceased after May as the comet
entered the sun's glare, but on December 4 Roemer recovered it at a nuclear

magnitude of 18.1.

The comet reached opposition in early April 1976 and was then near a nuclear magnitude of 17.5. Three weeks earlier Chernykh had obtained longer photos which gave the total magnitude as 15.5 and showed a tail extending 15 to 20 arcsec to the west. This gave great support to the belief that this object could be an annual comet and observations continued at each of the oppositions up through the perihelion passage in November 1979.

	T	q	P	Max. Mag.	Max. Tail
1975 VII	Aug. 6	3.57	8.53	15.0	1 arcmin

Spitaler

Discovered: November 17, 1890   Discovery Magnitude: 11.0

On the night of November 16-17, 1890, Rudolf Spitaler (Vienna Observatory) received a telegram from T. Zona (Palermo Observatory) which gave the position of a new comet discovered on the 15th. Interested in viewing the bright object, Spitaler turned the 68.6-cm refractor to the approximate location and found a comet, though he was a little disappointed since it appeared fainter than Zona's telegram had led him to believe. Nevertheless, Spitaler measured the position and 30 minutes later he repeated the measurement as part of his routine program for comet observations. Spitaler compared the positions and immediately noted that the rate of motion was less than the figure quoted by Zona; after a short deliberation he wondered if he was observing a different comet. Upon moving the telescope he came upon a brighter comet--Zona's--roughly one degree away.

Having already passed perihelion on October 27, Spitaler's comet faded very slowly during November and December as the distance from Earth changed very slightly from 0.95 AU. Subsequently, Spitaler still indicated a total magnitude near 12 on December 29. Observations continued throughout January 1891 and finally ceased after February 4 when it had faded to near magnitude 13[*].

By the end of 1890 Comet Spitaler was recognized as a new short-period object with a period near 6.4 years. The first definitive orbit was calculated in 1897 and this included a prediction for the comet's return in that year--but conditions were not particularly favorable and nothing was found. The comet passed close to Jupiter in 1899 which acted to increase the period to 6.8 years and the perihelion distance to 2.1 AU. The 1903 apparition was only slightly more favorable than in 1897, and the few searches made were unsuccessful. Later, in 1910, Dr. F. Hopfer remarked that had a proper ephemeris been available in 1903 the comet might have been found.

Several predictions were made for the comet's return in 1910 as conditions appeared to be very favorable. The most extensive set of ephemerides was published by Dr. Hopfer. He computed nine different ephemerides, each of which used a different perihelion date in the period September 12 to November 15. Despite the care taken by astronomers in both calculations and searches, nothing was found.

The most recent predictions for this comet were published in 1972 and 1978. The first was by Brian Marsden, who stated that he "felt that this comet was a rather good candidate for possible recovery." He recomputed the 1890 orbit and found a period of 6.37 years. This orbit was then advanced, taking planetary perturbations into account, and the resulting perihelion date was September 8, 1972. A number of ephemerides were distributed to selected observers. Of these observers, Charles Kowal (Palomar Observatory) made the most extensive photo-

graphic searches, of which none showed the comet. In 1978, R. J. Buckley up-
dated Marsden's 1972 orbit and found the perihelion date to be August 4, 1979.
Again nothing was found.

	T	q	P	Max. Mag.	Max. Tail
1890 VII	Oct. 27	1.82	6.37	11.0*	none

Stephan-Oterma

Discovered: January 22, 1867   Discovery Magnitude: 8.5

This comet was discovered on January 22, 1867 by 18-year-old Jerome E.
Coggia, a newly hired assistant at Marseilles Observatory. This was the first
of several comet discoveries in Coggia's career, but, at that time, he was an
unknown in the field and credit for the discovery was taken by E. J. M. Stephan,
the observatory's director, who obtained the first accurate position of the
comet on January 25. On January 28 William Tempel (Marseilles) independently
found the comet.

Observers during the first days following discovery indicated the total
magnitude was between 8 and 9* and the comet was described as circular, with a
distinct nucleus and a possible fan-shaped tail. Thereafter, a steady fading
set in as the distances from the sun and Earth increased and when last seen on
April 4, 1867 it was near magnitude 12*.

Comet Stephan was not immediately recognized as a periodic comet, primarily
due to the small observed arc of 73 days, but a few years following its discovery, A. Searle computed an orbit which gave the period as 33.62 years. In
1891, L. Backer recomputed the orbit and arrived at a period of 40.09 years,
with an uncertainty of 2 years. This discrepancy in the period deterred attempts
to recover the comet at future returns--the first of which came in 1904--and
the comet was considered to be lost.

On November 6, 1942 Liisi Oterma (Turku, Finland) discovered a comet moving
slowly northward through southern Taurus. A few days later, Fred L. Whipple
(Harvard College Observatory) found a prediscovery image on a photographic plate
exposed on November 5. By mid-November Whipple and J. Bobone had independently
calculated elliptical orbits with periods of 41.4 and 38.3 years, respectively.
The former astronomer also noted a great similarity in the orbits of Comet
Oterma and Comet Stephan and further calculations proved them to be identical.

Although Oterma had estimated the photographic magnitude as 13 at discovery,
observations by George van Biesbroeck (Yerkes Observatory) beginning November
11 showed it to be near magnitude 10.5. The comet brightened to 10 by December
2 and reached a maximum brightness of 9.2 on the 14th. Perihelion came on
December 19, and the comet slowly faded thereafter as its distances from the sun
and Earth increased. In early January 1943 observers indicated a total magnitude
near 10 and by the end of the month it was near 10.5. At the beginning of March
the comet had faded to magnitude 13 and when last observed on May 2 it was near
16.

The comet was next expected to arrive at perihelion on December 5, 1980 and
on June 13, 1980 H.-E. Schuster (European Southern Observatory) recovered it
near magnitude 18. The indicated correction to a prediction made by D. K.
Yeomans was only -0.07 day. The comet brightened fairly rapidly as it neared
both the sun and Earth and reached a maximum brightness of 8.5 during the first
half of December 1980. Thereafter, it steadily faded to magnitude 9 by January
1, 1981 and 11 by February 15. The last observation came on April 4, 1981.

	T	q	P	Max. Mag.	Max. Tail
1867 I	Jan. 21	1.58	37.12	8.5[*]	"short"
1942 IX	Dec. 19	1.60	38.88	9.2	"short"
1980 X	Dec. 5	1.57	37.71	8.5	20 arcmin

Swift

Discovered: August 21, 1895   Discovery Magnitude: 10.5

On the morning of August 21, 1895, Dr. Lewis Swift (Echo Mountain, California) turned his telescope to the spot of the last nebula discovered at his old
observatory in Rochester, New York. His intentions were to secure a more accurate position and brightness estimate so its presence could be announced, but
upon peering through the telescope Swift was astonished to find "a beautiful
comet instead of the expected nebula." The comet was described as a circular
object 5 arcmin across, with a "bright elliptical condensation" and a minute
star-like nucleus.

At discovery, Comet Swift was located 0.34 AU from Earth--nearly its closest
possible approach--and proceeded to slowly fade thereafter as it moved away from
both the sun and Earth. By early September the total magnitude had declined to
11 and by mid-November it was near 12. Observations continued up to February 6,
1896 when observers at Lick Observatory indicated a magnitude near 15.

The comet was first recognized as a new short-period object near mid-September 1895, when Dr. A. Berberich obtained an uncertain period of 3.22 years.
A couple of weeks later more reliable periods of 7.22 and 7.06 years were independently obtained by Professor Lewis Boss and Berberich, respectively. At this
time Comet Swift became Comet Swift 2, to distinguish it from Swift's earlier
periodic comet found in 1889 (the "2" was later dropped, when Comet Swift 1 was
accidentally rediscovered by Tom Gehrels in 1972). Finally, in early November
1895, Dr. Schulhof computed a more precise orbit which gave a period of 7.19
years. At this time, the comet's history began to be uncovered as calculations
showed the orbit to have been shaped by Jupiter during a close approach (0.4 AU)
in 1886. It was also discovered that the comet's future returns would get pro-
gressively more unfavorable and that a recovery would probably not be possible
before 1924. Several predictions were made for the 1902 return, but the fears
of astronomers were upheld--no comet could be found due to unfavorable conditions.
As the comet returned to perihelion in January 1910 searches were again conducted, but, again, conditions were too unfavorable. Few, if any, searches were
conducted in 1917. In February 1921, the comet passed 0.7 AU from Jupiter,
causing changes which made searches in 1924 nearly hopeless. A series of very
unfavorable returns again followed and in August 1955 another close approach
(0.28 AU) to Jupiter caused further unmeasurable perturbations due to the un-
uncertainty of the original period.

	T	q	P	Max. Mag.	Max. Tail
1895 II	Aug. 21	1.30	7.20	10.5	5 arcmin

Swift-Gehrels

Discovered: November 16, 1889   Discovery Magnitude: 10.5

Lewis Swift (Warner Observatory, Rochester, New York) discovered this comet
on November 16, 1889 while searching for new nebulae. He described it as a
faint, round, nebulous mass, without a tail. The comet was then located 0.63
AU from Earth and steadily moved away in the weeks that followed; however, its

brightness changed little as November progressed, due to an approaching perihelion date of November 30. Thereafter, fading began and by mid-December the
total magnitude had dropped to 12. The comet was last detected on January 22,
1890, when near magnitude 13.

Despite an observed arc of 68 days, several elliptical orbits were calculated which gave periods ranging from 8.5 to 8.9 years. When the probable
uncertainties of up to one year were added, astronomers realized that a future
recovery would have to rely on an accidental discovery. On February 8, 1973
that accident occurred while Tom Gehrels (Palomar Observatory) was using the
122-cm Schmidt telescope to photographically search for Apollo-type minor
planets. Upon examination of the plate, he noted a sharply condensed object
which appeared to be a comet and this was confirmed on the following night when
a further photo was taken of the region. The 19th-magnitude comet was again
photographed on February 10, when R. E. McCrosky and C.-Y. Shao (Harvard College
Observatory's Agassiz Station) obtained the last observation prior to the interference of moonlight. Calculations were quickly made using the three observations so that a recovery following the moon's exit from the sky could be made
and astronomers noted that the comet seemed to be of short period. This was
confirmed on February 25 when McCrosky and Shao recovered the comet. New calculations revealed that Comet Gehrels was identical to the lost comet Swift 1
(so designated after Swift's discovery of a second periodic comet in 1895).
It was also found that the comet's true period in 1889 had been 9.15 years and
that minor approaches to Jupiter in 1912 and 1957 had acted to increase the
period to 9.23 years.

Comet Swift-Gehrels had passed perihelion on August 31, 1972 and an intense
search for prediscovery images was undertaken. A flurry of excitement came
when it was found that in August 1972 the comet should have passed less than 1
degree from P/Kearns-Kwee and been as bright as magnitude 15; however, searches
of photographic plates revealed no trace. Some searches were also conducted on
plates exposed in 1935 and 1944 when it was found that the comet would then have
been well placed for recovery; however, no traces were found. The comet was
kept under observation until March 26, 1973 when Roemer (Steward Observatory)
estimated the magnitude to be 21.0. A further exposure on April 8 revealed
nothing. The comet was next recovered on July 31, 1981 by Shao and G. Schwartz.
Then near a nuclear magnitude of 18.5, the comet brightened to a maximum total
magnitude of 10.0 one month after passing its November 27 perihelion.

	T	q	P	Max. Mag.	Max. Tail
1889 VI	Nov. 30	1.36	9.15	10.5	none
1972 VII	Aug. 31	1.35	9.23	19.0	none
1981 XIX	Nov. 27	1.36	9.26	9.6	4 arcmin

Swift-Tuttle

Discovered: July 16, 1862   Discovery Magnitude: 7.5

Lewis Swift (Marathon, New York) discovered this comet in Camelopardalis
on July 16, 1862, while examining the northern sky with his 11.4-cm Fitz refractor. He described the comet as a somewhat bright telescopic object; however,
the finding was not immediately reported because Swift thought he was viewing
Comet Schmidt (1862 II) which had been found two weeks earlier. Subsequently,
the comet was independently discovered on July 19 by Horace Tuttle (Harvard
Observatory, Massachusetts) and Swift immediately made his announcement to get

some credit for his first comet discovery.

With the comet approaching an August 23 perihelion, its total magnitude brightened rapidly from the initial value of 7.5. At the end of July, the brightness was near 5.5*, with a tail extending 1 degree, and by mid-August the magnitude was near 4*, while the tail extended about 10 degrees. Telescopically, the comet was an interesting object in early August. First of all, a faint anomalous tail was observed during the first week by G. V. Schiaparelli (Italy) and was estimated as 30 arcmin long. Secondly, rapidly shifting luminous jets were observed in the coma near the nucleus. The comet was most impressive during the last days of August and the first days of September when the total magnitude attained 2 and the tail was estimated as between 25 and 30 degrees long. Thereafter, the comet faded, with the magnitude dropping to 5* by late September. Having been moving southward since passing 7 degrees from the north celestial pole in mid-August, the comet dropped below the horizon for Northern Hemisphere observers at the beginning of October; however, observers in the Southern Hemisphere continued to observe the comet until October 31.

During the next few years several astronomers found the comet to travel in an elliptical orbit with a period between 120 and 125 years and in 1866 Schiaparelli announced that the comet was responsible for the Perseid meteor shower seen shortly before mid-August of each year.

In 1889 F. Hayn (University of Göttingen, Leipzig, Germany) computed a definitive orbit for Comet Swift-Tuttle which gave a period of 119.64 years and this indicated the next return would be expected in 1982. In the early 1970s interest in this comet began to increase and in 1973 Brian Marsden recomputed the 1862 orbit and found a period of 119.98 years--very close to that obtained by Hayn. He then predicted the next perihelion date to be September 17, 1981. Minor searches for the comet began in 1980, which was within the error range given by calculations, and more rigorous searches have been conducted in 1981 and 1982, but, to date, nothing has been found. It should be noted that Marsden has found that comet 1737 II (Kogler) seemed to move across the sky much as Comet Swift-Tuttle would be expected to. If this comet is actually a previous appearance of 1862 III then the resulting period would be near 130 years and perihelion would probably not occur until late 1992.

	T	q	P	Max. Mag.	Max. Tail
1862 III	Aug. 23	0.96	119.98	2.0*	30 degrees

Taylor

Discovered: November 24, 1915   Discovery Magnitude: 10.0

Clement J. Taylor (Herschel View, South Africa) discovered this comet on November 24, 1915 near 31 Orionis. He described it as near magnitude 10 and determined the motion as slowly northward. On December 2 he reported his find to the Union Observatory (Johannesburg, South Africa) where it was observed later that evening near Delta Orionis.

The comet began to be widely observed around December 8 when the total magnitude was estimated as 9.5. At the same time, a tail was first reported and was estimated as 2 arcmin long. As December progressed the comet slowly brightened as it neared both the sun and Earth and by month's end it was near magnitude 9. The comet was brightest during the first week of January 1916 when reported to be near magnitude 8.7. It was then at perigee (0.64 AU) and, thereafter, faded very slowly as it approached a January 31 perihelion. During

the first week of February the total magnitude was near 9.5, but shortly thereafter an odd series of events began to occur. First, a secondary nucleus was
discovered on February 10 by E. E. Barnard. This nucleus was at first brighter
than the primary nucleus, but by late February it was fainter than the primary
and was not detected after March 21. Secondly, the comet began to fade more
rapidly than predicted with the magnitude dropping to 12 on March 1--1.5 magnitudes fainter than expected. When last seen on May 28, 1916 the comet was between magnitude 16 and 16.5--over 3 magnitudes fainter than predicted.

The comet was first recognized as a short-period object in January 1916,
when J. Braae and J. Vinter-Hansen determined the period to be 6.29 years. From
this orbit it was found that the comet had passed close to Jupiter in 1854 (0.06
AU) and 1912. This was confirmed in March when Braae recomputed the orbit and
found a period of 6.37 years. In June, Jeffers and Neubauer also found a
period of 6.37 years.

Comet Taylor was next expected to arrive at perihelion in June 1923 and,
during the previous year, Jeffers computed a definitive orbit for 1916 (period
equals 6.37 years) and provided an ephemeris for April 1923, when the comet
would be closest to Earth. Unfortunately the expected magnitude on April 13,
1923 was given as 17.5 and searches proved fruitless. On June 20, 1925 the
comet passed 0.24 AU from Jupiter which acted to increase the period. Y. C.
Chang and George van Biesbroeck predicted it to arrive at perihelion on October
22, 1928 and computed the new period to be 6.76 years. With the magnitude
expected to be between 13 and 14 in early October 1928, numerous photographic
searches were made at Yerkes, Heidelberg and other observatories, but no comet
was found. Astronomers began to fear that the comet had become lost for the
same reasons that Comet Biela had--the splitting later led to a complete break-
up.

Little additional work on Comet Taylor was carried out until 1973 when N.
A. Belyaev, V. V. Emel'yanenko and N. Yu. Goryajnova recomputed the 1916 orbit.
In 1974, Emel'yanenko computed more precise orbits for components A and B. These
orbits became the basis for predictions of the comet's return to perihelion in
January 1977, and on January 25, 1977 Charles T. Kowal (Hale Observatories,
California) identified images of the comet's B component on plates he had exposed with the 122-cm Palomar Schmidt on December 13 and 14, 1976. The total
magnitude was then near 16 and the published prediction was found to require
a correction of only -1.4 days. Searches for the A component were fruitless.
The comet slowly faded thereafter and was last seen on April 16, when near
magnitude 18. Pre-recovery images were identified by both G. V. Zhukov (Alma-
Ata Observatory, Russia) and T. Seki (Geisei Station, Japan) on plates exposed
on November 23.85, 1976 and November 23.68, 1976, respectively. The astronomers
gave the total magnitude as between 16.5 and 17.

	T	q	P	Max. Mag.	Max. Tail
1916 I	Feb. 1	1.56	6.37	8.7	2 arcmin
1977 II	Jan. 11	1.95	6.97	16.0	none

Tempel 1

Discovered: April 3, 1867   Discovery Magnitude: 9.0

William Tempel (Marseilles, France) was conducting a routine search for
comets on April 3, 1867 when he detected this 9th-magnitude[*]object moving slowly
southeastward through Libra. He described it as faint and diffuse, with a coma

4 to 5 arcmin across.

The comet was then heading towards both the sun and Earth and slowly brightened in the following weeks. Perigee came around May 6 when the comet was 0.56 AU from Earth and the magnitude was then near 8.5[*]. The brightness changed little thereafter as the decreasing distance from the sun countered the effects of an increasing distance from Earth; however, after perihelion on May 24 (r= 1.56 AU), fading became more rapid and when last seen on August 27 the comet was near magnitude 12.

Comet Tempel was soon recognized as a new short-period comet and the orbital period was established as about 5.7 years. As the 1873 return neared, several predictions were made--one of which allowed Stephan (Marseilles) to recover the comet on April 4, 1873. The comet was then between magnitude 11 and 12[*] and was moving southward. Perihelion came on May 10 and the comet reached magnitude 10 a few days later. Observations ended after July 1, when the comet had faded to about magnitude 13[*].

During the interval between the apparitions of 1873 and 1879, Gautier computed definitive orbits for the previous two returns and provided an ephemeris for 1879. On April 25, 1879, Tempel (Arcetri, France) recovered the comet close to Gautier's prediction. The comet was then between magnitude 10 and 11[*] and changed little until after its May 8 perihelion. Thereafter, it steadily faded until last detected at magnitude 12 on July 8.

The comet's orbit underwent a drastic change in 1881 when it passed 0.55 AU from Jupiter. The period increased to 6.5 years, which made only alternate returns favorable for observation, and the perihelion distance increased to 2.1 AU, making the comet a very faint object even when favorably placed. Subsequently, no further recoveries were made despite photographic searches in 1898 and 1905.

An investigation into the loss of this comet was finally carried out in 1963 by B. G. Marsden. He found that approaches to Jupiter in 1941 (0.41 AU) and 1953 (0.77 AU) acted to decrease both the period and the perihelion distance to values smaller than when the comet was discovered. Predictions were supplied for the comet's return in 1967 and 1972--the latter of which was expected to be very favorable. Several photographs were made during mid-1967, but all seemed to show nothing; however, in late 1968, Elizabeth Roemer (Catalina Observatory) identified a possible image on a plate exposed on June 8, 1967. The magnitude was estimated as 18 and the measured position was very close to predictions. Proof of the recovery had to await the comet's 1972 return and on January 11 of that year Roemer and L. M. Vaughn (Steward Observatory) recovered the comet very close to predictions. This confirmed the 1967 observation and, since the comets of 1967 had already received their perihelion designations, the comet was designated 1966 VII, even though it was the first comet to pass perihelion in 1967, so as to keep the perihelion passages in chronological order.

The 1972 apparition was very favorable and the comet attained a maximum magnitude of 11 in late May. Observations ceased after July 10. The 1978 return was unfavorable for observations; however, the comet was recovered on April 12, 1977 by Shao, Schwartz and McCrosky (Harvard College Observatory's Agassiz Station). Then near magnitude 20.4, the comet was lost in the sun's glare for several months before and after the January 1978 perihelion, but was again recovered later in 1978. Observations ended after September 27, 1978.

	T	q	P	Max. Mag.	Max. Tail
1867 II	May 24	1.56	5.65	8.5*	2 arcmin
1873 I	May 10	1.77	5.98	10.0*	none
1879 III	May 8	1.77	5.98	10.5*	none
1966 VII	Jan. 13*	1.50	5.51	18.0	none
1972 V	Jul. 15	1.50	5.50	10.7	2 arcmin
1978 II	Jan. 11	1.50	5.50	18.0	none

*Denotes a 1967 perihelion passage.

Tempel 2

Discovered: July 3, 1875   Discovery Magnitude: 9.5

William Tempel (Milan, Italy) discovered this comet in Cetus on July 3, 1873. He described it as faint, with a slow motion to the southeast. Tempel next observed the comet on July 5 and commented that it was much brighter than expected, with a coma diameter of 5 arcmin. During the remainder of July, observers continually indicated the total magnitude was near 8.5*. The comet began fading in early August as its distances from the sun and Earth increased. By the end of August, the total magnitude was near 9.5* and in late September it was near 10.5*. The comet was last seen on October 20, 1873, when near magnitude 12*.

The comet was recognized as a new short-period object and several independent calculations gave periods between 5 and 5.5 years. A definitive orbit was finally computed by Schulhof which gave the period as 5.16 years and his prediction for the comet's return in 1878 led to its recovery by Tempel on July 19. Then near magnitude 10*, the comet brightened to 8* by mid-October and was followed until December 21.

Comet Tempel 2 has been observed at every favorable apparition since its discovery. The unfavorable apparitions of 1883, 1889, 1910, 1935 and 1941 were missed due to the lack of powerful telescopes and the acquisition of such instruments prevented the comet from going unobserved at similar apparitions in 1957 and 1978. The comet's most favorable apparitions have always come during the summer months and it is not uncommon for the total magnitude to become brighter than 8. The comet's best apparition to date was that of 1925, when perihelion came on August 7. During the first days of August the total magnitude reached 6.5, while the tail was extending 20 arcmin. This was very near the date of the comet's perigee (0.35 AU).

The comet has consistently shown a very predictable physical appearance at each return, although occasional jumps in brightness are not uncommon. These jumps usually occur about a week or two after perihelion and can amount to 2 or 3 magnitudes. The most pronounced instances of this phenomenon occurred in 1873 and 1967.

Almost as consistent as the physical appearance has been the comet's orbit, which has hardly changed during the last 300 years. During this time, the aphelion distance has remained within the orbit of Jupiter and occasional approaches to within 1 AU have caused the period to fluctuate between 5.2 and 5.3 years, while the perihelion distance has varied from 1.3 to 1.4 AU.

	T	q	P	Max. Mag.	Max. Tail
1873 II	Jun. 26	1.34	5.21	8.5*	none
1878 III	Sep. 8	1.34	5.20	8.0*	none
1894 III	Apr. 24	1.35	5.22	10.5*	none
1899 IV	Jul. 29	1.39	5.28	8.5	16 arcmin

	T	q	P	Max. Mag.	Max. Tail
1904 III	Nov. 11	1.39	5.28	12.5	none
1915 I	Apr. 15	1.32	5.17	12.5	none
1920 II	Jun. 11	1.32	5.16	9.0	2 arcmin
1925 IV	Aug. 7	1.31	5.16	6.5	20 arcmin
1930 VII	Oct. 6	1.32	5.17	10.0	2 arcmin
1946 III	Jul. 2	1.39	5.31	8.0	2 arcmin
1951 VIII	Oct. 25	1.39	5.30	12.0	5 arcmin
1957 II	Feb. 5	1.37	5.27	19.0	none
1962 VI	May 13	1.36	5.26	16.0	2 arcmin
1967 X	Aug. 14	1.37	5.26	7.8	3 arcmin
1972 X	Nov. 15	1.36	5.26	13.5	12 arcsec
1978 V	Feb. 21	1.37	5.27	18.5	none

Tempel-Swift

Discovered: November 27, 1869   Discovery Magnitude: 9.0

William Tempel (Marseilles) discovered this comet in Pegasus on November 27, 1869 and indicated a total magnitude near 9[*]. Observers soon afterward described the comet as a diffuse circular object, with a coma diameter near 5 arcmin and some condensation. The comet passed perigee (0.25 AU) on December 7 and slowly faded thereafter as it moved away from both the sun and Earth. At the end of December the total magnitude was near 10 and observations ceased after January 3.

With observations spanning 39 days, astronomers could not accurately determine the orbit of Comet Tempel, though they did recognize it as a new member of the short-period family. Thus, the comet passed through its 1875 apparition (which happened to be unfavorable anyway) unobserved. On October 11, 1880, Lewis Swift (Warner Observatory, New York) discovered an 8th-magnitude[*] comet in Pegasus, which he described as diffuse and circular. The comet brightened slightly during the next few weeks as its distances from the sun and Earth decreased, but observations were not made in Europe, due to a telegraphic error, until November 7, when Lohse (Dun Echt Observatory, Scotland) made an independent discovery. The comet passed perihelion (1.07 AU) on November 8 and perigee (0.13 AU) on November 18. Shortly thereafter, orbital calculations showed Comet Swift to be of short period and it was then recognized as Comet Tempel.

Comet Tempel-Swift steadily faded during December 1880, and by month's end it was near magnitude 11.5[*]. When last seen on January 26, 1881, the magnitude was near 12.5[*]. The comet's orbital period of 5.5 years made the 1886 return very unfavorable, but the 1891 apparition was expected to be very favorable and the comet was recovered 4 degrees southwest of Bossert's computed position by E. E. Barnard (Lick Observatory) on September 28, 1891. An independent recovery was made by W. F. Denning (Bristol, England) on September 30. Observers in the following days described the comet as a large, shapeless nebula with a faint central condensation. The comet brightened to 10th magnitude by late October and finally began fading shortly after it passed perihelion on November 18. Observations continued until January 21, 1892.

The comet was again unfavorably placed for observations in 1897 and in May 1899 it passed 1.16 AU from Jupiter which acted to increase the orbital period to 5.68 years. The 1903 apparition was then missed due to unfavorable conditions, but the comet was again recovered on September 30, 1908, when Javelle (Nice

Observatory) found it less than 2 degrees from the position predicted by Maubant.
Then near magnitude 14, the comet reached a maximum brightness of 12 by late October and had faded to 16.5 when last seen on December 30.

The orbit of this comet was considerably altered after close approaches to
Jupiter in 1911 (0.61 AU) and 1923 (0.50 AU) with the resulting orbital period
of almost exactly 6 years bringing the comet to perihelion when in conjunction
with the sun. Subsequently, observations were impossible for the next few returns. The period was further lengthened in 1935 when the comet passed 0.56
AU from Jupiter. The period of 6.3 years made the 1944 return more favorable
than previous returns; however, wartime conditions prevented searches from being made. The comet passed 1.44 AU from Jupiter in 1946, and the orbital period
increased to 6.4 years. The 1950 apparition became the best since 1908 and S.
Kanda (Tokyo) provided an ephemeris. Searches in Russia and the United States
at the time of the comet's opposition with Earth in late November and early
December revealed nothing. The 1957 return was very unfavorable, but the 1963
return was considered "quite satisfactory" by Brian Marsden, who predicted the
perihelion date as August 29. Marsden's calculations showed Kanda's 1950 prediction to have been "some six weeks too early" and he felt observers would
have a good chance to recover the comet in 1963. During August, Kearns and
Kwee (Palomar Observatory) searched unsuccessfully and they were followed by
Roemer (Flagstaff) in September and October, and by Hendrie (Colchester) in
October, November and December. For the 1970 return Marsden predicted the perihelion date to be January 23, but conditions were not as favorable as in 1963
and searches revealed no comet.

The comet is feared lost, especially since Bossert and Maubant had to apply
a retardant to the daily motion for the returns of 1891 and 1908, respectively.
Maubant remarked that this correction "is too large to be accounted for by in-
accuracies in the computation of the perturbations." Today these corrections
are recognized as a necessary component of orbital calculations and are attributed to nongravitational forces--jet-like forces emanating from a comet's nucleus. Astronomers believe the recent failures in recovering Comet Tempel-Swift
are linked directly to their uncertainty as to how nongravitational forces have
affected the comet since the 1908 apparition.

	T	q	P	Max. Mag.	Max. Tail
1869 III	Nov. 19	1.06	5.48	8.5*	none
1880 IV	Nov. 8	1.07	5.50	7.5*	none
1891 V	Nov. 18	1.09	5.54	10.0	trace
1908 II	Oct. 5	1.15	5.68	12.0	none

Tempel-Tuttle

Discovered: October 25, 1366   Discovery Magnitude: 3.0

On December 19, 1865, William Tempel (Marseilles) discovered this comet in
the evening sky near Beta Ursae Minoris. He described it as a circular object,
with a central condensation and a tail 30 arcmin long. Other observers in the
next few days indicated a total magnitude near 6*, a nuclear magnitude near 12
and a coma diameter of 3 arcmin. On January 6, 1866 an independent discovery
was made by Horace Tuttle (Harvard College Observatory, Massachusetts).

The comet passed perihelion on January 12, and was then near its maximum
brightness of magnitude 5*. Thereafter, it steadily faded and when last detected
on February 9, 1866, it was near magnitude 10.5*. Shortly thereafter, orbital

calculations revealed the comet to be moving in an elliptical orbit with a period of 33 years and at the end of 1866, Oppolzer had computed a definitive orbit which refined the period to 33.17 years. Immediately following Oppolzer's publication of his orbit, a flurry of letters from noted astronomers suggested and later proved that this comet produced the Leonid meteor shower, which also moved in an orbit with a period of 33 years. Equally interesting was the theory brought forth by John Russell Hind a few years later. Hind suggested that Comet Tempel-Tuttle was identical to comets seen in 868 and 1366.

The comet was not searched for very diligently at the 1899 return, but the 1932 return acquired much interest. Several independent orbits were computed, but S. Kanda utilized observations of the comet of 1366 to hopefully upgrade his orbit. The result of his calculations was the prediction that perihelion would come on November 1.8. Searches were conducted at several observatories, but no trace of the comet was found on photographs. Despite this apparent failure, Kanda continued his investigations in 1933 to determine whether the comets of 868 and 1366 were really related to Tempel-Tuttle. His conclusion was that only 1366 was a past apparition. He first used Chinese observations made between October 25 and 30, 1366 and obtained an orbit very similar to that of Tempel-Tuttle. He also found that the comet's apparent brightness (about magnitude 3) was due to a close approach to Earth (0.06 AU).

The comet was next expected to return to perihelion in 1965 and J. Schubart undertook extensive calculations to try and arrive at the most precise prediction. He began by tracing the 1866 orbit back for over 500 years to the apparition of 1366, and used several hypotheses as to the motion. He found that one of the orbits represented a comet seen by Gottfried Kirch on October 26, 1699. Although the comet was not seen again, the position and direction of motion were so close to Schubart's orbit that identity with Tempel-Tuttle seemed certain. This orbit was used for the 1965 prediction and perihelion was expected on April 25. Several unsuccessful photographic searches were conducted, but in October 1965, Schubart announced his discovery of the comet on plates exposed on June 30 and July 1 by M. J. Bester (Boyden Observatory, South Africa). The total magnitude was then 16 and the positions indicated Schubart's prediction was 5 days too early. The recovery was confirmed later in the month when images of the comet were found on two Palomar Schmidt plates exposed on June 30 and July 26. A further photo was found on a plate exposed on September 1.

	T	q	P	Max. Mag.	Max. Tail
1366	Oct. 19	0.98	33.68	3.0*	--
1699 II	Oct. 11	0.96	33.94	4.0*	--
1866 I	Jan. 12	0.98	33.52	5.5*	30 arcmin
1965 IV	Apr. 30	0.98	32.92	16.0	none

Tritton

Discovered: February 11, 1978   Discovery Magnitude: 20.0

Keith Tritton (United Kingdom Telescope Unit) discovered this comet on photographic plates taken on February 11 and 13, 1978 with the 122-cm Schmidt telescope. He described it as near total magnitude 20 with a fuzzy nucleus and a tail. The comet was confirmed on February 15 by C.-Y. Shao (Harvard College Observatory's Agassiz Station), whose description of a 19th-magnitude comet possessing a very well condensed coma and a tail 10 to 15 arcsec long, led Brian Marsden to theorize that a possible outburst had occurred. Tritton

obtained one last photo on February 15 before moonlight began interfering and
M. P. Candy (Perth Observatory) used this and Tritton's two earlier positions
to compute an elliptical orbit with a period of 7.30 years.

The comet was recovered on March 10 by J. H. Bulgar (Agassiz Station),
who estimated the nuclear magnitude to be 20. On March 11, 13 and 14, Shao
obtained the last observations of the comet. Marsden took five precise positions obtained between February 11 and March 13 and confirmed Candy's earlier
elliptical solution, though the period was found to be only 6.33 years. The
comet had been fairly favorably placed for discovery in 1978 with the distance
from Earth being less than 0.9 AU at discovery. In 1982, Nakano recomputed
the orbit based on 7 precise positions and arrived at an orbital period of
6.35 years. The estimated uncertainty was only 10 days. Nevertheless, a
future recovery will be difficult--primarily because of the comet's faintness.

	T	q	P	Max. Mag.	Max. Tail
1977 XIII	Oct. 29	1.44	6.35	19.0	15 arcsec

Tsuchinshan 1

Discovered: January 1, 1965   Discovery Magnitude: 15.0

Comet Tsuchinshan 1 was the first of two comets discovered in China in
January 1965 and subsequently named after the observatory on Purple Mountain
(Nanking) where they were discovered. The 15th-magnitude comet was reported
to have been found on January 1 as it moved northward in Gemini. The announcement of the comet's discovery was delayed and the first observation outside
of China was made on February 9 by Philip Veron (Palomar Observatory). This
position, combined with two of the three Chinese positions, allowed Leland
Cunningham to compute two slightly different elliptical orbits (depending on
which of the three discordant positions were used).

On February 23, Elizabeth Roemer (U. S. Naval Observatory, Arizona) photographed a "practically stellar nuclear condensation of magnitude 16.8" which
was embedded in a faint symmetrical coma 18 arcsec across. Observations of
the fading comet continued until April 24, when the nuclear magnitude had dropped
to 19. A short time later, the Planetary Section of the Purple Mountain Observatory computed an elliptical orbit based on 9 observations obtained between
January 1 and March 4 which possessed a period of 7.22 years. Towards the end
of the year, G. Sitarski computed a more precise orbit based on 18 observations
obtained between January 1 and April 24. He calculated the period to be 6.62
years and found that the comet passed 0.145 AU from Jupiter on December 12.96,
1960. Prior to this encounter, the period had been about 7.15 years.

In 1968, Sitarski predicted the comet would return to perihelion on September 18, 1971 and using the prepared ephemeris, Roemer and L. M. Vaughn
(Steward Observatory) recovered the comet on December 20, 1971 at a nuclear
magnitude of 20.3. The position indicated Sitarski's prediction required a
correction of -1.53 days. The comet was then three months past perihelion.
Roemer continued to observe the fading comet until March 10, 1972, when the
nuclear magnitude had dropped to 21.4.

In 1976, the Planetary Section of Purple Mountain Observatory published a
prediction for the comet's return to perihelion in 1978. The computed perihelion date was May 7 and on February 4, 1978 J. H. Bulgar (Harvard College
Observatory's Agassiz Station) obtained very weak images of the comet very
close to the predicted position. The nuclear magnitude was given as between

20 and 20.5. Confirmation came on March 8, when G. Schwartz and C.-Y. Shao (Agassiz) obtained images of nuclear magnitude 19.5. Due to unfavorable conditions, no further observations were obtained.

	T	q	P	Max. Mag.	Max. Tail
1965 I	Jan. 29	1.49	6.62	15.0	none
1971 VIII	Sep. 16	1.49	6.63	20.3	none
1978 IX	May 7	1.50	6.65	19.5	none

Tsuchinshan 2

Discovered: January 11, 1965   Discovery Magnitude: 15.0

This comet was the second to be discovered in China during January 1965 and was named after the observatory on Purple Mountain (Nanking) where it was found. It was first detected on a photograph exposed on January 11 and was described as near magnitude 15 with a short tail. Then moving slowly southwestward in Cancer, the comet was again photographed in China on January 13, but observations outside that country could not immediately be made due to a delay in telegraphing the announcement. When precise positions for the January 11 and 13 observations arrived at the Central Bureau on February 3 no orbit could be calculated and no searches could be made. Finally, on March 4 Purple Mountain Observatory provided an orbit which indicated the comet had passed perihelion on February 10 and possessed a period of 6.69 years. Subsequently, on March 10, E. Roemer (U. S. Naval Observatory, Arizona) obtained a precise position and described the comet's nucleus as nearly stellar with a magnitude of 16.4. A fan-shaped tail extended 24 arcsec to the east. Observations continued until May 31, when the comet had faded to magnitude 19.

During 1966, G. Sitarski computed a precise orbit for this comet using 16 observations obtained between January 11 and May 31. The period was given as 6.79 years and the comet was found to have passed 0.46 AU from Jupiter on January 1.76, 1962. In 1968, Sitarski applied perturbations from Mercury to Neptune to this orbit and predicted the next perihelion passage would occur on November 29, 1971. On September 19, 1971, Roemer (Catalina Observatory) recovered the comet near magnitude 19.7. The position indicated Sitarski's prediction needed a correction of +0.97 day.

Roemer's confirmatory photo of September 22 indicated a magnitude of 19.4 and showed a tail 12 arcsec long. The comet brightened to magnitude 17.1 by November 22, but, despite perihelion coming one week later, the comet continued to brighten into early February as the Earth-comet distance decreased to 1.04 AU. Observers in February gave a total magnitude of 15 and estimated the tail length as 10 arcmin. Thereafter, a steady fading set in, and when the comet was last detected on July 9 the total magnitude was near 19.

In 1976, the Planetary Section of Purple Mountain Observatory predicted the comet would next arrive at perihelion on September 21, 1978. On October 19, 1978, T. Seki recovered the comet very close to the predicted position. Then near a total magnitude of 18, the comet was followed only until December 1.

	T	q	P	Max. Mag.	Max. Tail
1965 II	Feb. 9	1.77	6.79	15.0	24 arcsec
1971 X	Nov. 30	1.78	6.80	15.0	10 arcmin
1978 XVI	Sep. 21	1.78	6.83	18.0	none

Tuttle

Discovered: January 9, 1790   Discovery Magnitude: 6.0

This comet was first discovered on January 9, 1790 by Pierre F. A. Mechain
(Paris, France) during a routine search for comets. Observers in the following
days described the comet as resembling a nebula without a nucleus and the total
magnitude must then have been near 6.[*]

The comet was found at perigee (0.33 AU) and steadily faded thereafter, despite a nearing perihelion. Observations ended on February 1, when the total
magnitude had dropped to about 7.[*] The observed arc of only 24 days did not
allow astronomers to recognize the comet's periodic nature and the best orbit
available was a parabolic one computed by Mechain shortly after the comet's disappearance. This orbit gave a perihelion date of January 28, 1790 and a perihelion distance of 1.06 AU.

The comet was accidentally rediscovered on January 5, 1858 by Horace Tuttle
(Harvard College Observatory, Massachusetts), who indicated the total magnitude
was near 8. An independent discovery was made by C. Bruhns (Berlin, Germany)
on January 11 and the comet was widely observed after January 23. The comet
brightened to magnitude 7[*] when closest to Earth (0.75 AU) at the end of January,
and shortly before its February 24 perihelion it reached 6.5.[*] Thereafter fading
was rapid, and when last observed on March 24 it was near magnitude 10.5.[*]

Shortly after the first parabolic orbit was calculated, astronomers began
suggesting the comet was periodic. Tuttle and Pape suggested Comet Tuttle was
identical to Comet 1790 II, while Bruhns suggested it was a return of Comet
1785 I. Indeed, both of these comets possessed similar orbits, aside from the
inclination, but around the time of Comet Tuttle's last observation Pape and
Bruhns independently calculated a period near 13.7 years--proving the link with
1790 II to be correct. Further investigations into the orbit were conducted by
several astronomers and as the comet neared its next return, Tischler provided
a search ephemeris. On October 13, 1871 the comet was recovered by Borrelly
(Marseilles). Independent recoveries by Winnecke (Karlsruhe) on October 16
and Tuttle (Harvard Observatory) on October 22 indicated the total magnitude was
near 8.5. The comet passed perigee (0.69 AU) shortly after mid-November and
passed perihelion on December 2. Around the latter date it was near its maximum brightness of 7.5.[*] Observations ended on January 30, 1872.

Comet Tuttle has been observed at every return since 1871, except for the
very unfavorable apparition of 1953. During these returns it became brighter
than 10th magnitude six out of eight times, but a tail has been virtually nonexistent. The comet's most favorable returns are those which have perihelion
dates during December, January and February. The comet's most unfavorable
apparitions occur in June, July and August; however, since 1858 only the 1953
return came within this period. This comet is the parent comet of the Ursid
meteor shower of December 22, which attains a maximum rate of about 15 meteors
per hour each year.

	T	q	P	Max. Mag.	Max. Tail
1790 II	Jan. 31	1.04	13.88	5.5	none
1858 I	Feb. 24	1.03	13.74	6.5	"short"
1871 III	Dec. 2	1.03	13.82	7.5	none
1885 IV	Sep. 12	1.02	13.76	9.5	none
1899 III	May 5	1.01	13.62	10.0	none
1912 IV	Oct. 29	1.03	13.51	8.0	trace

	T	q	P	Max. Mag.	Max. Tail
1926 IV	Apr. 29	1.03	13.54	12.5	none
1939 X	Nov. 11	1.02	13.61	8.5	none
1967 V	Mar. 31	1.02	13.77	9.0	none
1980 XIII	Dec. 15	1.01	13.68	6.5	trace

Tuttle-Giacobini-Kresak

Discovered: May 3, 1858   Discovery Magnitude: 9.5

Horace Tuttle (Harvard College Observatory, Massachusetts) discovered this comet in Leo Minor on May 3, 1858. Then described as very faint, the comet was located only 0.36 AU from Earth. Fading was rapid as the comet moved away from the sun and Earth, and when last seen on June 2, the total magnitude must have been near 11.* Despite the short observational arc of 31 days, astronomers realized that the comet traveled in an elliptical orbit, but the period could not be determined more precisely than Schulhof's estimate of 5.8 to 7.5 years.

On June 1, 1907, Professor M. Giacobini (Nice Observatory, France) was conducting a routine search for comets with a 46-cm telescope. Suddenly he found a 13th-magnitude comet at the limit of visibility and measured its diameter as nearly 2 arcmin. During the following days, the comet moved southeastward through Leo and was observed at several other observatories after June 4. Observations ended after June 14, when the increasing distances from the sun and Earth had caused the magnitude to drop to 14. Dr. E. Stromgren (Berlin) computed a parabolic orbit for the comet which gave the perihelion date as May 31, 1907 and the perihelion distance as 1.24 AU. Several years passed before the possibility of this comet's identity to Comet Tuttle was first suggested by W. H. Pickering in 1914 and in 1928 A. C. D. Crommelin mathematically linked the two apparitions by assuming that nine revolutions had taken place. This indicated an average period of 5.44 years and Crommelin predicted the comet would return to perihelion on November 17, 1928. Unfortunately, the return was unfavorable and no searches were made.

In 1933, Crommelin reinvestigated the orbit of Comet Tuttle-Giacobini and found that the number of revolutions between 1858 and 1907 was probably ten. Perihelion was predicted to next occur on March 21, 1934, but searches by George van Biesbroeck (Yerkes Observatory) in March and April 1934 revealed nothing. The comet was then given up as lost.

Participating in Skalnate Pleso Observatory's routine searches for comets with 25x100 binoculars, Lubor Kresak discovered a 10th-magnitude comet in Cancer on April 24, 1951. The convenient location allowed numerous observations during the next few days and J. Brady and Miss N. Sherman were soon able to compute an orbit which proved to be elliptical with a period of 8.5 years. Further observations showed the period to be much shorter and the comet was soon identified as the recovery of Comet Tuttle-Giacobini. The comet remained near magnitude 10 until after its May 9 perihelion when a slow fading set in. In early June the total magnitude was near 11 and in early July it was 13. The comet was last seen on August 9, 1951.

The comet's 5.5-year period caused the 1956 return to be unfavorable and, thus, unobserved; however, on January 28, 1962, Elizabeth Roemer (U. S. Naval Observatory, Arizona) made the first recovery of this comet only one degree from Kresak's prediction. Then near magnitude 19, the comet passed within 0.3 AU of Earth in April 1962 and brightened to magnitude 10.0. Thereafter, it slowly faded and when last seen on August 5, the magnitude was near 20.

The comet was missed in 1967, but it was again recovered on January 8, 1973 by Roemer and J. Q. Latta (Steward Observatory) near magnitude 21.0. As with previous apparitions, the comet steadily brightened as it approached perihelion and on May 20 it was estimated as between magnitude 12 and 14 by F. Seiler (Munich); however, only six days later, the tranquil character of this comet suddenly changed. On May 26.5 S. Ako (Siraishi, Japan) photographed an unexpected 8th-magnitude comet and immediately reported his discovery. Similarly, K. Mameda (Kobe, Japan) independently detected the object on May 27.5, but his magnitude estimate was given as 5. The comet was actually Tuttle-Giacobini-Kresak and on May 27.9, Seiler photographed the comet near magnitude 4! The comet's outburst quickly subsided and by June 2 it was back to magnitude 10. M. Antal (Skalnate Pleso Observatory) took an interest in the comet as June progressed and his photographs showed the comet fading to magnitude 12.6 by June 13 and 13.2 by June 23. On July 4.1 McCrosky (Harvard College Observatory's Agassiz Station) estimated the total magnitude as between 14 and 15; however, on July 6.9, Antal found the comet at magnitude 5.6. The magnitude reached 4.5 on the 7th, but quickly dropped to 9 on the 10th, 10 on the 12th and 11 on the 19th. No further activity was noted and the comet was last observed on September 23, 1973 at a magnitude of 17.

No explanation currently exists for the dramatic flares of this comet in 1973 and careful attention during the rather unfavorable return of 1978 revealed no unusual activity.

	T	q	P	Max. Mag.	Max. Tail
1858 III	May 4	1.14	5.35	9.5*	none
1907 III	May 29	1.16	5.57	13.0	none
1951 IV	May 9	1.12	5.47	9.8	trace
1962 V	Apr. 24	1.12	5.49	10.0	trace
1973 VI	May 30	1.15	5.56	4.0[ƒ]	45 arcmin[†]
1978 XXV	Dec. 26	1.12	5.58	15.0	none

[†] Outburst in brightness. Normal brightness was near 13 and the normal tail was less than 1 arcmin.

Väisälä 1

Discovered: January 19, 1939   Discovery Magnitude: 15.0

When Professor Y. Väisälä (University of Turku, Finland) examined an asteroid patrol plate exposed on February 8, 1939, he located an object whose appearance incited him to announce it as a new minor planet--it was even given the provisional minor planet designation 1939 CB. The 15th-magnitude object was soon sought by other astronomers to assure the calculation of an accurate orbit; however, their photos revealed the short tail and small coma characteristic of a comet. An orbit calculated in March not only confirmed the cometary nature, but also helped Väisälä locate a prediscovery image on a plate he had exposed on January 19, 1939.

At the time of discovery, Comet Väisälä was a few days past perigee (0.86 AU), but the brightness continued to climb and it reached 14 shortly before the perihelion date of April 26. Thereafter, the comet slowly faded with magnitude estimates being near 15.5 on May 17 and 16.5 on June 5. H. M. Jeffers (Lick Observatory) obtained the last photograph of the comet on June 8 and estimated the magnitude as 17.

The comet was first recognized as periodic in early March, when Liisi Oterma and Väisälä computed a preliminary orbit with a period near 10 years. At the end of April observations obtained at Turku between January 19 and April 22 were used to refine the orbit and the period was found to be 10.58 years. In 1948, Oterma published a definitive orbit for the comet which also gave the 1939 orbital period as 10.58 years and from this orbit she predicted the comet would return to perihelion on November 10, 1949. On December 19, 1949 Antonin Mrkos (Skalnate Pleso Observatory, Czechoslovakia) recovered the comet and determined its magnitude to be 17. The measured position indicated the prediction was only one day early.

Astronomers have not missed one apparition of Comet Väisälä 1, as it was recovered in 1960 and 1971, but they realize that it is a small, intrinsically faint object. The comet's appearance has always followed predictions and it was brightest and showed the most signs of activity during the very favorable returns of 1939 and 1960, when it made its closest approaches to Earth (0.9 AU). At these apparitions the magnitude obtained 14, but the comet's large perihelion distance allowed only slight development of both the tail and coma.

An examination of the orbit by the Russian mathematician E. I. Kazimirchak-Polonskaya for the period 1660 to 2060 indicated the comet would experience a total of seven minor and moderate encounters with Jupiter and four major ones with Saturn. These produce only small variations in the orbit--the period varying by barely one year during the 400 years studied.

	T	q	P	Max. Mag.	Max. Tail
1939 IV	Apr. 26	1.76	10.58	14.0	1 arcmin
1949 V	Nov. 11	1.75	10.52	17.0	none
1960 IV	May 11	1.74	10.46	14.0	1 arcmin
1971 VII	Sep. 12	1.87	11.28	20.0	none
1982 V	Jul. 31	1.80	10.88	19.0	none

Väisälä 2

Discovered: March 11, 1942   Discovery Magnitude: 13.0

On routine minor planet photos exposed at the University of Turku (Finland) on March 11, 1742, Professor Y. Väisälä discovered an asteroid (1942 EC) and a comet (1942 c). News of the discovery was immediately sent to Copenhagen--the clearing house for important astronomical discoveries at that time--and the information was relayed to major observatories. Unfortunately, wartime conditions caused delay and confusion in the announcement and some observatories never received the message.

In the United States, Harvard College Observatory received the radiogram on March 17, but the message told only of a fast-moving stellar object of magnitude 13. On March 18 and 19, H. L. Giclas took photographs of the region and found a distinctly cometary object traveling at a rate of more than 2 degrees per day. Using the available positions acquired by Giclas and Väisälä, Miss Stahr and Mr. Salanave (Berkeley, California) tried to compute an orbit, but no meaningful result could be obtained. Giclas again photographed the comet on March 22, but his attempts thereafter were fruitless. In Europe, the comet was photographed on several occasions by G. Kulin (Budapest, Hungary), J. Stobbe (Poznan, Poland) and Krumpholz (Vienna, Austria) using preliminary orbits and ephemerides provided by Turku Observatory. These indicated the comet was near perigee (0.38 AU) when discovered and had passed perihelion in mid-February. The comet was last seen on April 17.

The comet was first recognized as a new short-period comet in late April, when Väisälä used positions obtained on March 11, 25 and April 12 to compute an elliptical orbit. The period obtained was 85.52 years. In 1971, Liisi Oterma recomputed the orbit and found a period of 85.42 years. She added that the uncertainty amounted to 2 years.

	T	q	P	Max. Mag.	Max. Tail
1942 II	Feb. 16	1.29	85.42	13.0	none

van Biesbroeck

Discovered: September 1, 1954   Discovery Magnitude: 15.0

George van Biesbroeck (Yerkes Observatory) discovered this comet on a pair of plates taken with the 61-cm reflector on September 1, 1954. The plates had been exposed to recover the minor planet 1953 GC--which was not found. The comet was described as fairly well condensed, with a total magnitude of 15.

Comet van Biesbroeck had passed perihelion on February 21, 1954 (r= 2.41 AU) and when discovered it was near opposition. Early computations by L. E. Cunningham using observations extending about two weeks met with great difficulty, but one month after the discovery, he found the object to be a new short-period comet and gave the period as 14.1 years. Van Biesbroeck continued to photograph the comet until November 28, when it had faded to magnitude 18.5, but Elizabeth Roemer (Lick Observatory) managed to again detect the comet on January 15, 1955, when it was near magnitude 19. Shortly thereafter, Cunningham revised the orbit and found the period to be 12.4 years. As the comet approached its 1955 opposition it was again recovered by Roemer on September 21. Then near magnitude 19.5, the comet changed little during the remainder of its visibility, which finally ended on November 13, 1955.

In 1958, van Biesbroeck published definitive elements for his comet which indicated the orbital period was 12.43 years and in March 1965, Brian Marsden applied perturbations by Earth, Jupiter and Saturn to this orbit and calculated the comet's next perihelion date to be July 19, 1966. Four months earlier, S. W. Milbourn and G. Lea recomputed the 1954 orbit and, after subjecting it to perturbations by Venus to Saturn, they predicted the 1966 perihelion date to be July 17. On May 1, 1965 Roemer (U. S. Naval Observatory) recovered the comet near magnitude 20. Her precise positions indicated Milbourn and Lea's orbit required a correction of +0.3 day, while Marsden's orbit required a correction of -1.8 days. A short time later K. Tomita (Dodaira, Japan) found pre-recovery images on plates he had exposed on March 27, April 5, and April 24.

Comet van Biesbroeck was again photographed by Roemer on May 2, 22 and 23, 1965, before entering the sun's glare; however, after conjunction with the sun it was detected on May 13, 1966 near magnitude 16. The comet was brightest during June and July, when near magnitude 15, and began fading thereafter as it moved away from both the sun and Earth. By October 16, Roemer estimated the magnitude as 17 and the comet was last seen on November 12.

In December 1976, Brian G. Marsden used observations obtained at the previous two apparitions, as well as perturbations by all nine planets to compute the comet's next perihelion date. He found it to be December 3, 1978, and his ephemeris allowed R. E. McCrosky (Harvard College Observatory's Agassiz Station) to recover the comet on December 17, 1977 at a nuclear magnitude of 20. The comet reached a maximum brightness of 15 during April and May of 1978 and then began fading soon afterward. It was last detected on September 24, 1979.

Orbital calculations reveal the comet to be moving near a 1:1 resonance with Jupiter and Marsden has expressed his opinion that the comet was captured into its present orbit fairly recently. An overall look at the physical appearance of the comet reveals the total lack of a tail and a small coma which rarely exceeds a diameter of 5 arcsec. Thus, the comet shows little activity in its present orbit as a result of the large perihelion distance.

	T	q	P	Max. Mag.	Max. Tail
1954 IV	Feb. 21	2.41	12.43	15.0	none
1966 III	Jul. 18	2.41	12.41	15.0	none
1978 XXIV	Dec. 3	2.41	12.39	15.0	none

van Houten

Discovered: September 24, 1960   Discovery Magnitude: 17.0

During the early months of 1966, C. J. van Houten and I. van Houten-Groeneveld (Leiden Observatory, The Netherlands) were measuring the positions of minor planets from plates exposed by Dr. Tom Gehrels (Palomar Observatory) during the Palomar-Leiden Survey, when they discovered a comet on eight plates exposed between September 24 and October 26, 1960. The comet was noticed during a blink examination of the plates and was described as possessing a hazy appearance, with a magnitude near 17. Precise measures of the comet's positions were sent to Dr. Paul Herget (Cincinnati Observatory), who proceeded to compute an orbit which was distinctly elliptical in nature. The period was found to be 15.75 years and the perihelion date was determined as April 29, 1961 (r= 3.94 AU). Since the perihelion designations had already been assigned for 1961, the comet was added to the end of the list, even though it had arrived at perihelion before Comet 1961 V.

The comet was expected to next come to perihelion during 1976-77 and various orbits were computed which gave perihelion dates ranging from December 29, 1976 to February 20, 1977. Charles T. Kowal (Palomar Observatory) conducted the most extensive search for the comet covering the range of 70 days either side of a January 1, 1977 perihelion date. Although a comet as faint as magnitude 19 could have been found, nothing was detected.

	T	q	P	Max. Mag.	Max. Tail
1961 X	Apr. 27	3.96	15.62	17.0	none

West-Kohoutek-Ikemura

Discovered: October 15, 1974   Discovery Magnitude: 12.0

Richard M. West (European Southern Observatory headquarters, Geneva, Switzerland) discovered this comet in January 1975, while examining a plate exposed by G. Pizarro and D. Billereau (European Southern Observatory, La Silla, Chile) on October 15, 1974. He described it as near magnitude 12, with a condensation and a short tail. Unfortunately, the comet's ambiguous motion, as well as the elapsed time of three months since the exposure was made, gave little hope for recovery.

On February 27, 1975, Lubos Kohoutek was attempting to recover a comet he had found on February 9, whose direction of motion was ambiguous. On photographs taken southwest of the earlier position he found a comet oriented in the proper direction and since it was too coincidental that another comet had been found, he immediately notified the proper authorities. On March 1, Toshihiko Ikemura (Shinshiro, Japan) was attempting to photograph Kohoutek's comet, but

could find nothing in the indicated position; however, he did find a comet to
the northeast of the February 27 position and additional images were found on
photos exposed by N. Kojima, T. Seki and K. Suzuki (all of Japan) on February
28.

Ikemura immediately reported his discovery and a further image was then
identified on a plate taken at Harvard College Observatory's Agassiz Station
on March 1. From the available positions it was soon realized that Ikemura's
comet was identical to Kohoutek's February 27 comet, which was actually moving
northeastward instead of southwestward, and Kohoutek's February 9 comet was
found to be totally different and was later identified on other photos.

Comet Kohoutek-Ikemura was further observed on March 3 by Seki and from
all available precise positions Brian Marsden was able to compute an elliptical
orbit which showed the period to be 6.07 years. In addition, an ephemeris
extending back to September 10, 1974 revealed that the comet was identical to
Comet West and the comet's name became West-Kohoutek-Ikemura.

Perihelion had been passed on February 26, 1975 and the comet slowly faded
after its discovery. By March 16, the magnitude was near 12.5 and this further
decreased to 14 by the beginning of April. At Woolston Observatory, R. H. S.
South estimated the total magnitude to be 15.3 on May 4 and at Steward Observatory on May 17, E. Roemer and L. M. Vaughn estimated the nuclear magnitude to
be 19.0. The comet was last detected on May 30.

In 1976, Marsden recomputed the comet's orbit using 36 positions obtained
between October 15, 1974 and May 30, 1975 and determined the period to be 6.12
years. This confirmed the earlier suggested theory that the comet had passed
0.01 AU from Jupiter in March 1972. The previous period was about 30 years,
while the previous perihelion distance was near 5 AU. The comet was next expected to arrive at perihelion in 1981 and in 1980 S. W. Milbourn calculated
the perihelion date to be April 12, 1981. On November 12, 1980, H.-E. Schuster
(European Southern Observatory) recovered the comet in a position which indicated a correction to Milbourn's prediction amounting to -1.37 days. Then
near a total magnitude of 18.5 the comet brightened to about 17 before becoming lost in the sun's glare after February 15, 1981.

	T	q	P	Max. Mag.	Max. Tail
1975 IV	Feb. 26	1.40	6.12	12.0	"short"
1981 VIII	Apr. 12	1.40	6.12	17.0	none

Westphal

Discovered: July 24, 1852   Discovery Magnitude: 7.5

Dr. Westphal, an assistant at Göttingen Observatory (Germany) discovered
this comet nearly two degrees south of 89 Piscium on July 24, 1852. With the
use of a comet-seeker, Westphal described the comet as a tolerably bright
nebula with a diameter of several arcmin. The total magnitude must then have
been near 7.5.[*] On August 3 an independent discovery was made by C. H. F. Peters.

During the next couple of months Comet Westphal steadily brightened as it
neared both the sun and Earth, with magnitude estimates being near 6.5 by mid-
August and 6.0 in early September. Perigee came on September 13 (0.60 AU),
but the comet continued to brighten as it neared the sun. On October 4, the
total magnitude must have been near 5[*] and on the 7th it may have been near 4.5.[*]
Perihelion came on October 13 and the comet began fading thereafter. On October
20 it was near magnitude 5.5,[*] but by mid-November it was near 7--one magnitude

fainter than predicted. By mid-December observers indicated a magnitude near
10--nearly 3 magnitudes fainter than predicted--and when last seen on February
9, 1853 it was near magnitude 11.

Within a few years following this comet's appearance, several orbits had
been computed which showed that it moved in an elliptical orbit. A. Marth provided one of the most precise orbits which possessed a period of 67.8 years;
however, as the next perihelion passage neared, astronomers revised the orbital
period to between 60 and 62 years and the first searches were begun in 1911.
Further attempts were made in 1912 and in 1913, but to no avail. Suddenly, on
September 27, 1913, Paul T. Delavan (National Observatory, La Plata, Argentina)
discovered an 8th-magnitude comet near the northern border of Aquarius. The
comet's motion in the next few days quickly identified it as the expected Comet
Westphal.

The comet was located 0.59 AU from Earth when found by Delavan and passed
perigee (0.57 AU) a few days later. Calculations showed the comet would remain near magnitude 8 throughout October and November as the decreasing distance
from the sun countered the effects of an increasing distance from Earth; however,
instead of keeping the same appearance, Comet Westphal began an unexplained
fading. By October 1 it was near magnitude 8.5 and on October 19 it was near
9.5. At the end of October, observers indicated the total magnitude was between 12 and 13 and in the first days of November it was between 14 and 15. No
nucleus was ever detected during November, as the comet continued to become
more diffuse, and the final observation came on November 22--four days short of
perihelion--when R. Schorr (Hamburg Observatory) estimated the total magnitude
as 16 or 17.

In the years following the 1913 appearance many questions existed as to
what happened to Comet Westphal, but the usual answer was that the comet broke
up. The only test for this hypothesis had to await the 1975-76 return and in
1973 several independent predictions were made for the comet's next perihelion
date. There were four predictions published and the perihelion dates varied
from December 26.9, 1975 to January 3.9, 1976. Photographic searches reaching
to magnitude 18 and 20 were conducted in August 1975 at Cerro El Roble and El
Leoncito, but no comet was found. Later in the year and early in 1976 several
visual and photographic searches were conducted in the Northern Hemisphere, but
without success. The comet is now believed to have broken up in 1913.

	T	q	P	Max. Mag.	Max. Tail
1852 IV	Oct. 13	1.25	61.20	4.5*	9 arcmin
1913 VI	Nov. 27	1.25	61.86	8.0	4 degrees

Whipple

Discovered: October 15, 1933   Discovery Magnitude: 13.0

Fred L. Whipple (Harvard College Observatory, Oak Ridge, Massachusetts)
discovered this comet on a photograph taken with the 41-cm Metcalf telescope on
October 15, 1933. He described it as a 13th-magnitude object with a short tail.
Whipple confirmed the comet on October 21 and by the 23rd enough positions had
been obtained to show that it was a new short-period comet, with a period near
8.2 years.

The comet had passed perihelion on August 1, 1933 (r= 2.50 AU), but the
small eccentricity of 0.35 caused fading to be very slow during the next several
months. The magnitude dropped to 14 by mid-October, 15 by late December and 16

by mid-February, 1934. The comet was lost in the sun's glare after mid-March, but it was recovered in October at a magnitude of 18. The comet was last detected on March 28, 1935 at a magnitude of 19.

By the end of 1933, M. Davidson had revised the orbit of this comet and computed the period to be 7.77 years. In 1934 further computations were conducted and Davidson and A. D. Maxwell independently arrived at periods of 7.51 and 7.50 years, respectively. By 1937, Maxwell had computed a definitive orbit which gave the period as 7.54 years.

Comet Whipple was recovered on September 1, 1940, on a pair of plates exposed by L. E. Cunningham (Harvard College Observatory). Then located very near the predicted positions, the comet was described as near magnitude 15 with a coma diameter of 10 arcsec. The comet was estimated as near magnitude 14.5 between September 3 and 5, but, thereafter, the increasing distance from Earth caused the comet to fade despite a nearing perihelion. By November 25 the total magnitude was near 16 and the comet was lost in the sun's glare after February 1941. On November 22, 1941 George van Biesbroeck succeeded in recovering the comet after its conjunction with the sun and estimated the magnitude as 17. No further observations were obtained.

Comet Whipple has never been missed at a return and with the small eccentricity it is not unusual for the comet to remain visible for 1.5 to 2 years at a particular apparition. The brightest apparitions occurred in 1933 and 1955 when perihelion came in August and November, respectively. The magnitude then reached 13. The apparitions of 1963 and 1978 have been the most unfavorable (maximum magnitude near 18), with perihelion dates of April 29 and March 28, respectively.

In 1967, E. I. Kazimirchak-Polonskaya examined the orbital evolution of Comet Whipple between 1660 and 2060. She found that the present orbit formed due to a close approach (0.25 AU) to Jupiter in June 1922. Before that encounter the period had been near 10 years, while the perihelion distance was near 3.9 AU. Prior to that an approach to within 0.22 AU in April 1852 caused the period to drop from about 19 years and the perihelion distance to decrease from 5.7 AU. In July 1981, an approach to within 0.59 AU increased the period to 8.5 years and the perihelion distance to 3.1 AU.

	T	q	P	Max. Mag.	Max. Tail
1933 IV	Aug. 1	2.50	7.50	13.0	3 arcmin
1941 III	Jan. 22	2.48	7.48	14.5	trace
1948 VI	Jun. 26	2.45	7.41	14.0	1 arcmin
1955 VIII	Nov. 30	2.45	7.41	13.0	short
1963 II	Apr. 29	2.47	7.46	18.0	trace
1970 XIV	Oct. 9	2.48	7.47	16.0	none
1978 VIII	Mar. 28	2.47	7.44	18.0	none

Wild 1

Discovered: March 26, 1960   Discovery Magnitude: 14.3

Mr. Paul Wild (Astronomical Institute of Berne, Switzerland) discovered this comet on a photograph exposed on March 26, 1960. Although the total magnitude was then estimated as 15, Wild later revised his estimate to 14.3. Wild secured further photos on April 5 and 6, and estimated the magnitude as 14.7 and 14.5, respectively. Thereafter, moonlight interfered for several days but the comet was finally recovered by Wild on April 13 (magnitude 14.8), and on the 14th,

H. L. Giclas (Lowell Observatory) succeeded in photographing it. A few days later, Wild provided a provisional parabolic orbit which indicated the comet had passed perihelion in mid-February at a distance of 2.2 AU.

On April 16, Elizabeth Roemer (U. S. Naval Observatory, Flagstaff, Arizona) photographed a "quite sharp, essentially stellar nucleus of magnitude about 18.0." On April 21 and 25, she estimated the nucleus as magnitude 17.5 and by May 1 it had faded to 17.8. Wild estimated the total magnitude as 15.0 on April 21 and by the beginning of May it was near 16. Thereafter, Roemer was the only observer of the comet and her estimates of the nuclear magnitude were near 18.0 on May 16, 19.0 on June 21 and 19.5 when last seen on June 27.

The comet was first recognized as a periodic comet at the end of April and the perihelion date was adjusted to mid-March. By October, Brian G. Marsden had computed a very precise orbit with a period of 13.19 years. Marsden refined the orbit in 1968 to prepare for the comet's return in 1973 and arrived at a period of 13.20 years. In August of 1971, Wild provided Marsden with revised measurements of his 1960 plates and the latter astronomer took special care in computing the comet's orbit. He found the 1960 period to have been 13.17 years and predicted the comet would next arrive at perihelion on July 2, 1973. Based on this prediction, Roemer succeeded in recovering the comet on January 8, 1973. Then between magnitude 19.5 and 20.0, the comet was found to be less than 1 day behind Marsden's prediction. Due to the rather unfavorable conditions, the comet never became brighter than 19 and was followed only until June 5.

	T	q	P	Max. Mag.	Max. Tail
1960 I	Mar. 17	1.93	13.17	14.3	trace
1973 VIII	Jul. 3	1.98	13.29	19.0	trace

Wild 2

Discovered: January 6, 1978   Discovery Magnitude: 13.5

This comet was discovered by Paul Wild (Astronomical Institute of Berne, Switzerland) on photos taken by him on January 6 and 8, 1978. It was described as between magnitude 13.5 and 14, and possessed a marked condensation. Wild confirmed the comet on January 25 and from the available precise positions B. G. Marsden computed a preliminary elliptical orbit which gave the perihelion date as June 15, 1978 (r= 1.49 AU) and the period as 6.15 years. This orbit also indicated that the comet had passed close to Jupiter in 1974.

At discovery, Comet Wild 2 was located about 1 AU from Earth. This distance steadily increased in the following months, but, as the comet was approaching the sun, the magnitude slowly brightened. Photographically the comet was being estimated as near magnitude 13 in the first days of February, but visually it was estimated as near 11.5. The comet brightened very slowly during February and March, and attained a visual magnitude of 11.0 in mid-April. By the end of May visual estimates placed the magnitude near 10.5 and when seen at low altitude on June 6, the comet was near 10.4, according to J. E. Bortle (Brooks Observatory, New York). By late June, observations could only be made in the Southern Hemisphere and Perth Observatory continued to provide precise positions until August 26. Thereafter, the comet was lost in the sun's glare as it approached conjunction. At the same time, fading had begun to increase rapidly and by the time the comet exited the sun's glare early in 1979 it was too faint to be seen.

In 1979, S. Nakano used 165 observations to calculate a more precise elliptical orbit. He determined the period to be 6.17 years and found that the comet had passed less than 0.2 AU from Jupiter in 1974. Prior to this encounter the period had been nearly 40 years and the perihelion distance was 4.9 AU.

	T	q	P	Max. Mag.	Max. Tail
1978 XI	Jun. 16	1.49	6.17	10.4	5 arcmin

Wild 3

Discovered: April 11, 1980   Discovery Magnitude: 15.5

In early May, Paul Wild (Astronomical Institute, Berne University, Switzerland) discovered a comet on photographs made with the 40-cm Schmidt telescope at Zimmerwald on April 11 and 12. It was described as diffuse, with a rather strong central condensation. On May 7, he recovered the comet in northern Virgo and estimated the total magnitude as 15.5.

Both Wild and Brian G. Marsden (Harvard-Smithsonian Center for Astrophysics) determined the comet to be of short period, with the latter astronomer determining a preliminary perihelion date of October 6 (q= 2.30 AU) and a period of 6.90 years. Marsden also added that the orbit indicated "that the comet passed 0.13 AU from Jupiter in 1976 Aug. Before then q=4.2 AU, e=0.12, P=10.3 years."

Although approaching the sun, the comet slowly faded after discovery as its distance from Earth increased. Beginning on June 11, observers at Perth Observatory (Australia) were the main observers of the comet and continued observations until August 11, when the total magnitude had dropped to about 16. The comet was lost in the sun's glare thereafter and was not recovered.

	T	q	P	Max. Mag.	Max. Tail
1980 VII	Oct. 5	2.29	6.89	15.5	none

Wirtanen

Discovered: January 15, 1948   Discovery Magnitude: 16.0

Carl A. Wirtanen (Lick Observatory, California) discovered this comet on a photographic plate exposed on January 15, 1948, in the course of the special proper motion program. He described it as a diffuse 17th-magnitude object possessing a tail 1.4 arcmin long. Having passed perihelion on December 3, 1974 (r= 1.63 AU), the comet slowly faded in the following weeks as it moved away from both the sun and Earth and when last seen on March 11 it was near magnitude 19.

The comet was recognized as a new short-period object in February 1948, when L. E. Cunningham used seven observations covering 13 days of visibility to obtain a period of 7.25 years. In 1954, G. Merton revised the 1947 orbit using 9 observations during a 54-day arc and found a period of 6.71 years. Using this orbit, W. E. Beart computed the perturbations by Jupiter and Saturn and predicted the next perihelion date to be August 13.6, 1954. On September 8, 1954, Wirtanen himself recovered the comet in a position which indicated a correction of -0.08 day to the predicted perihelion date. Then near magnitude 18, the comet faded to 19.5 by October 28 and was not seen thereafter, despite attempts at Lick Observatory on January 27, 1955.

Comet Wirtanen underwent very little orbital change during the next two apparitions in 1961 and 1967, when it reached maximum magnitudes of 18 and 15, respectively; however during April 1972 it passed only 0.28 AU from Jupiter,

which caused the period to decrease to 5.9 years and the perihelion distance
to decrease to 1.26 AU. The 1974-75 apparition was very unfavorably placed for
observation, but, nevertheless, the comet was detected between December 20,
1974 and February 6, 1975 at a magnitude of about 21.5. The 1980 return was
very similarly placed, but the comet was not observed. In 1984, another close
approach to Jupiter is expected to decrease both the orbital period and the
perihelion distance to values of 5.5 years and 1.08 AU, respectively.

	T	q	P	Max. Mag.	Max. Tail
1947 XIII	Dec. 3	1.63	6.71	16.0	2 arcmin
1954 XI	Aug. 14	1.63	6.69	18.0	none
1961 IV	Apr. 15	1.62	6.67	18.0	none
1967 XIV	Dec. 16	1.61	6.65	15.0	none
1974 XI	Jul. 6	1.26	5.87	21.5	none

Wolf

Discovered: September 17, 1884   Discovery Magnitude: 9.5

Dr. Max Wolf (Heidelberg, Germany) discovered this comet as it was moving
slowly through Cygnus on September 17, 1884. He described it as between magnitude 9 and 10, with a coma 2.5 arcmin across. Beginning on September 21, observers reported that Wolf's magnitude estimate seemed to have actually referred to the central condensation and that the total magnitude was near 7[*]. On
September 22, while making spectroscopic observations of the stars, R. Copeland
(Dun Echt, Scotland) independently discovered the comet as a gaseous body in an
objective-prism spectroscope.

At discovery the comet was only a few days from its closest approach to
Earth. During the next two months, as the comet approached perihelion, the
total magnitude faded very slowly and when closest to the sun on November 18
(r= 1.57 AU), the total magnitude was estimated as between 7 and 8. Thereafter,
fading became more rapid, with magnitude estimates being near 8 by mid-December,
and 9.5 by mid-January, 1885. When last detected on April 7 it was near magnitude 12[*].

Comet Wolf was recognized as a new short-period comet with a period near
6.7 years and an ephemeris by Thraen allowed R. Spitaler (Vienna, Austria) to
recover the comet on May 2, 1891. Then near magnitude 13, with a coma diameter
near 15 arcsec, the comet was independently recovered on May 3 by E. E. Barnard
(Lick Observatory). The comet steadily brightened as the year progressed and
as it passed perihelion and perigee in September and October, respectively,
the total magnitude reached a maximum value of 8. Orbital studies around this
time indicated that the comet had passed 0.12 AU from Jupiter in 1875, which
acted to decrease the period from 8.8 years to 6.8 years and the perihelion
distance from 2.7 AU to 1.6 AU.

Except for the unfavorable apparition of 1905, Comet Wolf has been observed
at every return since its discovery. Of special interest is the comet's return
to Jupiter's vicinity in September 1922, when its passage to within 0.13 AU
caused the orbit to nearly change back to its pre-1875 shape, i.e. a period of
8.3 years and a perihelion distance of 2.4 AU. Although the 1925 apparition
attained a maximum magnitude of 14.5, those following have never exceeded 18.
This change in brightness was primarily due to the increased perihelion distance;
however, astronomers who have studied the physical behavior of Comet Wolf have
noted one of the steepest decreases of absolute magnitude of any periodic comet.

This decrease was most rapid between 1884 and 1912 when it amounted to 3 magnitudes. Since there was no change in the perihelion distance or solar radiation during this time, astronomers have hypothesized that the comet's close approach to Jupiter in 1875 had caused an abnormal increase in brightness.

	T	q	P	Max. Mag.	Max. Tail
1884 III	Nov. 18	1.57	6.77	7.0*	trace
1891 II	Sep. 4	1.59	6.82	8.0	9 arcmin
1898 IV	Jul. 5	1.60	6.85	11.0	none
1912 I	Feb. 24	1.59	6.80	12.0	none
1918 V	Dec. 14	1.58	6.79	10.5	15 arcmin
1925 X	Nov. 8	2.43	8.28	14.5	1 arcmin
1934 I	Feb. 28	2.45	8.33	18.0	none
1942 VI	Jun. 24	2.44	8.29	18.6	16 arcsec
1950 VI	Oct. 24	2.50	8.42	18.0	none
1959 II	Mar. 22	2.51	8.43	20.3	none
1967 XII	Aug. 30	2.51	8.43	18.0	none
1976 II	Jan. 25	2.50	8.42	19.8	none

Wolf-Harrington

Discovered: December 22, 1924   Discovery Magnitude: 16.0

Dr. Max Wolf (Heidelberg, Germany) photographically discovered this comet on December 22, 1924. The comet was then moving slowly southward through Taurus and was described as near 16th magnitude. The comet faded thereafter as its distance from Earth increased and by December 26 van Biesbroeck estimated the total magnitude as 16.5. On January 23, 1925, van Biesbroeck indicated a total magnitude near 17.5 and the comet was last detected on February 14.

The comet's orbit was first recognized as elliptical at the end of December 1924, when A. Kahrstedt (Berlin, Germany) computed a period of 11.12 years based on Wolf's photographic positions obtained on December 22, 23 and 25. Shortly after Wolf's next observation in early January, Kahrstedt revised the period to 7.44 years and calculated the perihelion date as January 29, 1925. Comet Wolf possessed a perihelion distance of 2.4 AU and, with it being observed very near to the perihelion date, astronomers realized their orbits were far from precise. Subsequently, as 1925 progressed, van Biesbroeck provided a more precise orbital calculation and obtained a perihelion date of December 30, 1924 and a period of 7.66 years. One fact which all of the orbital calculations did have in common was the prediction that the comet's next return would be unfavorable and that a close approach to Jupiter in 1936 would alter the orbit--though the degree of altering was unsure due to the uncertainty of the orbit.

On October 4, 1951, Robert G. Harrington (Palomar Observatory) discovered a 16th-magnitude comet on plates exposed with the 122-cm Schmidt telescope for the National Geographic Society-Palomar Observatory Sky Survey. The comet was described as a diffuse object with a tail 2 arcmin long. The distances from the sun and Earth decreased during the remainder of October and into November, and the total magnitude reached 13. The comet was brightest in mid-December when the magnitude reached 12 and remained unchanged until late January 1952, when a slow fading began. After perihelion on February 7, fading became more rapid and in mid-March the magnitude reached 14. Observations ended after April 24.

Comet Harrington was recognized as a short-period comet by early November 1951, when L. E. Cunningham computed a period of 6.53 years. At this time

Cunningham remarked on the similarity between the orbits of this comet and
Comet Wolf (referred to as Wolf 2, since it was Wolf's second periodic comet),
but several astronomers disagreed, including Antoni Przybylski, who primarily
based his argument on the calculation of a definitive orbit for Comet Wolf 2.
After subjecting the orbit to the Jupiter perturbations of 1936, Przybylski
concluded that identity with Comet Harrington was unlikely. Meanwhile, in 1957,
W. Wisniewski recomputed the orbit of Comet Harrington 1 (so designated after
Harrington's discovery of a second periodic comet in 1953) and J. Kordylewski
advanced it and predicted a return to perihelion on August 12, 1958. Subsequently, Elizabeth Roemer recovered the comet on November 18, 1957 as a 20th-
magnitude object very close to the predicted position. Shortly thereafter, the
noted astronomer-mathematician M. Kamienski expressed his opinion that this
comet was identical to Wolf 2 and, with his estimation that the probability was
85 percent certain, identity was widely accepted.

	T	q	P	Max. Mag.	Max. Tail
1924 IV	Jan. 11*	2.43	7.60	16.0	trace
1952 II	Feb. 7	1.60	6.50	12.0	5 arcmin
1958 V	Aug. 11	1.60	6.51	16.5	1 arcmin
1965 III	Feb. 16	1.61	6.53	17.0	trace
1971 VI	Sep. 1	1.62	6.55	15.0	5 arcmin
1978 VI	Mar. 16	1.61	6.53	14.0	none

*Indicates a 1925 perihelion date.

NOTES ON COMET WATCHING

Charles Messier, an eighteenth century comet watcher, suffered the illness and death of his wife as deeply as most men. While he attended her, a contemporary discovered a comet. When a friend spoke to him about his loss, Messier grieved, "Alas I had discovered a dozen comets, and Montagne had to take away from me the thirteenth." Then, catching himself at his lapse, he recovered, "Ah, that poor woman."

Although the above story may be apocryphal, Messier's preoccupation was understandable. In former days, the discovery of a comet brought personal fame and some fortune. It still carries a certain prestige—only 300 to 400 persons have discovered comets in all history, and only 150 have found more than one. Not infrequently, the discoverer has been an amateur astronomer, among whom stands the king of comet hunters, Jean Louis Pons. This doordeeper at a Marseille observatory discovered 37 comets between 1801 and 1827, establishing a record for a single observer.

Contemporary amateurs carry on Pons' tradition all over the world. From 1959 to 1965, the English amateur George Alcock logged 735 hours of skywatching to find four comets. A Japanese piano polisher, Kaoru Ikeya, searched 109 nights for 335 hours before he discovered his first comet in 1963, then followed it with four more discoveries, one in each year through 1967.

Ikeya's visual discoveries were typical of most amateurs, who are successful because they combine persistence with the use of small and maneuverable instruments. For comet hunting, telescopic mirrors of four to six inches complement a low magnification of 25 to 40 power so that comets may be recognized quickly.

Despite their efforts, amateurs—at least in the Northern Hemisphere—face overwhelming competition from professional observers who discover comets on photographic plates exposed with the aid of high-power telescopes. Dr. Fred L. Whipple, former director of the Smithsonian Astrophysical Observatory, discovered six comets on photographic plates between 1933 and 1942; all nine comets of 1973 were discovered this way. The visual discovery of a comet on Feb. 12, 1974 by the Australian amateur William Bradfield was the first such sighting in two years, since the last one in March 1972, also by Bradfield. Because relatively few large telescopes operate in the Southern Hemisphere, an amateur there probably stands a better chance of finding his own comet.

Even if one doesn't find a new comet, bright comets can be photographed easily. A standard 35-mm camera with a f/1.4 or f/2 lens kept wide open should be mounted on a tripod. Expose each frame from 10 seconds to a minute using high-speed film, such as Tri-X, for black and white exposures, or high-speed Ektachrome for color pictures. The color film should then be processed at ASA 400 to enhance the comet.

—From "Comets," a publication of the
Smithsonian Astrophysical Observatory

REFERENCES

Years -371 to 1800

Books

    Ho Peng Yoke. The Astronomical Chapters of the Chin Shu. Paris: Mouten
        and Co., 1966.

    Marsden, Brian G. Catalog of Cometary Orbits. Hillside, N.J.: Enslow
        Publishers, 1983.

    Pingré, Alexandre. Cometographie. 2 Vols. Paris, 1783-84.

    Thorndike, Lynn, ed. Latin Treatises on Comets Between 1238 and 1368 A.D.
        Chicago: University of Chicago Press, 1950.

    Vsekhsvyatskii, S. K. Physical Characteristics of Comets. Moscow, 1958.
        (English Translation by Israel Program for Scientific Translations,
        Ltd., 1964).

Periodicals

    Ho Peng Yoke. "Ancient and Mediaeval Observations of Comets and Novae
        in Chinese Sources." Vistas in Astronomy, Vol. 5 (1962), pp.127-225.

    Nature, Vols. 13-17, 20, 22, 27-32.

    The Observatory, Vols. 23-49.

    Transactions of the Royal Philosophical Society, Vols. 1-86 (1665-1796).

Years 1800 to 1982

Books

    Lyttleton, R. A. The Comets and Their Origin. Cambridge, England:
        Cambridge University Press, 1953.

    Marsden, Brian G. Catalog of Cometary Orbits. Hillside, N.J.: Enslow
        Publishers, 1983.

    Middlehurst, B. M. and G. P. Kuiper, eds. The Moon, Meteorites and Comets.
        Chicago: University of Chicago Press, 1963.

    Vsekhsvyatskii, S. K. Physical Characteristics of Comets. Moscow, 1958.

Periodicals

    Astronomical Journal, Vols. 1-83.

    Astronomische Nachrichten, Vols. 2-277.

    Bulletin of the Astronomical Institute of Czechoslovakia, Vols. 10-32.

    Harvard Announcement Cards, Nos. 55-1676.

    I.A.U. Circulars, Nos. 300-3715.

    Icarus, Vols. 30-47.

    International Comet Quarterly, Vols. 2-5.

    Journal of the Association of Lunar and Planetary Observers, Vols. 18-29.

    Journal of Astronomical Society of Pacific, Vols. 52-70.

    Journal of Astronomical Society of South Africa, Vols. 1-2.

    Kiang, T. "The Past Orbit of Halley's Comet." Memoirs of the Royal
        Astronomical Society, Vol. 76 (1972), pp.27-66.

    Lunar and Planetary Laboratory Communications, Vols. 3 and 8.

    Minor Planet Circulars/Minor Planets and Comets, Nos. 4396-8324.

    Monthly Notices of Royal Astronomical Society, Vols. 49-119.

    Nature, Vols. 1-117.

    The Observatory, Vols. 23-70.

    Popular Astronomy, Vols. 1-59.

    Quarterly Journal of Royal Astronomical Society, Vols. 1-19.

    Sky and Telescope, Vols. 9-64.

    Soviet Astronomy, Vols. 11-24.

    van Houten, C.J. and I., "A New Periodic Comet, Observed in 1960." Bulletin
        of the Astronomical Institute of the Netherlands, Vol.18 (1966), p. 441.

NOTES

NOTES